WILLIAM F. MAAG LIBRARY
YOUNGSTOWN STATE UNIVERSITY

ELECTROANALYTICAL CHEMISTRY

ELECTROANALYTICAL CHEMISTRY

A SERIES OF ADVANCES

Edited by
ALLEN J. BARD

DEPARTMENT OF CHEMISTRY
UNIVERSITY OF TEXAS
AUSTIN, TEXAS

VOLUME 18

Marcel Dekker, Inc.   New York • Basel • Hong Kong

The Library of Congress Catalogued The First Issue of This Title as Follows:
Electroanalytical chemistry: a series of advances, v. 1
   New York, M. Dekker, 1966-
   v. 23 cm.
   Editor: 1966-   A. J. Bard
     1. Electromechanical analysis-Addresses, essays, lectures
     1. Bard, Allen J., ed.
QD115E499   545.3   66-11287
Library of Congress
ISBN 0-8247-9092-8 (v.18)

The publisher offers discounts on this book when ordered in bulk quantities. For more information, write to Special Sales/Professional Marketing at the address below.

This book is printed on acid-free paper.

Copyright © 1994 by Marcel Dekker, Inc. All Rights Reserved.

Neither this book nor any part may be reproduced or transmitted in any form or by any means, electronic or mechanical, including photocopying, microfilming, and recording, or by any information storage and retrieval system, without permission in writing from the publisher.

Marcel Dekker, Inc.
270 Madison Avenue, New York, New York 10016

Current printing (last digit):
10 9 8 7 6 5 4 3 2 1

PRINTED IN THE UNITED STATES OF AMERICA

INTRODUCTION TO THE SERIES

This series is designed to provide authoritative reviews in the field of modern electroanalytical chemistry defined in its broadest sense. Coverage will be comprehensive and critical. Enough space will be devoted to each chapter of each volume so that derivations of fundamental equations, detailed descriptions of apparatus and techniques, and complete discussions of important articles can be provided, so that the chapters may be useful without repeated reference to the periodical literature. Chapters will vary in length and subject area. Some will be reviews of recent developments and applications of well-established techniques, whereas others will contain discussion of the background and problems in areas still being investigated extensively and in which many statements may still be tentative. Finally, chapters on techniques generally outside the scope of electroanalytical chemistry, but which can be applied fruitfully to electrochemical problems, will be included.

Electroanalytical chemists and others are concerned not only with the application of new and classical techniques to analytical problems, but also with the fundamental theoretical principles upon which these techniques are based. Electroanalytical techniques are proving useful in such diverse fields as electro-organic synthesis, fuel cell studies, and radical ion formation, as well as with such problems as the kinetics and mechanisms of electrode reactions, and the effects of electrode surface phenomena, adsorption, and the electrical double layer on electrode reactions.

It is hoped that the series will prove useful to the specialist and nonspecialist alike—that it will provide a background and a starting point for graduate students undertaking research in the areas mentioned, and that it will also prove valuable to practicing analytical chemists interested in learning about and applying electroanalytical techniques. Furthermore, electrochemists and industrial chemists with problems of electrosynthesis, electroplating, corrosion, and fuel cells, as well as other chemists wishing to apply electrochemical techniques to chemical problems, may find useful material in these volumes.

A. J. B.

CONTRIBUTORS TO VOLUME 18

ALLEN J. BARD University of Texas at Austin, Austin, Texas

FU-REN F. FAN University of Texas at Austin, Austin, Texas

GYÖRGY INZELT Eötvös University, Budapest, Hungary

MICHAEL V. MIRKIN University of Texas at Austin, Austin, Texas

JAMES F. RUSLING University of Connecticut, Storrs, Connecticut

CONTENTS OF VOLUME 18

Introduction to the Series	iii
Contributors to Volume 18	v
Contents of Other Volumes	ix

ELECTROCHEMISTRY IN MICELLES, MICROEMULSIONS, AND RELATED MICROHETEROGENEOUS FLUIDS
James F. Rusling

II.	Historical Survey	2
II.	Historical Overview	2
III.	Surfactant Microstructures	4
IV.	Electrochemistry in Micellar Solutions	16
V.	Electrochemistry in Microemulsions	65
VI.	Electrochemistry in Lamellar and Vesicle Dispersions	74
VII.	Summary and Conclusions	79
	References	80

MECHANISM OF CHARGE TRANSPORT IN POLYMER-MODIFIED ELECTRODES
György Inzelt

I.	Introduction	90
II.	General Remarks and Scope	91
III.	Theories of Electron Transport in Polymer Film Electrodes	95
IV.	Results on Charge Transport in Polymer Films	124
V.	Effect of Film Morphology on Charge Transport in Polymers	218
VI.	Conclusion	228
	References	231

SCANNING ELECTROCHEMICAL MICROSCOPY
Allen J. Bard, Fu-Ren F. Fan, and Michael V. Mirkin

I.	Introduction	244
II.	Instrumentation	251
III.	Theory	268
IV.	Applications	310
V.	Conclusions	365
	Abbreviations	365
	List of Symbols	366
	References	370

Author Index   375

Subject Index   393

CONTENTS OF OTHER VOLUMES

VOLUME 1

AC Polarograph and Related Techniques: Theory and Practice, Donald E. Smith
Applications of Chronopotentiometry to Problems in Analytical Chemistry, Donald G. Davis
Photoelectrochemistry and Electroluminescence, Theodore Kuwana
The Electrical Double Layer, Part I: Elements of Double-Layer Theory, David M. Mohilner

VOLUME 2

Electrochemistry of Aromatic Hydrocarbons and Related Substances, Michael E. Peover
Stripping Voltammetry, Embrecht Barendrecht
The Anodic Film on Platinum Electrodes, S. Gilaman
Oscillographic Polarography at Controlled Alternating Current, Michael Heyrovksy and Karel Micka

VOLUME 3

Application of Controlled-Current Coulometry to Reaction Kinetics, Jiri Janata and Harry B. Mark, Jr.
Nonaqueous Solvents for Electrochemical Use, Charles K. Mann
Use of the Radioactive-Tracer Method for the Investigation of the Electric Double-Layer Structure, N. A. Balashova and V. E. Kazarinov
Digital Simulation: A General Method for Solving Electrochemical Diffusion-Kinetic Problems, Stephen W. Feldberg

VOLUME 4

Sine Wave Methods in the Study of Electrode Processes, Margaretha Sluyters-Rehbach and Jan H. Sluyters

The Theory and Practice of Electrochemistry with Thin Layer Cells,
A. T. Hubbard and F. C. Anson
Application of Controlled Potential Coulometry to the Study of Electrode
Reactions, Allen J. Bard and K. S. V. Santhanam

VOLUME 5

Hydrated Electrons and Electrochemistry, Geraldine A. Kenney and
David C. Walker
The Fundamentals of Metal Deposition, J. A. Harrison and H. R. Thirsk
Chemical Reactions in Polarography, Rolando Guidelli

VOLUME 6

Electrochemistry of Biological Compounds, A. L. Underwood and Robert
W. Burnett
Electrode Processes in Solid Electrolyte Systems, Douglas O. Raleigh
The Fundamental Principles of Current Distribution and Mass Transport
in Electrochemical Cells, John Newman

VOLUME 7

Spectroelectrochemistry at Optically Transparent Electrodes; I. Electrodes
Under Semi-infinite Diffusion Conditions, Theodore Kuwana and
Nicholas Winograd
Organometallic Electrochemistry, Michael D. Morris
Faradaic Rectification Method and Its Applications in the Study of
Electrode Processes, H. P. Agarwal

VOLUME 8

Techniques, Apparatus, and Analytical Applications of Controlled-
Potential Coulometry, Jackson E. Harrar
Streaming Maxima in Polarography, Henry H. Bauer
Solute Behavior in Solvents and Melts, A Study by Use of Transfer
Activity Coefficients, Denise Bauer and Mylene Breant

Contents of Other Volumes

VOLUME 9

Chemisorption at Electrodes: Hydrogen and Oxygen on Noble Metals and their Alloys, Ronald Woods

Pulse Radiolysis and Polarography: Electrode Reactions of Short-lived Free Radicals, Armin Henglein

VOLUME 10

Techniques of Electrogenerated Chemiluminescence, Larry R. Faulkner and Allen J. Bard

Electron Spin Resonance and Electrochemistry, Ted M. McKinney

VOLUME 11

Charge Transfer Processes at Semiconductor Electrodes, R. Memming

Methods for Electroanalysis In Vivo, Jiří Koryta, Miroslav Březina, Jiří Pradáč, and Jarmila Pradáčová

Polarography and Related Electroanalytical Techniques in Pharmacy and Pharmacology, G. J. Patriarche, M. Chateau-Gosselin, J. L. Vandenbalck, and Petr Zuman

Polarography of Antibiotics and Antibacterial Agents, Howard Siegerman

VOLUME 12

Flow Electrolysis with Extended-Surface Electrodes, Roman E. Sioda and Kenneth B. Keating

Voltammetric Methods for the Study of Adsorbed Species, Etienne Laviron

Coulostatic Pulse Techniques, Herman P. van Leeuwen

VOLUME 13

Spectroelectrochemistry at Optically Transparent Electrodes, II. Electrodes Under Thin-Layer and Semi-infinite Diffusion Conditions and Indirect Coulometric Iterations, William H. Heineman, Fred M. Hawkridge, and Henry N. Blount

Polynomial Approximation Techniques for Differential Equations in Electrochemical Problems, Stanley Pons
Chemically Modified Electrodes, Royce W. Murray

VOLUME 14

Precision in Linear Sweep and Cyclic Voltammetry, Vernon D. Parker
Conformational Change and Isomerization Associated with Electrode Reactions, Dennis H. Evans and Kathleen M. O'Connell
Square-Wave Voltammetry, Janet Osteryoung and John J. O'Dea
Infrared Vibrational Spectrosopy of the Electron-Solution Interface, John K. Foley, Carol Korzeniewski, John L. Dashbach, and Stanley Pons

VOLUME 15

Electrochemistry of Liquid-Liquid Interfaces, H. H. J. Girault and D. J. Schiffrin
Ellipsometry: Principles and Recent Applications in Electrochemistry, Shimson Gottesfeld
Voltammetry at Ultramicroelectrodes, R. Mark Wightman and David O. Wipf

VOLUME 16

Voltammetry Following Nonelectrolytic Preconcentration, Joseph Wang
Hydrodynamic Voltammetry in Continuous-Flow Analysis, Hari Gunasingham and Bernard Fleet
Electrochemical Aspects of Low-Dimensional Molecular Solids, Michael D. Ward

VOLUME 17

Applications of the Quartz Crystal Microbalance to Electrochemistry, Daniel A. Buttry
Optical Second Harmonic Generation as an In Situ Probe of Electrochemical Interfaces, Geraldine L. Richmond
New Developments in Electrochemical Mass Spectroscopy, Barbara Bittins-Cattaneo, Eduardo Cattaneo, Peter Königshoven, and Wolf Vielstich
Carbon Electrodes: Structural Effects on Electron Transfer Kinetics, Richard L. McCreery

ELECTROCHEMISTRY IN MICELLES, MICROEMULSIONS, AND RELATED MICROHETEROGENEOUS FLUIDS

James F. Rusling
University of Connecticut
Storrs, Connecticut

I. Introduction 2
II. Historical Survey 2
III. Surfactant Microstructures 4
    A. Surfactants and micelles 4
    B. Microemulsions 10
    C. Vesicles and lamellar dispersions 11
    D. Surfactant structure controls system architecture 12
    E. Adsorption of surfactants on solid surfaces 13
IV. Electrochemistry in Micellar Solutions 16
    A. Adsorption of surfactants 16
    B. Diffusion in micellar solutions 28
    C. Heterogeneous electron transfer in micellar solutions 41
    D. Electrochemical reactions in micelles 46
    E. Electrochemical catalysis in micellar solutions 59
V. Electrochemistry in Microemulsions 65
    A. Diffusion in conductive microemulsions 65
    B. Diffusion studies with microelectrodes 68
    C. Electrochemical reactions in microemulsions 69
    D. Electrochemical catalysis 71
VI. Electrochemistry in Lamellar and Vesicle Dispersions 74
    A. Lamellar dispersions 74
    B. Vesicle dispersions 76
    C. Detecting photochemistry at CdS particles in vesicles 77
    D. Immunological detection: marker release from liposomes 78
VII. Summary and Conclusions 79
    References 80

I. INTRODUCTION

Solutions of micelles and microemulsions depend on surfactants for their stabilities and microheterogeneous structures. Surfactant is short for "surface-active agent." Surfactants have polar or charged head groups and nonpolar regions in the same molecule [1]. Being surface active means that these molecules adsorb at the interface between two bulk phases, such as air and water, oil and water, or electrode and solution. The driving force for adsorption is the lowering of interfacial tension, i.e., minimization of interfacial free energy [2].

Micelles are perhaps the most widely studied aggregates of surfactants. In water, micelles form when the concentration of a water-soluble surfactant exceeds a characteristic value called the critical micelle concentration (CMC). Micellar structure and properties are discussed in Sec. III.

The aim of this chapter is to review modern uses of surfactants in electrochemistry and applications of electroanalytical methods to characterization of fluid surfactant systems. A major focus is on the use of fluid surfactant media such as micellar solutions, microemulsions, and other microheterogeneous fluids to purposely influence the outcome of electrochemical reactions. Surfactant concentrations used for such purposes are generally above the CMC for micelles, or at relatively high bulk or surface concentrations in other surfactant systems. The use of electroanalytical methods to help characterize aggregates in micelles and microemulsions is also reviewed.

Section II briefly summarizes the history of surfactants in electrochemistry. Section III gives an overview of the current understanding of structure and dynamics of surfactant aggregates. These concepts are essential to understanding and using electrochemistry in organized surfactant media. Section IV discusses recent research on electrochemistry in micellar systems; Sec. V does the same for microemulsions. Section VI discusses results of electrochemical studies in surfactant vesicles and lamellar dispersions.

II. HISTORICAL SURVEY

Surfactants have been employed in electrochemistry for over half a century [2–5]. They are widely used as brighteners in the electroplating industry [4] and have promising roles in advanced battery design [6]. Surfactants such as gelatin and Triton X-100 began to be used routinely in electroanalytical chemistry to suppress so-called streaming maxima at the dropping mercury electrode (DME) shortly after the invention of polarography by Héyrovský in 1922 [3]. Surfactants adsorb at the mercury-solution interface and retard streaming of solution near the DME, suppressing unwanted convection currents [3,7,8]. It was also recognized many years ago that adsorption of surfactants on the electrode can have large effects on the kinetics of heterogeneous electron transfer reactions at the DME [3,7].

A large fraction of the research on controlling electrochemical reactions with surfactants as well as aggregate characterization by electrochemical methods has been published within the past 15 years. Such studies gained popularity about the same time that novel work was being done with surfactant microstructures to control chemical and photochemical reactions [9–11]. However, several seminal publications prefigured many present-day activities in this field. Colichman, working in Meites's lab at Yale, reported in 1950 that CMC values of surfactants could be determined by their effects on polarographic limiting currents and half-wave potentials of reducible metal cations [12]. He showed that the CMC also corresponded approximately to the minimum amount of surfactant needed to suppress polarographic maxima. Meites [13] extended this approach to other systems and showed that CMCs could be determined from the influence of increasing surfactant concentration on the drop time of the DME. The decrease in drop time is caused by the surfactant-induced decrease in interfacial tension at the mercury-solution interface.

Early landmark papers featuring effects of surfactants also appeared in the literature of organic electrochemistry. In 1952, Holleck and Exner [14] reported that surface-active agents inhibited follow-up reduction of nitroaniline anion radical formed at the DME by reduction of nitroaniline. They showed that reversible reduction of aromatic nitro compounds can be effected in certain surfactant solutions, enabling the properties of stabilized arylnitro radicals to be studied by electron spin resonance (ESR), for example. This spawned much further research and contributed significantly to the unraveling of detailed reduction mechanisms for aromatic nitro compounds [15].

A related discovery led to perhaps the greatest commercial success of industrial organic electrochemistry, the electrolytic production of adiponitrile, the precursor to hexamethylenediamine in the manufacture of Nylon 66 [16]. Manuel Baizer, working at Monsanto Co., reported in 1964 that acrylonitrile can be electrochemically dimerized to adiponitrile at low cost in concentrated aqueous solutions of tetraethylammonium p-toluenesulfonate (TEATS) [17]. The surface-active tetraethylammonium ions form an adsorbed layer at Hg and Pb cathodes, favoring dimerization over a competing two-electron reduction. This type of electrolytic process is used to produce more than one billion pounds of adiponitrile worldwide per year [16].

Proske in 1952 first reported using surfactants at concentrations well above the CMC to solubilize nonpolar organic compounds in water for electroanalytical measurements [18]. Most of Proske's surfactant media were not simple aqueous micelles. They were macroscopically homogeneous mixtures of water, organic solvent, and surfactants Aerosol MA or Aerosol AY (dihexyl and diamyl sodium sulfosuccinates).

Thus, there was evidence as early as the 1950s that electroanalytical methods could be used to help characterize micellar systems, and that surfactants could be used to control electrochemical reactions and solubilize organic compounds for electrochemical studies in water. Progress in this area occurred at a relatively slow pace through the 1960s and early 1970s. A number of studies reported the polarographic behavior of some organic compounds solubilized in micelles [19].

Research since the late 1970s has demonstrated that coulombic and hydrophobic interactions with surfactants can stabilize various electrochemically produced ion radicals [20], that micelles in solution or adsorbed on electrodes can significantly enhance [21,22], inhibit, or control [22,23] reactions between an electrochemically generated reductant and a reducible substrate, and that electrochemical methods can be used to characterize diffusion of micelles [20,21]. Research during this more recent period is the main focus of Sec. IV to VI.

III. SURFACTANT MICROSTRUCTURES

A. Surfactants and Micelles

Surfactants owe their unique ability to form different types of aggregates to their amphiphilic nature and their structural diversity. The unifying structural features of surfactants, or detergents as they are sometimes called, are their distinct hydrophobic and hydrophilic regions. The hydrophilic region of the molecule is called the head group and may be positive, negative, neutral, or zwitterionic. The hydrophobic region is called the tail and consists of one or more hydrocarbon chains, usually with 6–22 carbon atoms. Chains may be linear or branched. Some typical surfactants referred to in this chapter are shown in Fig. 1. Phospholipids, major components of membranes in living cells, are also surfactants.

The properties of solutions containing surfactants of a homologous series, e.g., $CH_3(CH_2)_xR$, change in a regular fashion with increasing x. In 1897, Traube reported that the surfactant concentration needed to give a specific value of interfacial tension at the air-water interface decreased threefold for

Micelles and Microemulsions

$CH_3(CH_2)_{11}SO_4^-Na^+$

sodium dodecylsulfate (SDS)

$CH_3(CH_2)_{15}N(CH_3)_3^+Br^-$

cetyltrimethylammonium bromide (CTAB)
hexadecyltrimetylammonium bromide

$CH_3(CH_2)_{11}(OCH_2CH_2)_{23}OH$

polyoxyethylene(23)dodecyl ether (Brij-35)

$[CH_3(CH_2)_{11}]_2N(CH_3)_2^+Br^-$

didodecyldimethylammonium bromide (DDAB)

$[CH_3(CH_2)_3CHCH_2OOC]_2CHSO_3^-Na^+$
         $|$
       CH_2CH_3

sodium bis(2-ethylhexyl)sulfosuccinate aerosol AT (AOT)

$CH_3(CH_2)_{15}Py^+Br^-$

cetylpyridinium bromide
hexadecylpyridinium bromide

$4\text{-}CH_3(CH_2)_8]C_6H_4\text{-}$
$(OCH_2CH_2)_{11}CH_2CH_2OH$

Igepal CO-720

$[CH_3(CH_2)_{15}]_2PO_4^-$

dihexadecylphosphate

$CH_3(CH_2)_{14}OCOCH_2$
              $|$
$CH_3(CH_2)_{14}OCOCH$
              $|$
          $CH_2OPO_2CH_2CH_2\overset{+}{N}(CH_3)_3$

3-sn-dipalmitoylphosphatidyl-1'-sn-choline (lecithin)

FIG. 1. Molecular structures of some commonly used surfactants.

each additional methylene group in the chain [5]. Such regular changes in surface activity with increasing chain length follow what has become known as Traube's rule. Other examples are encountered throughout this review.

1. Structures of Micelles

When concentrations of soluble surfactants such as SDS or CTAB are increased in water, physical properties such as conductivity and surface tension show sharp discontinuities at a characteristic concentration called the critical micelle concentration (CMC). The discontinuities indicate the formation of dynamic aggregates of surfactant molecules called micelles. CMC values for commonly used surfactants range from about 10^{-4} to 10^{-2} M [9,10]. Addition of salt to the solution usually decreases the CMC.

Just above the CMC, micellar structure is considered to be roughly globular or spherical [9,10]. A schematic representation of such a structure is given in Fig. 2. However, the exact structure of micelles is still somewhat controversial. There is evidence, for example, that micelles have a rough surface with considerable penetration of water between head groups [10].

Although an oversimplification, Fig. 2 is a useful model for qualitative understanding of experimental results. Hydrophobic cores of micelles have diameters of about 10–30Å. The charged coat of ionic micelles, called the Stern layer, is usually 60–90% neutralized by counterions in aqueous surfactant solutions without added salt [10]. The surface charge of ionic micelles results in an electrical potential on the order of 100 mV at the micelle-water interface with the same sign as the surfactant head group [11]. If salt is added to the solution, the surface potential is partly neutralized. This decreases coulombic repulsion between adjacent head groups and allows the formation of larger micelles. A solution having a single, very narrow, distribution of micellar sizes is often called monodisperse.

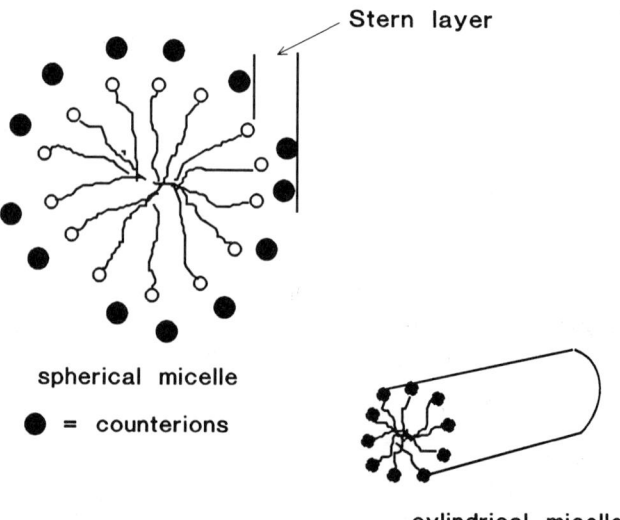

FIG. 2. Oversimplified conceptual structures of spherical and cylindrical micelles. Circles are head groups or counterions; wiggly lines are hydrocarbon tails.

Micelles and Microemulsions

The radius of a micelle corresponds roughly to the length of the extended hydrocarbon chain. The average number of monomers (n_{ag}) in micelles in a given population distribution is called the aggregation number and is typically 30–200 in water. If n_{ag} is known, the micellar core volume (V_c) in $Å^3$ can be estimated from [24]:

$$V_c = m'(27.4 + 26.9n'_c) \qquad (1)$$

where m' is the total number of hydrocarbon chains in the micelle and n'_c is one less than the number of carbon atoms per chain. For a single-chain surfactant, $m' = n_{ag}$; for a double-chain surfactant $m' = 2\,n_{ag}$.

As concentrations of surfactant or salt (or both) in water are increased, globular micelles gradually turn into larger, rodlike micelles. Under some experimental conditions, spherical and rodlike micelles coexist in the same solution. An example is a solution of 0.1 M CTAB and 0.1 M KBr in water [25]. Such systems containing two distinct distributions of micellar sizes are called polydisperse. At higher concentrations of surfactant or salt, generally above 0.1 M, rodlike micelles begin to predominate. Finally, at very high surfactant concentrations, lamellar liquid crystal phases may be formed [10].

For common water-soluble surfactants like SDS and CTAB, micellization is spontaneous; its standard free energy ($\Delta G°$) is negative. Recall that

$$\Delta G° = \Delta H° - T\Delta S° \qquad (2)$$

Up to 90% of the free energy of micellization is contained in the $T\Delta S°$ term [2]. The entropy of micellization is large and positive. This has been interpreted in terms of minimizing entropically unfavorable interactions between hydrogen-bonded water structures and the hydrocarbon chains of the surfactant by bringing the chains together in the core of the micelle. This association of alkyl tails has been termed the hydrophobic effect [24], or hydrophobic interaction.

Not all surfactants form micelles in water. Depending on structure, some surfactants disperse in water as lamellar liquid crystal phases or vesicles. Structural factors governing the types of aggregates formed will be discussed in more detail later. The practical result is that water-soluble, single-chain surfactants such as SDS, CTAB, and polyoxyethylene alcohols form micelles in water. Double-chain surfactants such as didodecyldimethylammonium bromide (DDAB), dihexadecylphosphate, and many phospholipids are insoluble in water and do not form micellar structures.

2. Micellar Dynamics

An important consideration for electrochemistry is that micelles exist in a highly mobile equilibrium with free surfactant monomers, present at concentrations nearly equal to the CMC. Ejection of one monomer from a micelle occurs on the µs time scale, and its recapture by micelles occurs at close to diffusion-controlled rates (i.e., $k = 10^8$–3×10^9 M/sec) [26,27]. Recapture rates are nearly independent of the type of surfactant, but exit rates decrease with the length of the hydrocarbon tail. Residence time of a monomer in a micelle increases about threefold with the addition of each methylene group to the tail [27], another illustration of Traube's rule.

A slower relaxation process is associated with equilibria between micelles and submicellar aggregates. Just above the CMC, this occurs on the millisecond time scale. The slower relaxation time (υ_2) depends on surfactant concentration and ionic strength [27]. As concentration of ionic surfactants is increased up to 0.1 M, $1/\upsilon_2$ goes through a minimum, then increases. Addition of salt has a similar effect. For nonionic surfactants, $1/\upsilon_2$ increases continually with concentration. These observations suggest that interactions between submicellar aggregates to give larger aggregates become faster at higher concentrations.

Equilibria between solutes and micelles are of primary importance in electrochemical studies. Rates of solute entry into and exit out of micelles have been studied extensively by fluorescence decay, flash photolysis, and pulse radiolysis [9,10,26,27]. A highly dynamic picture similar to micelle-monomer equilibria has emerged. For solutes such as aromatic hydrocarbons, halogenated aromatic hydrocarbons, ketones, and aliphatic dienes, entry rate constants from SDS and CTAB micelles are very fast, essentially at diffusion control (ca. 10^{10} M/sec) [26,27]. Entry rates do not depend much on molecular structure. Exit rate constants range from about 10^3 to 10^7 per sec. These decrease with decreasing solubility of the solute. The addition of hydrocarbon tails to solutes tends to decrease the exit rate by about threefold for each CH_2 group. Since the recapture rate is relatively independent of structure, longer-chain solutes are likely to have higher binding constants $K = k_{entry}/k_{exit}$ and longer residence times in micelles [27].

On average, nonpolar solutes like aromatic hydrocarbons are considered to reside close to the hydrophobic side of the Stern layer [10,26] near the surface of the micelle. Microenvironments of solutes like pyrene are considerably more polar in ionic micelles than in hydrocarbon solvents, but less polar than in water [29].

Multiply charged ionic solutes can have strong coulombic attraction to surfaces of oppositely charged micelles. For example, tris(2,2'-bipyridyl)-cobalt(II), $Co(bpy)_3^{2+}$, binds strongly to SDS micelles. However, charge repulsion between ions and ionic micelles of the same charge sign can be overcome by strong hydrophobic interactions. This is illustrated by significant binding of $Co(bpy)_3^{2+}$ to CTAB micelles in 0.1 M KBr [28]. In general, both hydrophobic and coulombic interactions are important for solubilization.

3. Reverse Micelles

Surfactants dissolved in apolar solvents form aggregates with the tails facing out towards the solvent. These are called reverse micelles. Such systems usually have small amounts of water present from impurities in the solvent or surfactant. These water molecules will be strongly associated with the head groups of the surfactant [30]. The head groups are in the interior of the reverse micelle and are effectively shielded from unfavorable interactions with the apolar solvent. Electrochemical experiments in reverse micellar solutions require ultramicroelectrodes because of high resistance of the medium [23].

4. Kinetics in Micellar Systems

The use of surfactant structures to alter or enhance reaction rates has been known for several decades [9]. More recently, surfactant structures have been used to control reaction pathways. An example of wide interest is the use of ionic surfactants to facilitate charge separation and retard back reactions in sensitized photochemical generation of hydrogen from water [11].

Rates of chemical reactions (R_{obs}) in micellar solutions are usually considered to be the sum of rates in the continuous aqueous phase (R_w) and the micellar "pseudophase" (R_m):

$$R_{obs} = R_w + R_m \tag{3}$$

For a pseudo-first-order decomposition of reactant A, Eq. (3) yields

$$k_{obs} = f_w k_w + f_m k_m \tag{4}$$

where f_w and f_m are fractions of A in the water and micellar phases, respectively. Rate constants k_w and k_m also refer to the respective phases. Since $f_w = 1 - f_m$, we have:

$$k_{obs} = (1 - f_m)k_w + f_m k_m \tag{5}$$

Expansion of f_m in terms of binding equilibria of A with micelles gives expressions for k_{obs} as a function of micelle concentration. This approach provides a way to get rate constant k_m in the micelles [10,26].

Consider a second-order reaction of A and B in aqueous micelles. In the simplest case, where both reactants are bound completely to micelles, the rate of the reaction in the water phase is negligible and

$$R_{obs} = [A][B]k_{obs} = [A]_m[B]_m k_m \tag{6}$$

where observed rate constant k_{obs} is found on the basis of the moles of A and B in the total system volume V_t. However, k_{obs} estimated in this fashion will depend on the volume available to the reactants. Actual concentrations in the micelles are approximately $[A]_m = [A]/\phi_m$ and $[B]_m = [B]/\phi_m$, where $\phi_m = V_c/V_t$, the volume fraction of the micellar core [see Eq. (1)]. Thus,

$$k_{obs} = \frac{k_m}{\phi_m^2} \tag{7}$$

Eq. (7) reflects the fact that the observed rate is enhanced by compartmentalization of reactants into the smaller reaction volume $V_t\phi_m$, leading to an apparent enhancement of reaction rate. This effect on bimolecular reactions is sometimes called micellar catalysis. In most cases, it is mainly a concentration effect. Usually, k_m is found to be the same order of magnitude as rate constants for the same reaction in isotropic homogeneous solutions. However, bimolecular rate enhancement in micelles is real and of considerable practical importance. Although much more sophisticated treatments of kinetics in micellar systems are available [26], the simple concepts described above are quite useful for understanding chemical reactions coupled to charge transfer at electrodes in micellar solutions.

B. Microemulsions

Pure water and pure oil don't mix. However, addition of surfactant, sometimes with a cosurfactant, to oil/water systems can lead to thermodynamically stable, optically clear fluids called microemulsions (2,5,10). The term "microemulsion" is usually attributed to Schulman, who found in 1955 that addition of medium-chain alcohols such as hexanol to coarse emulsions of oil in water stabilized by ionic surfactants produced clear, stable emulsions with colloidal-sized droplets that appeared to the eye to be homogeneous solutions [5]. Microemulsions are important for practical applications, including detergency and secondary oil recovery [31]. These fluids with intimately mixed water and oil phases are also useful for bringing together reactants of widely different polarities [22] and for making colloidal particles [31c].

Oil-in-water (o/w) microemulsions have a continuous water phase, with surfactant surrounding the oil in globular aggregates resembling swollen micelles (Fig. 3). These microheterogeneous fluids are conductive when made with ionic surfactants and can be used for all sorts of electrolytic experiments. O/W microemulsion droplets are larger than micelles and often have a larger capacity for solubilizing nonpolar solutes [30–33].

Water-in-oil (w/o) microemulsions have a continuous oil phase, with surfactant surrounding the water in microdroplets. The structures of these so-called water pools are similar to inverse micelles, with surfactant head groups

Micelles and Microemulsions

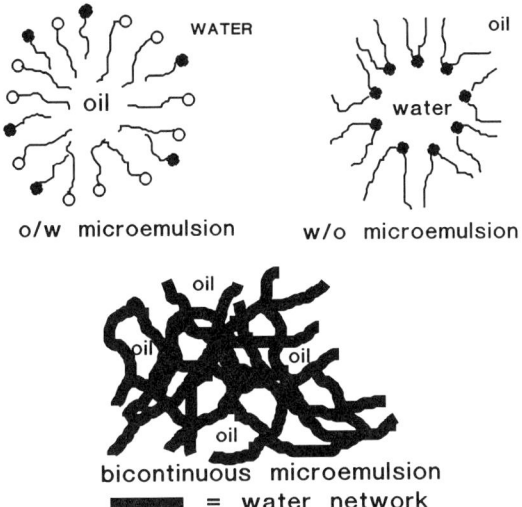

FIG. 3. Oversimplified conceptual structures of three types of microemulsions.

facing toward the water and hydrocarbon tails extending out into the continuous oil phase [30]. These fluids are nonconducting, and ultramicroelectrodes must be used for voltammetric measurements [22].

Bicontinuous microemulsions are a third structural variant. These systems have both oil and water as continuous phases with surfactant residing at extended oil-water interfaces. Because of the continuous water phase, bicontinuous microemulsions of ionic surfactants conduct electricity. Three component bicontinuous microemulsions have recently attracted considerable attention because of their relative simplicity. They are microheterogeneous two-phase networks with oil and water separated by a monolayer of surfactant [32]. In principle, they should be usable in electrochemical studies with electrodes of any size.

Similar to micelles, surfactant-coated droplets in w/o and o/w microemulsions are in dynamic equilibria with their components in the continuous phase. Solutes also exist in dynamic equilibria with microemulsion droplets. Time scales of solute entry and exit for droplets in microemulsions are similar to those in micellar solutions [27].

C. Vesicles and Lamellar Dispersions

Some water-insoluble surfactants that do not form micelles, usually those with two or more hydrocarbon tails, can be suspended in water as vesicles or

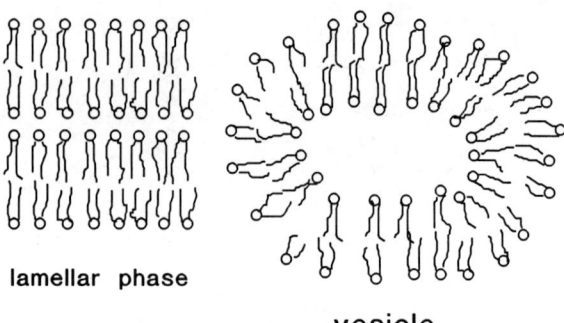

FIG. 4. Idealized structures of lamellar phase and unilamellar bilayer vesicle formed from insoluble surfactants.

lamellar liquid crystal phases. Vesicles are closed bilayer structures which often have several nested compartments, resembling the skin of an onion. Sonication of these systems often gives single compartment structures (Fig. 4). Vesicles made from phospholipids are called liposomes and are important models for biological membranes.

Lamellar liquid crystal phases formed from insoluble surfactants such as DDAB (see Fig. 1) are layered structures composed of interleaved surfactant bilayers and water. Like micelles, dispersions of these systems can be useful for solubilization.

Vesicles are much larger than micelles or microemulsion droplets. They are typically on the order of 50 nm long and 5 nm wide. Vesicles are stable upon dilution. Equilibria involving surfactant monomers are much slower than for micelles. For this reason, water-soluble solutes can be incorporated in the large inner water pools of the vesicles, and they escape only slowly. Photochemically induced electron transfer across vesicle bilayers has been studied extensively [10,11,26].

D. Surfactant Structure Controls System Architecture

The surfactant packing parameter is $v/a_o l_c$, where a_o is to a first approximation the area per head group in a bilayer configuration, l_c is the optimal chain length, and v is the volume of the hydrocarbon tail region per surfactant [32]. The last two quantities are found from the equations:

$$l_c = 1.5 + 1.26n \text{ Å}$$
$$v = 27.4 + 26.9n' \text{ Å}^3$$

Micelles and Microemulsions

where n' is one less than the number of carbon atoms in the chain. Of course, v for a double-chain surfactant is twice that of a single-chain surfactant with the same length tail.

The following rules have been derived for predicting the dependence of structure on the surfactant packing parameter [30,32]:

$v/a_o l_c$	System Architecture
< 1/3	spherical micelles
1/3 to 1/2	rod-shaped micelles
1/2 to 1	vesicles or bilayers, 3-component o/w and bicontinuous microemulsions
> 1	reverse micelles, w/o microemulsions

Thus, the packing parameter allows the choice of a surfactant for a desired type of organized fluid system from its molecular dimensions.

Here are some examples of how the rules work. For SDS, a micelle-forming surfactant, $v/a_o l_c < 1/3$, indicating spherical micelles. The area per head group of ionic surfactants can be decreased by adding salt, neutralizing head groups on the micelle's surface, and allowing them to get closer together. When enough salt is added to an SDS solution, a_o decreases, $v/a_o l_c$ increases above 1/3, and rod-shaped micelles form. The double-chain surfactant DDAB has $v/a_o l_c = 0.82$; it does not form micelles. DDAB exists in lamellar dispersions in water and readily forms three-component o/w and bicontinuous microemulsions. These rules provide predictions only; exceptions may occur.

E. Adsorption of Surfactants on Solid Surfaces

As will be seen, adsorption of surfactants on electrodes can have a profound influence on electrochemistry in fluids organized by surfactants. Since electrode surfaces are charged, we briefly review related work on adsorption of surfactants on charged surfaces. Two aspects are of prime importance: (1) adsorption isotherms, i.e., dependence of the amount of surfactant adsorbed on its concentration in solution, and (2) the structure of the adsorbed film on the surface.

The Langmuir isotherm is usually considered the "ideal" model for adsorption. It assumes that the adsorbate on the surface acts as an ideal two-dimensional solution, that there are no interactions between adsorbed molecules, and that the surface is smooth and homogeneous. For a dilute solution of adsorbed species i, the Langmuir isotherm can be written [5].

$$\theta = b_i C_i / (1 + C_i) \tag{8}$$

where $\theta = \Gamma_i / \Gamma_t$, the fractional coverage of adsorbate on the surface, Γ_i is the surface concentration of i at solution concentration C_i, and Γ_t is the surface concentration for full coverage of the surface. The constant b_i is directly related to the equilibrium constant, which is in turn related to the negative exponent of the standard free energy of adsorption [2]. The larger the b_i, the more negative the free energy of adsorption or the lower the interfacial tension. Experimentally, the amount of surfactant adsorbed onto polar solid surfaces from aqueous solutions at a given concentration increases strongly as chain length increases [5] according to Traube's rule. This reflects an increased surface activity per CH_2 group.

Most surfaces are microscopically heterogeneous. Further, interactions between adsorbate molecules become significant as fractional coverage of the surface approaches unity. Thus, the Langmuir equation often fails to fit experimental data. Many other adsorption isotherms have been developed to account for these nonidealities, often with the inclusion of an interaction parameter. General discussions of these isotherms can be found in standard texts [5,34].

Adsorption of surfactants on charged surfaces is of particular relevance to electrochemical applications. Much work has been done on adsorption of surfactants onto inorganic oxide surfaces such as alumina, silica, and titanium dioxide. The charge at the solid-water interface is controlled by pH, making use of the surface processes:

$$MOH_2^+ = MOH + H^+ \tag{9}$$
$$MOH = MO^- + H^+ \tag{10}$$

For amphiphilic ions of opposite charge to the surface in aqueous solution, the first stage of adsorption is thought to result from neutralization of surface charge on the oxide. A commonly observed isotherm for adsorption of anionic surfactants on positively charged alumina surfaces has at least three distinguishable regions (Fig. 5) [35–37]. The shape of this isotherm is clearly not described by the Langmuir equation. Results can be explained in terms of the hydrophobic effect, the heterogeneity of the surface, and aggregation of surfactant on the surface. At extremely low concentrations (region I), surfactants adsorb as individual ions and have little interaction with one another. In region II, where adsorption increases sharply with concentration, surface aggregates begin to form.

Micelles and Microemulsions

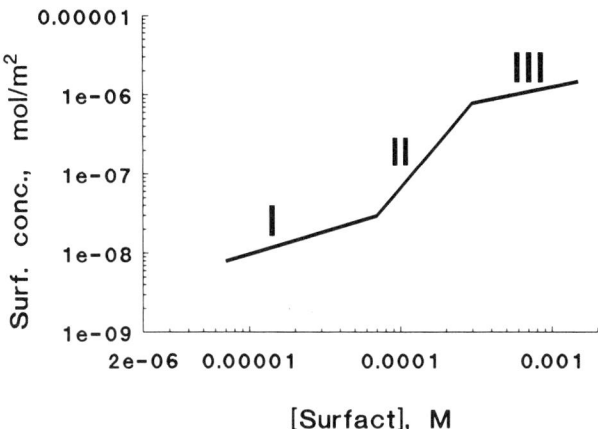

FIG. 5. Typical adsorption isotherm of amount absorbed vs. concentration in solution for anionic surfactants on positively charged metal oxides.

Scamehorn et al. [38] and Harwell et al. [36] suggest that in region II tail-to-tail bilayers form on patches of the surface with the highest charge densities. The outer layer of surfactant has its charged head groups facing the water phase. Other regions of the surface may be covered with individually adsorbed surfactant ions. When the high charge density patches are covered with bilayers, sometimes called admicelles, the other regions of the surface begin to get coated until complete bilayer coverage is reached at the plateau of region III. The initial positive surface charge (zeta potential) of alumina particles changes to a negative surface charge during transition from region I to III of the isotherm [35]. The negative zeta potential in region III is consistent with anionic bilayers of surfactant adsorbed to the alumina surface.

Cetyltrimethylammonium bromide (CTAB) adsorbed on colloidal silica forms bilayers at concentrations above the CMC, yielding a hydrophilic surface [39]. Neutron reflection results of adsorbed CTAB on an amorphous silica plate at concentrations just below the CMC (0.92 mM) were best fit by a bilayer model [40]. At 0.3 mM CTAB, 35% of the interface was covered by bilayer, while at 0.6 mM the value increased to 80%. The mean size of the surface aggregates was 1 μm. The measured adsorbate thickness of 28 Å is considerably smaller than twice the surfactant chain length, 43 Å, suggesting that surfactant tails were intermingled or tilted at an angle to the surface.

Results discussed above suggest that the same hydrophobic effect, which is the driving force for forming micelles, is also operative for bilayer formation on solid surfaces. Surfactant molecules can avoid thermodynamically unfavorable interactions of their tails with water by forming a bilayer on the surface, with head groups facing the water phase. Bilayer structures on surfaces are called admicelles [36]. Aggregates with head groups down on the surface and hydrophobic tails facing the water phase are sometimes called hemimicelles. Exact details of molecular arrangements of these surface structures are still controversial.

IV. ELECTROCHEMISTRY IN MICELLAR SOLUTIONS

A. Adsorption of Surfactants

Research on the influence of surface-active agents on the kinetics of electron transfer reactions at electrodes spans half a century [3–5,8,12–15,41]. Most early work on this topic used very low concentrations of surface-active agents and did not address solubilization of reactants. Moreover, older work often lumped together all species that adsorbed strongly onto electrodes. Thus, effects of gelatin, medium-chain-length alcohols, β-naphthol, and camphor were grouped in the same category with amphiphilic surfactants defined in the modern sense, such as CTAB, Triton X-100, and SDS. Much early work was done on mercury electrodes, especially the DME. Historical examples were given in Sec. II. Until recently, few studies trying to directly elucidate the supramolecular structures of surfactant adsorbate films were made.

A good part of the research on the influence of adsorbates on heterogeneous electron transfer rates addressed the problem in terms of inhibition of electron transfer and electrostatic interactions. The effect of adsorbed surfactants on electrodes on the apparent heterogeneous rate constant k_{app}^o has been expressed by [41, 42]:

$$k_{app}^o = k_o(1 - \theta) + \theta k_1 \tag{11}$$

where k_o is the rate constant on the bare electrode, k_1 is the rate constant on a fully covered electrode, and θ is fractional surface coverage (see Eq. (8)). However, many experimental systems deviate from Eq. (11). Work prior to 1978 on this topic has been summarized [41].

More recent work concerning surfactants adsorbed from micellar solutions has focused on elucidating, or utilizing, aggregate structures formed on the electrode. In principle, if structures, dimensions, and polarities of interfacial aggregates and the positions of electroactive centers within them are known, the effects on electrochemical kinetics can be predicted by using modern theories of electron transfer [43]. Such predictions would take into account the influence of distance of ET and the environment surrounding the

Micelles and Microemulsions

reactants [44]. It is quite possible that the dynamics of the surfactant aggregates also play a role.

We are concerned here mainly with adsorption from a surfactant solution in which the electrochemical reaction takes place. However, a parallel line of research is being developed that employs chemisorption of special functional groups on surfactant molecules to form ordered, chemically bound, mono- and multilayer films of surfactantlike molecules on electrodes. Examples include chemisorption of long-chain thiols, disulfides [45], and alkyltrichlorosilanes [46] onto gold, platinum, and carbon electrodes. The chemisorption is usually done from dry organic solvents, and the coated electrode is then studied in a different electrolyte solution. Although structural studies on these films are related to those of films adsorbed from surfactant solutions, an in-depth discussion of this rapidly growing research area is beyond the scope of this review.

As mentioned previously, interfacial tension (free energy) is minimized by adsorption of surfactants. The drop time (t_d) of dropping mercury electrodes is directly proportional to interfacial tension (γ). Thus, γ is easily found at the mercury-solution interface by measuring t_d of a DME. When this is done as a function of applied potential, the resulting plot of t_d vs. E is called an electrocapillary curve. This method has been used extensively to study adsorption at the DME [3,7,34].

Differential capacitance at the electrode-solution interface is also lowered upon adsorption of surfactants. On mercury electrodes, the interfacial capacitance can be obtained from electrocapillary data via Lippman's equation [34,47,48]:

$$\sigma^M = -(\delta\gamma/\delta E)_\mu \tag{12}$$

where σ^M is the excess charge on the electrode at constant chemical potential μ of the solute. σ^M is the slope of the electrocapillary curve at any E. The differential capacitance (C_d) is the slope of a plot of σ^M vs. E at any E [34]. Surface concentrations can be obtained as the negative slope of γ vs. concentration plots [47]. C_d can also be measured directly vs. potential by ac bridge methods or by phase selective ac voltammetry [47,48]. In principle, differential capacitance can be measured at any electrode surface. In practice, for electrodes other than mercury, reproducibility and cleanliness of surfaces becomes quite problematic [49].

For solid electrodes, spectroscopic techniques such as raman, UV-VIS, FT-IR, and fluorescence, as well as ellipsometry, have been used for in situ studies of surfactant adsorption, as discussed below. Useful results can sometimes be obtained from ultra-high-vacuum (UHV) spectroscopy, although in situ use of UHV methods is not yet feasible.

1. Nonelectroactive Surfactants: Electrochemical Studies

A major problem with electrocapillary and differential capacitance measurements in surfactant solutions is the interpretation of the results in terms of adsorbate structures. Classical Gouy-Chapman theory rarely holds for electrodes in surfactant solutions [47]. C_d and γ contain no structural or molecular information in themselves; surface structures must be inferred from models for the data. Also, many of these types of studies have been done for very low surfactant concentrations, near and below the CMC, and may be of limited use in extrapolating to the higher concentrations necessary for many practical applications of surfactant solutions in electrochemistry [4,22].

Having pointed out the capabilities and limitations of differential capacitance measurements, we now discuss their use in studies of surfactant adsorption. In some cases, results can be compared with in situ spectroscopy.

Schuhman et al. [47] summarized results on differential capacitance at mercury interfaces in solutions containing anionic surfactants, specifically sodium octylbenzene sulfonate and sodium alkylsulfates. Experiments were done at concentrations of surfactant up to 0.012 M. A common feature of the data was a peak attributed to desorption of the anionic surfactant at a σ^M considered to reflect the surface charge on micelles in solution. Surface charges estimated in this way for alkylsulfate micelles changed by about -15 mV per CH_2 group in salt-free solutions [47]. Alcohols changed the desorption potential. At zero σ^M (i.e., the electrode's point of zero charge, or PZC), the authors proposed a micellelike structure on the electrode surface. Similar results were discussed for SDS on Hg by Shinozuka and Hayano [41], who interpreted shapes of C_d vs. E plots in terms of reorientation of surfactant on the electrode in response to changes in applied potential.

Besio et al. [50] used ellipsometry to measure the thickness of adsorbed SDS films on platinum electrodes at different applied potentials. SDS concentrations below the CMC were used. Below about 0.4 V vs. Ag/AgCl, the thicknesses measured suggested the approximate dimensions of SDS lying flat on the surface. Above 0.6 V vs. Ag/AgCl, at concentrations well below the CMC, thicknesses corresponded to that of a surface bilayer or hemimicelle [50]. At 1 mM SDS, still below the CMC and the CHMC (critical hemimicelle concentration), multilayer adsorption was induced at the higher potentials.

Kaisheva et al. [51] studied adsorption of nonionic dodecylhexaoxyethyleneglycol monoether at concentrations $\le$ 1 mM by differential capacitance at a stationary Hg electrode. Peaks at positive and negative potentials were interpreted in terms of desorption of surfactant because C_d returned to values characteristic of the electrolyte at potentials beyond these peaks.

Micelles and Microemulsions

Results suggested that full coverage of the mercury at the PZC occurred at a surfactant concentration of 1 µM, slightly below the micellar CMC. Shinozuka and Hayano [41] discussed structure of differential capacitance plots in solutions of nonionic surfactants in terms of adsorption of monomers and "hemimicelles" and structural reorientations driven by potential changes.

Adsorption of cationic alkyltrimethylammonium bromides and chlorides on Hg has also been studied by differential capacitance. C_d vs. E data in the medium potential range were again interpreted in terms of reorganizations of surface aggregates [41]. Peaks at high and low potentials were suggested to indicate desorption of surfactant at low concentrations (< 1 mM) [52–55]. However, C_d and ellipsometric delta values for 3–4 mM decyltrimethylammonium chloride in aqueous KCl did not return to those of electrolyte alone beyond the more negative peak. Hayter and Hunter [52] interpreted C_d values in surfactant solutions that were smaller than those with electrolyte alone at very negative potentials as suggesting bilayer or multilayer coverage of the electrode.

Aqueous solutions of tetraalkylammonium surfactants have been shown to provide excellent micellar media for electrochemical studies. The positive limit in CTAB and CTAC solutions on solid electrodes is about +0.8 V (vs. SCE), presumably where halide is oxidized. The negative potential window extends to values of about −2.3 V on Hg and glassy carbon [21,25,56,57]. Octadecyltrimethylammonium perchlorate was found to adsorb strongly to Pt and iodine-coated Pt electrodes and to shift potentials for hydrogen evolution to more negative values [58].

Voltammetry on hydrophobic solutes in micellar solutions has also provided information about surfactant adsorption in electrodes. One cautionary note is that solubilization of some water-insoluble solutes, for example, ferrocene, anthracene, and their nonpolar derivatives, may take considerable time. Equilibrations of 3 or more days at 30°C may be needed to solubilize 1 mM or more of such water-insoluble solutes [25,56]. Ultrasonication can be used to increase the rate of solubilization. Equilibration can be followed by measuring voltammograms as a function of time on a scale of hours until no further changes are seen. Solutions should be equilibrated at the temperature of experimentation, since rapid temperature changes may result in changes in sizes and shapes of surfactant aggregates in solution, which also require time to equilibrate.

Adsorption of nonelectroactive surfactant on electrodes can be detected via the influence on the voltammetry of a hydrophobic electroactive probe. Examples are the one-electron reductions of anthracene, 9-phenylanthracene (9-PA), and 9,10-phenylanthracene, occurring at about −2.2 V vs. SCE at 30°C in 0.1 M CTAB/0.1 M TEAB on Hg electrodes. Cyclic voltammograms

(CV) of fully equilibrated solutions showed surface features at very low scan rates ($\leq$ 0.002 V/sec [56], but reversible diffusion-controlled CVs were observed at scan rates (v) between 0.01 and 51 V/sec. Apparent diffusion coefficients measured by CV and chronocoulometry were on the order of 10^{-3} cm^2/sec for these probes in the CTAB system. These anomalously large D' values coupled with the observation of surface voltammetry at $v \leq 2$ mV/sec suggested that the electrochemistry took place in a thick film of CTAB containing a high concentration of the phenylanthracene. Comparison of calibration curves for 9-PA in 0.1 M CTAB and in N,N-dimethylformamide (D = 9×10^{-6} cm^2/sec) at scan rates where the electrode reaction is diffusion controlled shows a much larger slope in the surfactant solution (Fig. 6) [57]. This reflects the preconcentration of 9-PA in the surfactant film on the electrode.

The micellar hydrophobic core volume in 0.1 M CTAB estimated from Eq. (1) is 2.6% of the total volume of the solution. Assuming that all the 9-PA resides in this hydrophobic volume, an effective concentration of 9-PA in the CTAB film on the electrode can be obtained. This was used with D' to estimate a perfectly reasonable effective diffusion coefficient of about 10^{-6} cm^2/sec for 9-PA in the CTAB film on the electrode [56].

Similar voltammetry reflecting preconcentration of 9-PA in films of nonionic polyoxyethylene surfactants were also found on Hg electrodes at about -2.17 V vs. SCE. In contrast, electroactive probes such as 1,2-dicyanobenzene [21], perylene, and tetracene [59], all with E°s well positive of -2 V, gave CV peaks in 0.1 M CTAB solutions suggesting apparent diffusion coefficients considerably smaller that in homogeneous solvents. No

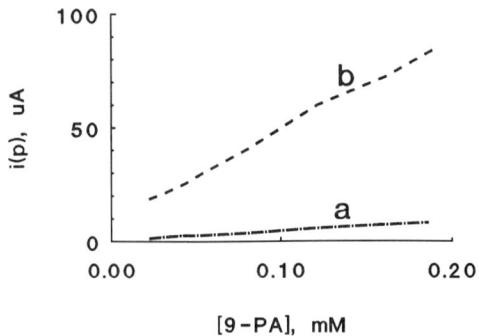

FIG. 6. Calibration curves for 9-phenylanthracene by CV at 0.1 V/sec at a HDME: (a) in 0.1 M TEAB in N,N-dimethylformamide; (b) in 0.1 M CTAB/0.1 M TEAB. (Adapted from Ref. 57.)

Micelles and Microemulsions

evidence for strong adsorption of these species was found. In addition, perylene and tetracene gave chemically irreversible CVs in 0.1 M CTAB. The above species are not preconcentrated in a surfactant film on the electrode at potentials near their E^os, presumably because the correct type or thickness of surfactant film does not form at potentials more positive than about -2 V.

The influence of medium-chain-length alcohols on surface electrochemistry in CTAB solutions has been investigated by voltammetric methods [59]. Sharp peaks in capacitance current vs. potential curves were observed at mercury and silver electrodes in aqueous micellar CTAB solutions containing straight-chain alcohols of two to five carbons (Fig. 7a). These peaks were attributed to rapid structural reorganization of a mixed alcohol/CTAB layer on the electrode surface. Longer-chain alcohols gave capacitance peaks at more positive potentials and at lower concentrations, following Traube's rule. When type and concentration of alcohol were adjusted so that capacitance peak potentials were matched with the reduction potentials of 1,2-dicyanobenzene, perylene, or tetracene, cyclic voltammograms in alcohol/CTAB/water were symmetric and reversible (Fig. 7). This suggested reversible reductions in a CTAB/alcohol layer adsorbed on the electrode surface. Anion radical products of these one-electron reductions were much more stable in the CTAB/alcohol film on the electrode than in aqueous CTAB micelles. This is consistent with a decrease in polarity of the microenvironment for the electroactive solutes by incorporation of alcohol in the surfactant layer on the electrode. This decrease in polarity compared to CTAB solutions without alcohol may reflect exclusion of water from solubilization sites on the electrode surface.

2. Nonelectroactive Surfactants: In Situ Surface Spectroscopy

Surface-enhanced raman spectroscopy (SERS) was used by Sun et al. [60] to study adsorption of surfactants on rough silver electrodes. Interpretation of the C-C stretching region of the spectra suggested that CTAB was adsorbed from aqueous micellar solutions on Ag electrodes in an ordered, solidlike phase with hydrocarbon chains in all-*trans* configurations. SERS spectra of CTAB/Ag at -0.9 V suggested a surface-bound head group and an extended, close-packed tail region. An orientation with head groups down was indicated even at potentials positive of the potential of zero charge (PZC) of silver. The presence of a bilayer was not inferred or ruled out from the spectral data. The second layer of a bilayer might be too far away from the Ag surface to contribute to SERS.

SERS intensities showed a dramatic dependence on applied potential (Fig. 8). Intensities for CTAB/Ag reached a large maximum at about -1.3

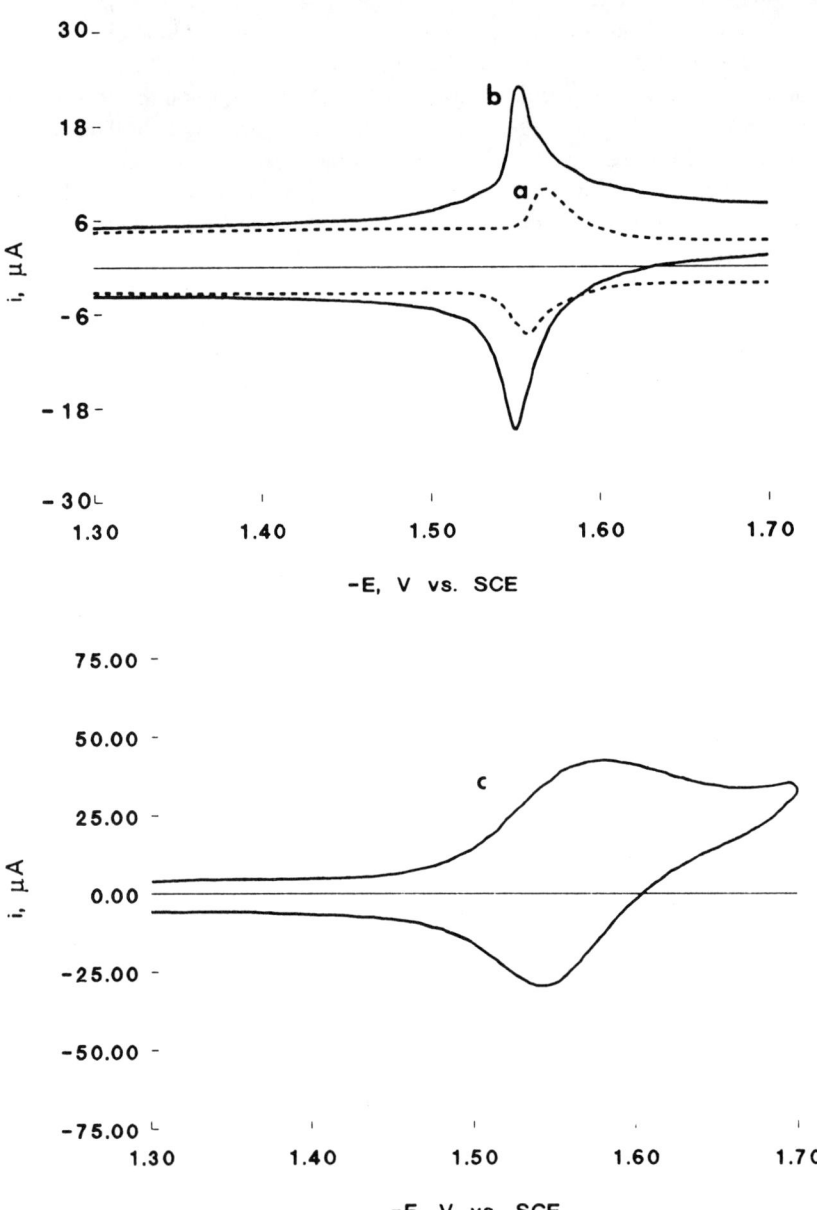

FIG. 7. CVs at HDME in 0.1 M CTAB/0.1 M TEAB at 10 V/sec for (a) 0.2 M 1-butanol; (b) 0.1 mM 1,2-DCB and 0.2 M 1-butanol; (c) 1 mM 1,2-DCB, no butanol added. (From Ref. 59.)

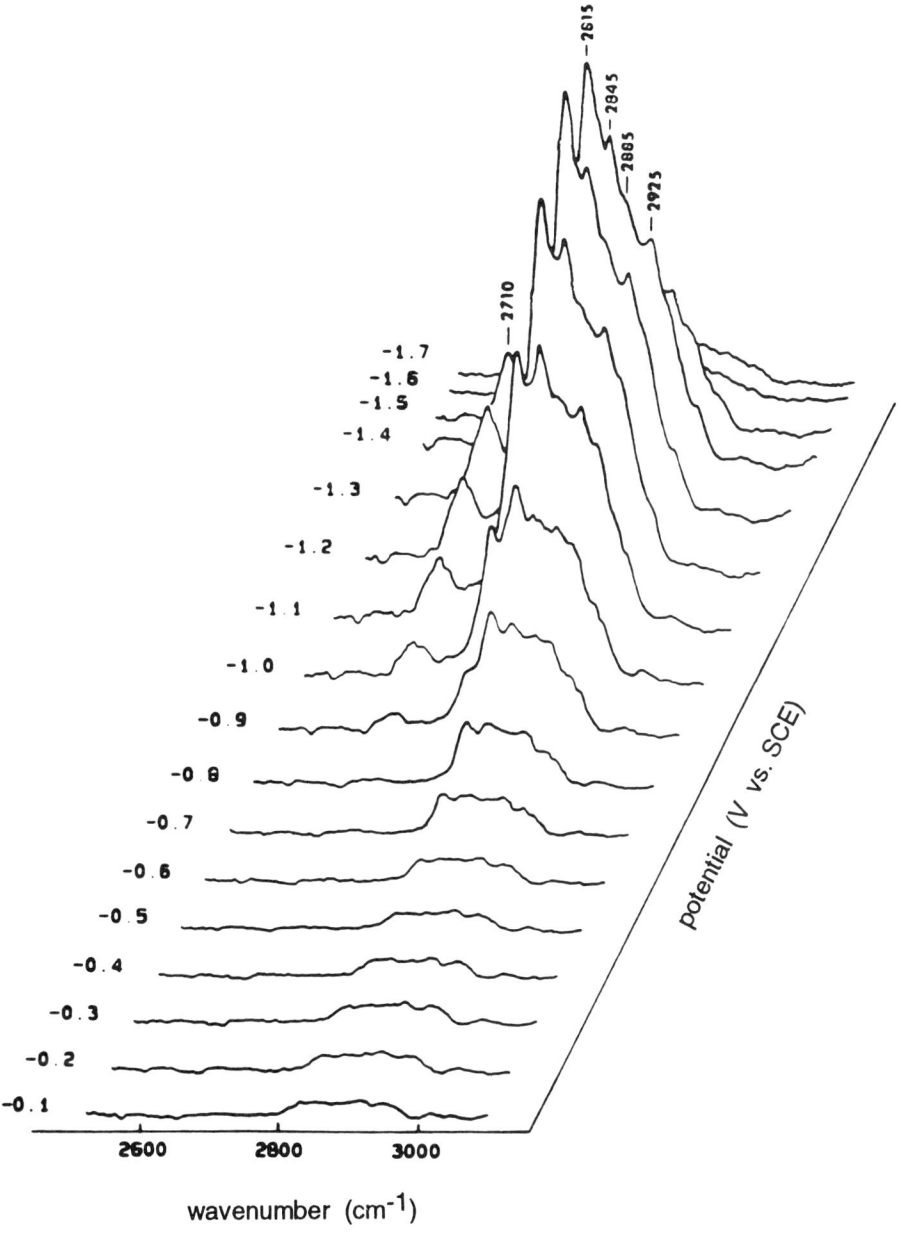

FIG. 8. SERS spectra of 10 mM CTAB on a Ag electrode in the 2700–3100 cm^{-1} region at different potentials. (From Ref. 60.)

V, then decreased at more negative potentials. A reorientation of structure of the CTAB adsorbate was suggested as the possible cause for this maximum [60].

Adsorption of CTAC, the chloride salt of CTA, was studied by surface fluorescence probe spectroelectrochemistry [61]. In this method, a fluorescent probe molecule that binds strongly to surfactant aggregates is introduced into a micellar solution in a thin layer spectroelectrochemical cell. Fluorescence spectra are measured by directing incident light normal to the electrode and collecting emitted light at a 22.5° angle of detector to the incident beam. In aqueous micellar solutions of 0.1 M CTAC, the probe pyrene is strongly associated with CTAC. The maximum fluorescence of pyrene in 0.1 M CTAC/0.05 M NaOH on Ag electrodes was found at potentials of about -1.3 V vs. SCE, slightly positive of those causing electrolytic evolution of hydrogen gas. This is about the same as the potential for the maximum in SERS intensity discussed above. The potential dependence of pyrene fluorescence was interpreted in terms of the potential dependence of the amount of adsorbed surfactant on the electrode and/or to structural reorganization of surfactant aggregates on the surface.

In both SERS and fluorescence probe studies, changes in the adsorbed halide or other anions on the Ag surface may influence the spectral intensities. Also, the decrease in fluorescence intensity at potentials negative of -1.3 V was associated with copious hydrogen evolution on the silver electrode. This process may physically remove surfactant from the surface.

Sun et al. also used SERS to study adsorption of cetylpyridinium chloride (CPC) and nonionic polyoxyethylene surfactants on silver electrodes [60]. Very strong SERS intensity for CPC suggested interaction between the pyridinium head group and the electrode surface. Even at potentials positive of the PZC of Ag, SERS data were consistent with an end-on orientation of the pyridinium ring on the surface. Weaker spectra were observed for polyoxyethylene glycol surfactants Brij-35 and Triton X-100. These nonionic surfactants were also found to have a "head-down" orientation of (-OCH$_2$CH$_2$)OH head groups at potentials positive of the PZC. Brij-35 SERS intensities showed a strong dependence on applied potential similar to that of CTAB. Intensities reached a maximum at -1.3 V, then decreased at more negative potentials. As discussed above, this maximum may be due to structural reorganization of surfactant aggregates on the surface and/or to potential dependence of the amount of surfactant adsorbed.

Dong et al. used absorbance spectroelectrochemistry in a long-pathlength thin-layer cell to measure the adsorption isotherm of CPC on glassy carbon at open circuit [62]. They found that orientation of the surfactant depended on concentration, much the same as for surfactant adsorption on charged metal oxides. At $c < 3 \times 10^{-5}$ M, the pyridinium ring and tail were considered to adsorb parallel to the surface. At $C > 6 \times 10^{-5}$ M, a head-down orientation of the pyridinium ring with tails extending away from the surface was suggested.

3. Adsorption of Electroactive Surfactants

Adsorption of electroactive surfactants from solution can be studied by measuring faradaic currents from reduction and oxidation of the adsorbed species. The amount of adsorbed surfactant can be found by integrating symmetric surface cyclic voltammograms or by controlled potential coulometry [34], providing a way to obtain adsorption isotherms in the potential region of the surfactant's redox reactions. In this section, studies of electroactive surfactants adsorbed from aqueous solutions are reviewed. The discussion is restricted mainly to adsorbates with head groups related to those of common nonelectroactive surfactants. Not included are molecules with reactive sulfur- or silicon-containing substituents chosen specifically to chemisorb to electrode surfaces from dry organic solvents.

Facci studied adsorption of ferroceneylmethyldimethyloctadecylammonium hexafluorophosphate (FODAH) at platinum electrodes coated with iodine atoms [63]. FODAH was more strongly adsorbed than its oxidized, dicationic form. Adsorption isotherms were dramatically different depending on whether FODAH was adsorbed from solutions of sulfuric or perchloric acid (Fig. 9). The plateau of the isotherm at low concentrations (Γ_1) corresponded to an area/molecule value suggesting a flat orientation of surfactant on the electrode. These low concentration data fit the Frumkin isotherm, giving negative interaction parameters suggesting attractive intermolecular interactions consistent with a hydrophobic driving force.

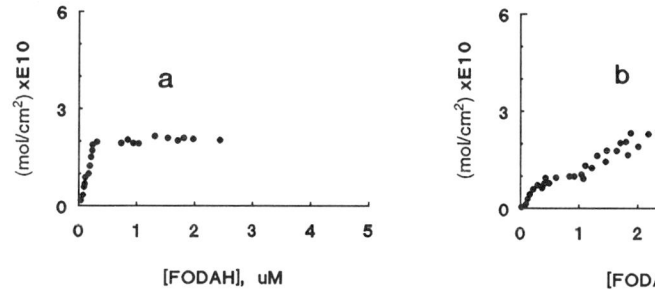

FIG. 9. Adsorption isotherms for ferroceneylmethyldimethyloctadecylammonium hexafluorophosphate (FODAH) on iodine-coated Pt electrodes: (a) from 1 M H_2SO_4; (b) from 1 M $HClO_4$. (Adapted from Ref. 63.)

The second plateau (Γ_2) of the isotherm in sulfuric acid at about $2\Gamma_1$ was interpreted as a second "flat" layer of surfactant. However, the second plateau in perchloric acid occurred at much higher surface and solution concentrations than in sulfuric acid. This was thought to correspond to surface micelles, or hemimicelles, formed on the surface by adsorption from the perchloric acid solution. Contact angles indicate a hydrophobic surface for the bare Pt/I electrode and for electrodes covered with $0.65\Gamma_1$ FODAH, but suggested a hydrophilic, wettable surface for Γ_2 coverage from either electrolyte. The latter results are consistent with exposure of head groups of adsorbed electroactive surfactant to the aqueous phase.

Donohue and Buttry investigated adsorption of C10, C12, C14, C16, and C18 homologs of FODA+ (shown below) on gold by using simultaneous cyclic voltammetry and quartz crystal microbalance (QCM) measurements [64]. The CV/QCM method can measure mass changes and voltammetry

$$(CH_3)_2 \overset{|}{\underset{CH_2Fc}{N^+}} - CH_2(CH_2)_x CH_3$$

simultaneously for electrochemically driven redox reactions by using gold electrodes deposited directly on the QCM crystal. This method showed an immediate loss in mass on the electrode upon oxidation of ferrocene surfactants with alkyl chains $<$ C14, consistent with desorption of the ferrocinium oxidation product. Oxidation yields a dication, which desorbs from the electrode for alkyl chains of $\leq$ 14 carbons, but remains adsorbed for C18. Thus, desorption rate is controlled by chain length. Association of anions with the oxidation products occurred before desorption.

Strength of adsorption of the reduced forms of the above ferrocene surfactants decreased with chain length, following Traube's rule. Adsorption isotherms were obtained by both CV and QCM, and these approximately fit the Langmuir model [64]. For C14 and C16 derivatives, the area per molecule at the plateau of the isotherm corresponded to the area of the head group. Although this would seem to suggest head-down orientations, the authors favored a tail-down orientation on the grounds that the electrode was positive of its PZC. Considering the spectroscopically determined head-down orientations for cationic surfactants positive of the Ag PZC [60], the weight such arguments carry is unknown.

Multilayer coverage (5 layers) of the C18 ferrocene surfactant 1 was measured at solution concentrations as low as 25 μM. Area/molecule values slightly larger than head group areas were found for the C12 and C10 derivatives at full saturation, but these values were too small to imply a flat

Micelles and Microemulsions

orientation. A partly disordered film was suggested [64]. Similar behavior was found for the C12 derivative in 0.2 M Li_2SO_4, pH 3. The C12 derivative was also used to probe the properties of electrodes coated with functional organic monolayers [65].

Diaz and Kaifer reported a brief CV study of adsorption of octadecylethylviologen ($C_{18}VE^+$) on gold and platinum electrodes. They concluded that a head-down orientation prevailed at saturation coverage, presumed to be a monolayer [66]. Widrig and Madja investigated adsorption of a similar electroactive surfactant, octadecylmethylviologen ($C_{18}VM^{2+}$), on gold [67]. From adsorption isotherms obtained by CV, they concluded that monolayer coverage occurred at 2 mM $C_{18}VM^{2+}$ with viologen head groups down and perpendicular to the surface and hydrocarbon tails away from the surface. The free energy of adsorption of about 22 kJ/mole was consistent with the hydrophobic effect as the driving force for adsorption. A mediated coulometric determination of surface coverage enabled detection of adsorbed surfactant that was not directly electroactive [67]. Results of these latter experiments showed that bilayer coverage with an electroactive amphiphile may go undetected by CV. The authors suggested coverage by an intercalated bilayer at concentrations of about 0.01 mM $C_{18}VM^{2+}$. A structure was suggested in which head groups of the outer layer of the bilayer facing the solution are too far away from the electrode to accept electrons directly [67].

4. Summary—Surfactants Adsorbed on Electrodes

The above discussion reveals that structures of surfactant films adsorbed on electrodes from aqueous solutions are as yet poorly understood. A more self-consistent picture seems to have evolved for adsorbates on charged metal oxide surfaces (see Sec. II.E), whose structures may be expected to be similar to those on electrodes. For oppositely charged amphiphilic ions on metal oxides, adsorbed molecules begin to form patches of bilayers even at concentrations well below the CMC. At saturation of the surface, complete bilayers or hemimicelles form. Bilayer-type structures are intellectually satisfying because they are consistent with the hydrophobic effect [24] as the driving force for adsorption, as in formation of micelles and vesicles.

Results above and just below the CMC for ionic surfactants on Pt and Hg electrodes [41,47,50,52,55] are in agreement with formation of bilayers or hemimicelles. Studies with electroactive FODAH led to a similar conclusion at saturation coverage in perchloric acid solutions [63]. Furthermore, multilayers of surfactants formed in solutions at concentrations well above the CMC, especially at extreme potentials of opposite sign of that of the surfactant head group, are suggested by voltammetry of hydrophobic probes

[56,57]. Multilayer coverage was also found for the C18 version of amphiphile 1.

Spectroscopic studies add to the picture. SERS on Ag electrodes [60] and surface absorbance studies on open-circuit glassy carbon [62] showed that cationic surfactants adsorbed in close-packed, head-down orientations at concentrations above the CMC. On Ag electrodes, the head-down orientations of adsorbed cationic and nonionic surfactants were retained even at potentials slightly positive of the PZC [60]. Both SERS and surface fluorescence spectroelectrochemistry revealed a rich dependence of adsorption characteristics on potential and the resulting condition of the electrode surface. However, the relation between adsorption of anions on Ag and the adsorption of cationic surfactants is poorly understood. Maxima in intensities were found in both experiments in the potential range just positive of the onset of evolution of hydrogen, signaling profound changes in the aggregate structure or amount of adsorbed surfactant.

On the other hand, studies with some electroactive surfactants seem to suggest that saturation coverage corresponds to a monolayer [64–67]. Structures of such monolayers are uncertain. A monolayer that presents extended hydrocarbon chains to the aqueous phase would be thermodynamically unfavorable and is not in agreement with results for charged metal oxides. The mediated coulometry experiments of Widrig and Madja [67] reveal that CV experiments may not measure all the adsorbed electroactive surfactant on the electrode, especially in structures with more than one adsorbed layer. This is reasonable for molecules with electroactive head groups, since for hydrocarbon chains of C12 to C18, the outer row of head groups in a bilayer may be 25–50 Å away from the inner row on the electrode surface. The same authors [67] point out the possibility that outer surfactant layers could be washed away during handling.

Although the work discussed above has provided significant insights into structures of surfactant adsorbates on electrodes, there is clearly a need for additional electrochemical, in situ spectroscopic, CV/QCM, and surface dynamic information to provide a detailed picture of adsorbate structure and its influence on electrochemical properties.

B. Diffusion in Micellar Solutions

Diffusion plays a fundamental role in electroanalytical measurements as well as in applications of surfactant media. Diffusion coefficients (D) measured for electroactive probes in micellar solutions can provide diffusion coefficients of the micelles themselves [21,25,41,68–70]. If the system contains only spherical micelles, the average micelle radius (r) can be obtained from the well known Stokes-Einstein equation:

$$D = \frac{kT}{6\pi\eta r}$$

where k is Boltzmann's constant, η is viscosity, and T is temperature in kelvins.

In using electrochemical probes to estimate micellar diffusion coefficients, it is advantageous for the probe to have a high equilibrium constant for binding to the micelle and to have reversible or nearly reversible electrochemistry. Generally, the measured response is proportional to some power of D. Some of the methods used are listed with relevant equations in Table 1.

1. Models for Diffusion

The diffusion coefficient measured by an electroanalytical method in a surfactant solution is really an apparent value, D', because of equilibrium of the free probe in solution with probe bound to micelles. For systems with a single distribution of micelles, D' can be expressed as [21,25,69,71,75–78]:

$$D' = f_a D_0 + f_b D_1 \tag{13}$$

where f_a is the fraction of free probe, f_b is the fraction of probe bound to micelles, and D_1 and D_0 are the diffusion coefficients of the micelle and the free probe in the bulk phase, respectively. The fraction of free probe in the bulk is $f_a = 1 - f_b$. Evans [78] pointed out that Eq. (13) holds when equilibrium between free and micelle-bound probe is fast compared to the experimental time scale.

In Sec. III.A, the very rapid entry and exit rates of solutes in micellar systems were discussed. These rates place kinetics of probe-micelle equilibria in the µsec time range. The methods in Table 1, when used to measure D', usually have characteristic time scales on the order of 10 msec to several seconds. Although the use of very tiny ultramicroelectrodes could possibly make the time scale of the experiment comparable to exit rates, steady-state microelectrode voltammetry gives a limiting current directly proportional to D'. Thus, unless quite slow probe-micelle kinetics can be demonstrated, it seems safe to use Eq. (13) for the majority of micellar diffusion studies.

On the other hand, if the kinetics of the particular system are truly slow with respect to the experimental time scale, probe-micelle equilibrium is fixed during the measurement. Then the expression for D' reflects the relation between the measured quantity and D (Table 1). For cyclic or square wave

TABLE 1
Methods and Equations for Micellar Diffusion Studies

Method	Equation	Ref.
Voltammetry[a]		
cyclic, large electrode	$i_p = 0.4463nFAC^*(nF/RT)^{1/2}v^{1/2}D^{1/2}$	34
ultramicroelectrode[b]	$i_l = 4nFDC^*r$ (disk)	72
square wave	$i_p = nFAD^{1/2}C^*\psi_p/(\pi t_p)^{1/2}$	73
rotating disk[b]	$i_l = 0.620nFAD^{2/3}C^*\omega^{1/2}v^{-1/6}$	34
Polarography[b]	$i_l = 708nD^{1/2}C^*m^{2/3}t_d^{1/6}$	3
Chronocoulometry[b]	$Q = 2nFAD^{1/2}C^*t^{1/2}/\pi^{1/2} + a'/r + Q_{dl}$	34,74

[a] Expressions for peak currents based on reversible charge transfer under the experimental conditions used.
[b] For mass transport–controlled electrode reactions.

Symbols:

n	number of electrons in reaction
i_p	peak current
i_l	limiting current
F	Faraday's constant
A	electrode area
C^*	total concentration of probe
D	apparent diffusion coefficient
R	gas constant
T	temperature in Kelvins
v	scan rate
r	radius of electrode
ψ_p	peak current function
t_p	square wave pulse width
ω	angular frequency of electrode rotation
ν	kinematic viscosity
m	Hg flow rate of DME
t_d	drop time of DME
t	time
a'	edge or spherical correction parameter
Q_{dl}	double layer charge

voltammetry, polarography, and chronocoulometry on conventional-sized electrodes, the following equation pertains for the slow equilibrium case:

$$D' = (f_a D_0^{1/2} + f_b D_1^{1/2})^2 \qquad (14)$$

As early as 1975, Hayano and Fujihira [69a] concluded that Eq. (14) was inappropriate based on a study of diffusion and absorption spectra of dyes in micellar solutions of tetradecyldimethylbenzylammonium chloride and decaoxyethylene nonylphenyl ether. Several authors [70,79,80] have used Eq. (14), but without confirming that equilibria in their systems were slow.

2. *Single Size Distribution of Micelles*

Starting from Eq. (13), equilibrium of the probe (X) with micelles (M) can be analyzed to obtain expressions for f_a and f_b. This has been done in terms of a formation equilibrium constant [25,75–77]:

$$M + X = MX \qquad K = \frac{[MX]}{[M][X]} \tag{15}$$

Equation (15) represents the reaction of one probe molecule per micelle. Substituting equilibrium expressions for f_a and f_b into Eq. (13):

$$D' = \frac{D_0}{1 + K[M]} + \frac{D_1 K[M]}{1 + K[M]} \tag{16}$$

At low probe concentrations, [M] may be replaced by the total concentration of micelles (C_M), giving the approximate relation:

$$D' = \frac{D_0}{1 + KC_M} + \frac{D_1 KC_M}{1 + KC_M} \tag{17}$$

Alternatively, using the so-called pseudophase model, solute equilibria can be expressed in terms of a distribution coefficient [69,81]:

$$K_D = \frac{[MX]}{[X]} \tag{18}$$

i.e., the ratio of bound to free probe. From Eq. (15), $K_D = [M]K$. Thus, the distribution coefficient depends on the concentration of micelles. The pseudophase expression for D' is

$$D' = \frac{D_0}{1 + K_D} + \frac{D_1 K_D}{1 + K_D} \tag{19}$$

It has been known for some time that D' for probes decreases with increasing micelle concentration [41,70,71,81]. Typical data for a probe such as ferrocene, which binds strongly to micelles, clearly shows this decrease (Fig. 10a). This is described by Eq. (17), but not explicitly by Eq. (19). In 1988, we reported that for several probes, D' also decreased with increasing probe concentration in micellar solutions (Fig. 10b) [25]. Similar decreases in D' with concentration of anthraquinone dyes measured by polarography in nonionic surfactant solutions were reported previously [41] but were attributed to solute-induced increases in micelle size. However, it was clearly

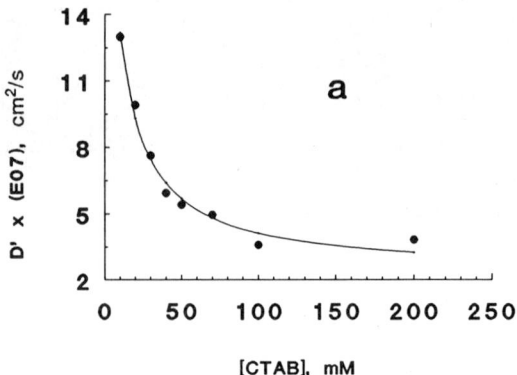

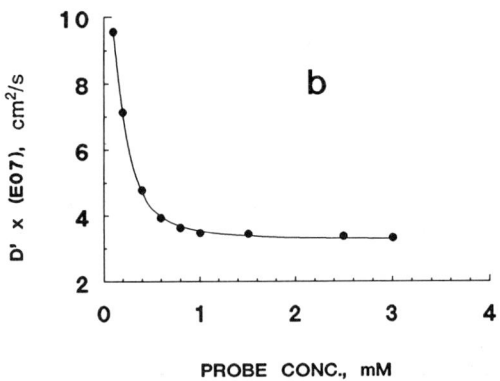

FIG. 10. (a) Influence of concentration of CTAB in 0.1 M KBr on D' measured by voltammetry at 10 mV/sec at 12.5 μm radius Pt microdisk electrode using 0.3 mM ferrocene. Points are experimental; line is best fit to Eq. (22) (n = 3) by nonlinear regression. (b) Influence of concentration of ferrocene on D' measured by cyclic voltammetry (0.05-1 V/sec) at glassy carbon (A = 0.071 cm^2) in 0.15 M CTAB/0.1 M tetraethylammonium bromide. Points are experimental; line is best fit to Eq. (22) (n = 3) by nonlinear regression. (From Ref. 71.)

demonstrated for ferrocene and methylviologen (MV^{2+}) in micellar SDS and CTAB that decreases in D' with C_X were a consequence of multiple occupancy of probes in the micelles [25,71].

Typical probe concentrations vary from about 0.1 to 5 mM, with surfactant concentrations from the CMC to well above it. A 0.05 M solution of CTAB in 0.1 M KBr has an average aggregation number of about 100 [28], so

Micelles and Microemulsions

that $C_M = 0.5$ mM. If $C_X \geq 0.5$ mM, the excess X would have no micelles to occupy if only one X per micelle were allowed. Obviously, there can be more than one probe bound per micelle.

Clearly, Eqs. (16) to (19) do not predict the observed dependence of D' on C_X. It is usually assumed that on an instantaneous time scale, solutes occupy micelles according to a Poisson distribution [11]. However, since diffusion is slow with respect to micelle-probe equilibria, electroanalytical experiments should be insensitive to the type of solute distribution. All that is required is a model that accounts for multiple occupancy and the relevant equilibria.

Consider a micelle capable of binding n probe molecules in the overall equilibrium:

$$M + n X = MX_n \qquad K^n = \frac{[MX_n]}{[M][X]^n} \qquad (20)$$

Making the assumption that $[X] \ll C_X$, i.e., that the probe is almost entirely bound to micelles, the following approximate relation was derived [25]:

$$f_b = \frac{C_M K^n C_X^{n-1}}{1 + C_M K^n C_X^{n-1}} \qquad (21)$$

In Eq. (21), K is now an "apparent" binding constant per bound probe molecule. The following approximate model for D' was obtained by substituting expressions for f_b and f_a into Eq. (16):

$$D' = \frac{D_0}{1 + C_M K^n C_X^{n-1}} + \frac{D_1 C_M K^n C_X^{n-1}}{1 + C_M K^n C_X^{n-1}} \qquad (22)$$

Eq. (22) holds approximately for micellar systems with a single size distribution and correctly describes the dependence of D' on both C_M and C_X. Examples of both types of data fit to Eq. (22) show good agreement of experimental and computed results (see Fig. 10). Equation (22) predicts that for tightly bound probes D' approaches D_1 as C_X becomes large. Also, if the apparent binding constant K is very large, a dependence of D' on C_X is not observed experimentally [71].

We now return briefly to the question of fast [Eqs. (13) and (22)] vs. slow probe-micelle equilibrium [Eq. (14)]. A slow equilibrium analog of Eq. (22) can be found by substituting Eq. (21) for f_b, and $f_a = 1 - f_b$, into Eq. (14) [71,79,80]. We previously concluded that most probe-micelle systems should follow Eq. (13). What happens if the incorrect expression is used? Consider the data in Fig. 10a. Using nonlinear regression analysis onto the fast (Eq. (22)) and slow models [Eqs. (14) and (21)] for D', we compared goodness of fit and values of parameters (Table 2). Values of D_0 and D_1 from the two

TABLE 2

Results of Nonlinear Regression Analysis of D' vs. C_X Data onto Fast and Slow Equilibrium Models for a Monodisperse System[a]

System	Model	$10^6 D_0$ (cm^2/sec)	$10^6 D_I$ (cm^2/sec)	$C_M^{1/n}K$ (mM$^{1/n-1}$)	SD[c] (cm^2/sec)
0.15 M CTAB/0.1 M TEAB ferrocene probe	Fast	1.12 ± 0.14	0.33 ± 0.03	3.00 ± 0.05	0.06
	Slow	1.11 ± 0.12	0.33 ± 0.03	2.70 ± 0.04	0.06
Simulated data:					
Theor. parameter values:		*1.500*	*0.700*	*3.000*	
Theor. data	Fast	1.502	0.701	3.013	0.00113
0.1% noise[b]	Slow	1.498	0.701	2.820	0.00117
Theor. data	Fast	1.496	0.694	2.91	0.0160
1.0% noise[b]	Slow	1.492	0.695	2.73	0.0160
Theor. parameter values:		*10.00*	*0.800*	*3.00*	
Theor. data	Fast	10.01	0.805	3.004	0.0099
0.1% noise[b]	Slow	9.70	0.862	2.453	0.0408
Theor. data	Fast	9.96	0.761	2.96	0.047
1.0% noise[b]	Slow	9.64	0.820	2.42	0.039

[a] Data from Fig. 10a; parameters reported with std. errors.
[b] Absolute normally distributed noise added as percent of largest D' value.
[c] Standard deviation of the regression.

different models are within experimental error. However, differences on the order of 10% are found between the two models in the binding constant parameter $C_M^{1/n}K$. Similar conclusions were reached when theoretical data generated from Eq. (22) were analyzed by both models. When the slow equilibrium model was used, error in $C_M^{1/n}K$ was larger at larger D_0/D_1, and errors in D_1 also began to appear (Table 2). This shows that values of the micelle diffusion coefficient are relatively insensitive to the model used, although small errors result if the slow equilibrium model is used to analyze data for which fast equilibria pertain.

3. Model for Two-Micelle Distribution

Equation (22) successfully predicts the dependence of D' on C_X and C_M and fits very well when the assumptions of the derivation hold. However, the quality of fits for a number of micellar systems was poorer than expected. These systems had been shown to contain two types of micelles by light scattering and NMR [25]. Thus, a model for binding of the probe to two micelles was developed.

Consider the following equilibria of X with micelles M_1 and M_2:

$$M_1 + n X = M_1 X_n \qquad K_1^n = \frac{[M_1 X_n]}{[M_1][X]^n} \tag{23}$$

$$M_2 + m X = M_2 X_m \qquad K_2^m = \frac{[M_2 X_m]}{[M_2][X]^m} \tag{24}$$

K_1 and K_2 are apparent binding constants for equilibria with M_1 and M_2, respectively. For fast probe-micelle equilibria, the expression for D' is:

$$D' = f_a D_0 + f_{b1} D_1 + f_{b2} D_2 \tag{25}$$

where f_{b1} and f_{b2} are fractions of X bound to micelles M_1 and M_2 with diffusion coefficients D_1 and D_2, respectively. Expressing these fractions in terms of equilibria in Eqs. (23) and (24), and assuming $[X] \ll C_X$:

$$\begin{aligned} D' &= \frac{D_0}{1 + C_{M1} K_1^n C_X^{n-1} + C_{M2} K_2^m C_X^{m-1}} \\ &+ \frac{D_1 C_{M1} K_1^n C_X^{n-1}}{1 + C_{M1} K_1^n C_X^{n-1} + C_{M2} K_2^m C_X^{m-1}} \\ &+ \frac{D_2 C_{M2} K_2^m C_X^{m-1}}{1 + C_{M1} K_1^n C_X^{n-1} + C_{M2} K_2^m C_X^{m-1}} \end{aligned} \tag{26}$$

where C_{M1} and C_{M2} are the total concentrations of M_1 and M_2, respectively [25].

It's important to realize that, like Eq. (22), Eq. (26) is *approximate*. The Ks in these equations are conditional binding constants; *they are not equilibrium constants*. They depend on the experimental conditions under which D' is measured, and may contain a kinetic component [71]. In addition, if the probe is strongly adsorbed to the electrode, the above models are invalid [82].

Figure 11a shows that data for methyl viologen in micellar SDS solution give a poor fit to the monodisperse model, Eq. (22). These data also did not fit a model that accounted for the three individual steps in the equilibria for a monodisperse micellar system [25]. However, the data give an excellent fit to Eq. (26) (Fig. 11b), which considers two micellar size distributions. Equation (26) was shown with $>$ 99% confidence to fit better than Eq. (22) by an extra sum of squares F-test, which accounts statistically for the additional two parameters in Eq. (26) compared to Eq. (22) [25].

In general, probe-micelle systems were found where one-micelle [Eq. (22)] or two-micelle (Eq. (26)) models gave statistically better fits. Micellar diffusion coefficients found by this approach (Table 3) were in very good agreement with those previously measured by nonprobe methods such as pulsed field gradient spin echo NMR, Taylor dispersion tube, or quasi-elastic light scattering (QELS) [83–87]. In addition to demonstrating the validity of the approach, this agreement indicates that solute-induced aggregation is not the cause of the D' vs. C_X dependence.

Systems following Eq. (26) have been shown by alternative methods to have coexisting spherical and rod-shaped micelles [25]. The few comparative data available seem to be self-consistent. For example, 0.15 M CTAB/0.1 M TEAB was found to be monodisperse and had a D_1 very close to that of the D_2 of 0.1 M CTAB/0.1 M KBr, in which spherical and rod-shaped micelles coexist. The value is also similar to D' to 6 mM ferrocene in 0.22 M CTAB/0.1 M NaCl found by RDV [69b]. This suggests that the larger micelle in polydisperse 0.1 M CTAB/0.1 M KBr, presumably rod-shaped, is the same as that in the $\geq$ 0.15 M CTAB systems, which are likely to have been fully converted to rods by the increase in surfactant concentration.

Table 3 also shows that with the pseudophase distribution model [Eq. (19)], values of D_1 were significantly smaller that those found by nonprobe methods. This may result partly from not considering polydispersity. Low-biased values of D_1 also were found when the incorrect monodisperse model [Eq. (22)] was used to analyze data for systems where spherical and rod-shaped micelles coexist [25]. Similar low average values were found when ultramicroelectrodes insensitive to probe-micelle equilibria were used to measure D' [71].

Even when voltammetric measurements are made at a single probe concentration, reasonable agreement with alternative techniques can be obtained

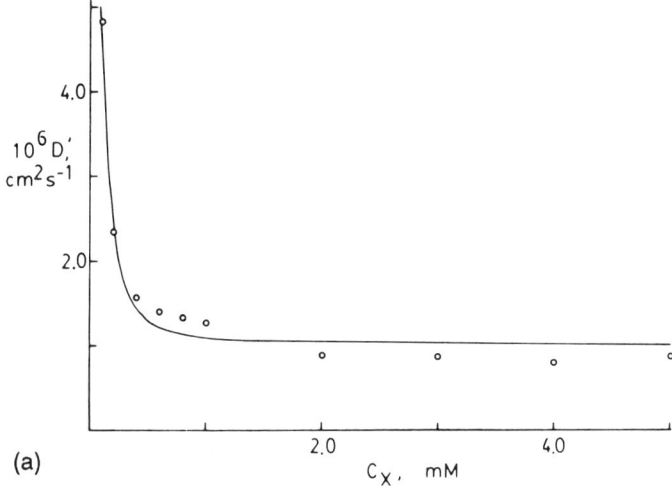

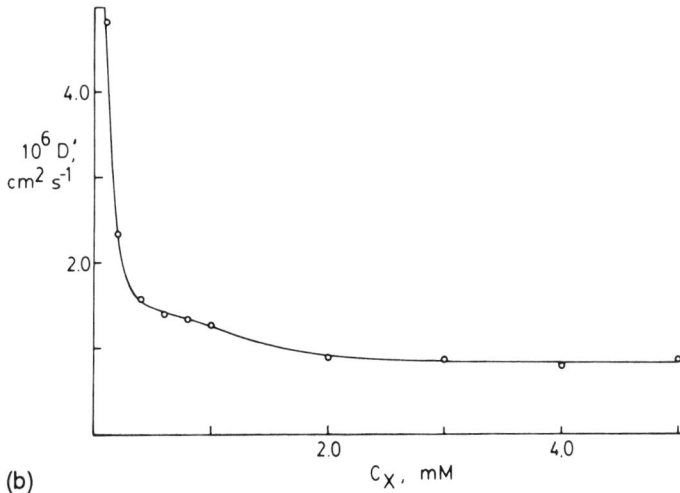

FIG. 11. (a) Influence of concentration of methyl viologen on D' in 0.1 M SDS/0.1 M NaCl. Circles are experimental data; line is best fit from nonlinear regression onto Eq. (22) with n = 3. (b) Data from plot in "a" with best fit (line) from nonlinear regression onto Eq. (26) with n = 4, m = 8. (From Ref. 25.)

TABLE 3

Micellar Diffusion Coefficients from Electrochemical Probe Techniques[a]

System	Probe[a]	Method [Ref.][b]	Model	$10^6 D_1$ (cm^2/sec)	lit[Ref.][c] (cm^2/sec)	$10^6 D_2$ (cm^2/sec)
Anionic surfactants:						
0.1M SDS/0.1M NaCl	MV^{2+}	CV[25]	Eq. (26)	1.41	1.40–1.45 [83]	0.84
0.1M SDS/0.1M NaCl	MV^{2+}	CC[25]	Eq. (26)	1.35		0.70
0.1M SDS/0.1M NaCl	Fc	CV[25]	Eq. (26)	1.45		0.99
0.069M SDS/0.1M NaCl	Fc	RDV[69b]	Eq. (19)	0.63		
0.14M SDS/0.1M NaCl	Fc	RDV[69b]	Eq. (19)	0.64		
0.07M SDS/0.05M NaCl	MV^{2+}	CC[77]	meas.[e]	0.95		
Cationic surfactants:						
0.1M CTAB/NaCl	Fc	CV[25]	Eq. (26)	1.01	0.5–1.0 [84–86][d]	0.77
0.11M CTAB/NaCl	Fc	RDV[69b]	Eq. (19)	0.67		
0.1M CTAB/KBr	Fc	CV[25]	Eq. (26)	0.73	0.5–0.83 [85][d]	0.38
0.15M CTAB/0.1M TEAB	Fc	CV[69b]	Eq. (22)	0.33	—	none
0.22M CTAB/0.1M NaCl	Fc	RDV[69b]	Eq. (19)	0.27		
0.04M CTAB:						
+0.02M NaBr	C$_{16}$P	Pol[87]	meas.[e]	1.5	2.1 [87]	
+0.05M NaBr	C$_{16}$P	Pol[87]	meas.[e]	1.2	1.4 [87]	
+0.08M NaBr	C$_{16}$P	Pol[87]	meas.[e]	0.99	1.1 [87]	
+0.10M NaBr	C$_{16}$P	Pol[87]	meas.[e]	0.66	0.44 [87]	
+0.20M NaBr	C$_{16}$P	Pol[87]	meas.[e]	0.43	0.24 [87]	
				D_m^o found	lit. [ref.]	
SDS	PyNO$_2$	Pol[88a]	Eq. (27)	1.1	1.53 [89][d]	
SDS/0.05M NaCl	PyNO$_2$	Pol[88a]	Eq. (27)	0.7	0.98 [89][d]	
TTAB	PyNO$_2$	Pol[88a]	Eq. (27)	0.72		
TTAB	C$_{16}$P	Pol[88a]	Eq. (27)	0.72		
TTAB/0.1M KBr	PyNO$_2$	Pol[88a]	Eq. (27)	0.72	0.92 [90]	
TTAB/0.1M KBr	C$_{16}$P	Pol[88a]	Eq. (27)	1.0		

[a] Probes: MV^{2+} = methylviologen; Fc = ferrocene; PyNO$_2$ = 1-nitropyrene; C$_{16}$P = N-octadecyl-4-cyanopyridinium bromide.
[b] Methods: CV = cyclic voltammetry; CC = chronocoulometry; RDV = rotating disk voltammetry; Pol = polarography.
[c] By alternative nonelectrochemical techniques.
[d] Conditions similar but not exactly the same as in electrochemical measurements.
[e] D' at fixed probe concentration.

Micelles and Microemulsions

if the probe is strongly bound to the micelles. This is illustrated by a comparison of polarographic and QELS measurements on CTAB micelles at different NaBr concentrations by Mackay et al. (Table 3) [87]. It is likely that the system chosen, 0.04 M CTAB, has only one micellar size distribution. Also, the choice of the long-chain N-octadecyl-4-cyanopyridinium bromide as probe may result in a D' that is invariant with probe concentration, as reported for this probe in other surfactant systems [70]. According to Traube's rule, the long chain should enhance the binding constant with micelles, making D' a good direct estimator for D_1 [71].

It may be concluded that when the proper models are used, reliable values of micellar diffusion coefficients can be obtained by the electrochemical probe method. For unknown systems, testing Eq. (22) with a series of values of n and testing Eq. (26) with various combination of n and m is recommended before deciding on the model of best fit [25]. The extra sum-of-squares F-test for comparing models with unequal numbers of parameters can be used to help decide between the best one- and two-micelle models.

4. Microelectrode Voltammetry

Electrodes with one dimension in the μm range are less sensitive to coupled chemical reactions than larger conventional electrodes [72,91]. The limiting steady-state current, usually measured at low scan rates, is directly proportional to D' (see Table 1). Using ferrocene as a probe, diffusion in several SDS and CTAB micelles was studied with disk ultramicroelectrodes with radii between 4 and 12.5 μm. In these experiments, D' for the aqueous micellar systems was *independent* of C_X but decreased with C_M [71]. The lack of dependence on C_X may result because the conditional binding constant K in not a true thermodynamic quantity and may also reflect dissociation kinetics [71]. Increasing K [Eq. (20)] removes the dependence on C_X in the observable concentration range. Very tiny electrodes are much less sensitive to fast dissociation reactions preceding electron transfer at the electrode. Thus, ultramicroelectrodes may result in a larger conditional K and no C_X dependence.

5. Diffusion Coefficients at the CMC

The dependence of D_1 on surfactant concentration C_s can give the micelle self-diffusion coefficient in the absence of intermicellar interactions (D_m^o) from the relation:

$$D_1 = \frac{D_m^o}{1 + k_f(C_s - CMC)} \tag{27}$$

where k_f is the interaction coefficient. Zana and co-workers [70,88a] obtained D_1 from dc polarographic data assuming that the probes were totally bound to micelles. They plotted $1/D_1$ vs. C_s to obtain D_m^o. Results for a few systems for which values from nonprobe methods are available are listed in Table 3. Agreement is surprisingly good. These workers also suggested that adsorption of cationic surfactants at the DME may invalidate such measurements at high surfactant concentrations [88a].

6. Detection of Micelles

As mentioned in Sec. II, electrochemical probe methods can also be used for determining CMC values [12,13]. Cyclic voltammetry has been used recently to obtain CMCs of various surfactants by monitoring adsorption peaks of methyl viologen cation radical [77] and the peak current of ferrocyanide [79]. Rotating disc voltammetry (RDV) was also used for CMC determinations [88b]. The latter two papers reported detection of "second" CMC values at higher concentrations than the first, presumably due to formation of a second type of micelle.

7. Consequences of Solute Binding

From the above discussion, it is evident that strong binding of a electroactive solute to micelles can be immediately recognized by a diminished current and D'-value compared to homogeneous solutions. For example, ferrocene at concentrations of several mM in 0.11 M CTAB/0.1 M NaCl is almost totally bound and gives a D' of about 0.7–0.8×10^{-6} cm^2/sec [69b]. Ferrocinium ion, its oxidation product, it not bound to CTAB micelles and gives a D' of 6×10^{-6} cm^2/sec. An approximate 10-fold decrease in D' is thus characteristic of binding of small solutes to micelles. Binding constants may be derived from electrochemical data, but they are conditional and dependent on the model used to analyze the data.

An interesting variation to the above scenario occurs in the oxidation of tetrathiofulvalene (TTF) in aqueous CTAC solutions [81]. TTF is oxidized in two separate one-electron steps to TTF$^{\cdot+}$ and TTF^{2+}. The second oxidation wave, which would usually be the same height as the first, had an anomalously large peak current in micellar CTAC. This was caused by:

$$TTF + TTF^{2+} = TTF^{\cdot+}$$

This coupled chemical reaction has a large effect on the size of the second peak if diffusion coefficients of the two reactants are very different. In this case, D' for TTF bound to CTAC micelles is much smaller than that of TTF^{2+}, which is mainly free in the solution [81a]. However, this reaction was

Micelles and Microemulsions 41

considered to take place after dissociation of TTF from the micelle. Similar results were found in nonionic Brij 35. In SDS, all the reactants and products are bound to micelles and equal wave heights are observed [81c].

C. Heterogeneous Electron Transfer in Micellar Solutions

1. Electron Transfer Processes

We first consider simple heterogeneous charge transfer processes at electrodes, where both forms of a given redox couple are stable. In order to understand electron transfer reactions in micellar solutions and to interpret half-wave potentials, equilibria of micelles with both forms of the redox couple must be considered. This has been done via a "square scheme" for a reduction [69b,76,77,93]:

Scheme I

$$A_m + ne \rightleftarrows B_m$$
$$\Updownarrow (K_o) \qquad\qquad (K_r) \Updownarrow$$
$$A_w + M + ne \rightleftarrows M + B_w$$

Here, the pseudophase approach has been taken, with equilibrium constants written as:

$$K_o = \frac{[A_w]}{[A_m]}$$

and

$$K_r = \frac{[B_w]}{[B_m]}$$

Subscripts w and m refer to water and micelle phases, respectively. In the general case, equilibria of both A and B with M must be considered. The reversible half-wave potential $E^r_{1/2}$ in the micellar solution is [69b]:

$$E^r_{1/2} = E^{o'}_w + \frac{RT}{nF} \ln \frac{K_o(1 + K_r)}{K_r(1 + K_o)} - \frac{RT}{2nF} \ln [D_o/D_r] \quad (28)$$

where D_r and D_o are measured values in micellar solutions, expressed as:

$$D_r = \frac{D_m}{1 + K_r} + \frac{D_{rw}K_r}{1 + K_r} \quad (29)$$

$$D_o = \frac{D_m}{1 + K_o} + \frac{D_{ow}K_o}{1 + K_o} \quad (30)$$

D_m is the micelle diffusion coefficient, $E°'_w$ is the formal potential of A/B in water, and D_{ow} and D_{rw} are the diffusion coefficients of A and B in water, respectively. Equations (29) and (30) are equivalent to Eq. (19), but the Ks in the former are the reciprocals of K_Ds in Eq. (19). K_r and K_o are related to formation constants such as in Eq. (15) by the expression $K_r = 1/([M]K)$, assuming one X per micelle.

There are several important special cases of Eq. (28). If both A and B exist mainly in the micelles, K_r and K_o are small with respect to 1, D_o and D_r are nearly equal, and Eq. (28) simplifies to [77,93]:

$$E^r_{1/2} = E°'_w + \frac{RT}{nF} \ln[K_o/K_r] \tag{31}$$

Equation 31 was found to pertain to reductions of methylviologen and a series of iron, ruthenium, and osmium complexes having the ligands CN^-, 2,2'-bipyridine, 4,4'-dimethyl-2,2'-bipyridine, 5,5'-dimethyl-2,2'-bipyridine, 1,10-phenanthroline, and 5-chloro-1,10-phenanthroline in micellar solution of SDS [93a].

Equation 28 has also been applied to the ferrocene/ferrocinium (Fc/Fc+) couple in alkyltrimethylammonium surfactants [69,76]. In this case, Fc is almost entirely bound to micelles, while its oxidation product Fc+ is not bound at all. Equation 28 then becomes:

$$E^r_{1/2} = E°'_w + \frac{RT}{nF} \ln \frac{1 + K_r}{K_r} - \frac{RT}{2nF} \ln[D_o/D_r] \tag{32}$$

When applying Eqs. (28), (31), or (32), care must be taken that the reactions involve only electron transfer. No additional coupled chemical reactions should be present. Such reactions could occur at different rates in the micellar media and might influence peak potentials. An example is the oxidation of ferrocyanide ion, which undergoes a preceding reaction with water in aqueous systems shifting its peak potential well positive of that in dry organic solvents [92a]. Ferrocyanide ion in 0.1 M CTAB has a peak potential shifted 120 mV negative compared to the value in aqueous solution at bare PG [46e]. The above equations would suggest stronger interactions with cationic headgroups of CTAB for the ferricyanide ion with -3 charge than for ferrocyanide with -4 charge. This is opposite of what is expected from the relative strengths of the coulombic interactions between these ions and the micellar surface. However, analysis with Eq. (28) or (31) is invalidated in this case by the preceding reaction with water, which is likely to be influenced significantly by the micellar environment.

2. Heterogeneous Electron Transfer Kinetics

Although a great deal of past research was aimed at elucidating the influence of "surface-active substances" on heterogeneous charge transfer kinetics [3,7,41], the majority of early work was done only at very low concentrations and with substances that do not fit the modern definition of surfactants. Only a few studies report electron transfer kinetics of micelle-solubilized electroactive species (Table 4). Note also in Table 4 that D' data were used to obtain values for the fraction f_b of solute molecules bound to micelles.

Only a few studies have addressed the question of exactly how electrons are exchanged between electrode and reactants solubilized in micelles. The answer to this question probably involves solute-micelle equilibria and kinetics [81,93–95]. Also, considering our earlier discussion of adsorption of

TABLE 4
Apparent Electrode Kinetics, Diffusion Coefficients, and Fractions of Electroactive Species in Micelles

Reactant[a] (C, electrode)	electrolyte system	$10^6 D'$ (cm^2/sec)	f_b^b	$k°$ (cm/sec)	Ref.
Co(bpy)$_3^{2+}$	MeCN/0.1 M TMAP	12		>0.1	28
(2 mM, GC)	0.1 M KBr, aq.	8.6		0.012	28
	0.1 M SDS/0.1 M TMAP	0.8	0.99	0.0014	28
	0.1 M CTAB/0.1 M KBr	2.5	0.78	0.0035	28
Co(dmbpy)$_3^{2+}$	MeCN/0.1 M TMAP	8.0		>0.1	28
5 mM, GC)	0.1 M KBr, aq.	5.2		0.0001	28
	0.1 M SDS/0.1 M TMAP	0.1	≈1	0.0001	28
	0.1 M CTAB/0.1 M KBr	2.5	0.85	0.0013	28
Co(bpy)$_2$(bpyC$_{16}$)$^{2+}$					
(0.9 mM, GC)	0.1 M SDS/0.1 M TMAP	0.88	0.98	0.0006	28
DMPZ	0.1 M SDS/0.1 M Li$_2$SO$_4$	0.9	≈1	1.1	93a
(Pt)					
TTF	aq. CTAC/0.1 M KCL	0.75^c	≈1	0.0092	81a
(Pt)					
Ferrocene	MeOH/0.1 M TEAP			0.18	94
(Pt)	PrOH/0.1 M TEAP			0.029	94
	acetone/0.1 M TEAP			0.015	94
	THF/0.1 M TBAP			0.013	94
	0.1 M SDS/0.1 M NaCl	1.45^d	≈1	0.0134	94
	0.1 M CTAB/0.1 M NaCl	1.0^d	≈1	0.020	94

[a] Reactant and salt abbreviations: Co(bpy)$_2$(bpyC$_{16}$)$^{2+}$ = bis (2,2'-bipyridyl) (4,4'-dihexadecyl-2,2'-bipyridyl)cobalt(II); DMPZ = 5,10-dimethyl-5,10-dihydrophenazine; TTF = tetrathiofulvalene; TMAP = tetramethylammonium perchlorate; TEAP = tetraethylammonium perchlorate; TBAP = tetrabutylammonium perchlorate.
[b] Computed from Eq. (13) and measured values of D', D_1, and D_0.
[c] For 0.2 M CTAB.
[d] Values of D_1 from Ref. 25.

surfactants on electrodes (Section IV.A), it is possible that coupled equilibria between surfactant adsorbed on the electrode, electroactive solute, micelles, and surfactant monomers may also play a role.

We will first discuss kinetics of electron transfer in micellar systems with surfactant concentrations well above the CMC, then return to speculate on how electrons are exchanged between electrodes and reactants. An important fact concerning the electrical double layer at the electrode-solution interface is that surfactant molecules are likely to be adsorbed specifically (Sec. IV.A) throughout the entire range of potentials used.

Interpretations of electrode kinetics in surfactant media need to consider at least four factors: (1) the polarity of the average microenvironment of the reactant during electron transfer, (2) the effect of surfactant at the electrode-solution interface on the orientation of the reactant at the time of electron transfer, (3) the role of interfacial surfactant in possible adsorption or co-adsorption of electroactive species, and (4) the influence of micellar aggregates in solution on mass transport of electroactive species to the electrode. In the absence of evidence to the contrary, measured values of the standard heterogeneous rate constant (k^o) in micellar systems probably should be considered apparent values without absolute significance.

For all the Co(II) bipyridyl complexes studied [28], values of k^o were smaller in micellar solutions than in isotropic acetonitrile (Table 4). This may be attributed to an inhibition or blocking effect of the surfactant on the electrode surface similar to that observed for some species at lower concentrations of surface-active agents [41,42]. Apparent k^os for the Co(II)/Co(I) couples are 10-fold larger in CTAB than in SDS. Electron transfer from glassy carbon to tris(2,2'-bipyridyl)cobalt(II) is slower in micelle solutions than in water, but the electron transfer rate constant for tris(4,4'-dimethyl-2,2'-bipyridyl)cobalt(II) is the same in SDS as in water and 10-fold faster in CTAB. The reason for much slower electron transfer to the dimethylbipyridyl complex is unclear.

The k^o data for the Co(II) bipyridyl complexes do not fit any simple classical picture of the influence of double-layer structure on electrode kinetics, even when specific adsorption of CTAB and SDS is considered. For example, specific adsorption of a small anion tends to make the potential at the outer Helmholtz plane more negative, making the apparent k^o larger for an electroactive ion of $z = +2$, such as the Co(II) complexes under discussion. The reverse effect on k^o is predicted when small positive ions are adsorbed [34]. However, this is the opposite of what is observed for the Co(II)/Co(I) reductions. Values of k^o in CTAB, with the same sign head groups as the reactants, are larger than in SDS. Similar surfactant effects on electron-transfer rates were observed qualitatively for the reduction of the negatively charged (1,2-diaminocyclohexanetetraacetic acid)Bi(III) complex, where dodecylammonium perchlorate accelerated the electrode reaction and SDS retarded it. Other, sometimes conflicting, examples of such observations have been reviewed [20]. Clearly, amphiphilic surfactant ions influence k^o in a way not easily predictable by classical double-layer theory, so that electron transfer rate constants must currently be interpreted in a largely phenomenological manner.

We note in passing that many reactants show fast, reversible electron transfer kinetics in micellar system. These include one-electron reductions of anthracene, 9-phenylanthracene, and 9,10-diphenylanthracene in CTAB (30°C) solutions on Hg [56,57], DMPZ in 0.1 M SDS on Pt (Table 4), and nitrobenzene in SDS on Hg [75]. The anthracenes, reduced at potentials negative of -2 V vs. SCE, appear to exchange electrons with the electrode while in a thick layer of surfactant on the Hg electrode surface.

Abbott et al. [94] used the concept of solvent parameters to explore the influence of microenvironment on electron transfer kinetics of ferrocene in CTAB and SDS. They correlated k^os for ferrocene at Pt with polarities of a series of isotropic organic solvents. These data were compared with heterogeneous electron transfer rates of ferrocene at Pt in micellar solutions. Electron transfer rates for ferrocene in CTAB and SDS micelles were slightly smaller than in alcohols (Table 4) and much smaller than predicted for water from Taft solvent parameters. The electron transfer rate in SDS was somewhat smaller than in CTAB. Assuming equal electron transfer distances in the two micellar systems, the authors suggest average microenvironments for ferrocene during electron transfer in aqueous micellar CTAB and SDS that are more polar than in many apolar solvents, but less polar than in water. This is in agreement with microenvironmental polarities found for hydrophobic solutes in micelles [10,26,29].

3. *Can Reactants Exchange Electrons with Electrodes While in Micelles?*

Since rates of exit of solutes from micelles are usually very fast, such rates may not influence electrochemical experiments in the time range of milliseconds or greater. Ohsawa et al. [93] suggested that 5,10-dimethyl-5,10-dihydrophenazine (DMPZ) in SDS might undergo rapid electron transfer at Pt electrodes while still in the micelle. However, the same group reported that DMPZ in micellar solutions of alkyltrimethylammonium bromides dissociate from micelles before electron transfer [95]. Analysis of potential step chronoamperometry experiments gave exit rate constants which varied from 10^3 to 10^4 per second as surfactant chain length was decreased from C_{16} to C_{12}.

Eddowes and Gratzel used cyclic and rotating disk voltammetry to study the kinetics of oxidation of tetrathiofulvalene (TTF) in micellar CTAC [81]. They concluded that TTF dissociates from the micelles before accepting an electron from Pt electrodes. They used RDV to estimate an exit rate constant of 7×10^3 per second and an entry rate constant of 5×10^9 per M/sec for TTF. These values are consistent with exit and entry rates measured by alternative methods for other systems (see Sec. II.A).

The above reports present convincing evidence that in certain cases dissociation of the reactant from micelles occurs prior to electron transfer. It is not known whether this scenario is general. However, the situation may not be so simple as electron transfer involving a dissociated "free" reactant at the surface of the electrode. Recall that well above the CMC, charged surfaces tend to be fully coated with surfactant, at least a bilayer thick. With a C_{12} surfactant bilayer on the electrode, electrons would have to transfer across a distance of about 4 nm to reach the outside of the bilayer. This distance increases with increasing surfactant chain length. If the reactant resides outside this bilayer, outer sphere electron transfer across such distances should be very slow [44]. Thus, the possibility that reactants partition into a surfactant layer on the electrode before electron exchange needs to be considered.

D. Electrochemical Reactions in Micelles

1. Reduction of Aromatic Nitro Compounds

Since the pioneering work of Holleck and Exner [14] many studies have been done on polarographic reduction of aromatic nitro compounds in the presence of "surface active agents" (see Sec. II). Camphor, tylose, and gelatin adsorbed to Hg electrodes inhibit electron transfer to the initially formed nitrobenzene anion radicals. Work up to the late 1970s has been reviewed, and a general mechanism for reduction of aromatic nitro groups has been developed [15].

In the absence of inhibitors, nitrobenzenes ($ArNO_2$) in water undergo a four-electron reduction with coupled protonations in a single polarographic wave to yield hydroxylamines (ArNHOH). In the presence of the above surface active inhibitors at pH > 10, however, the first one-electron reaction [Eq. (33)] of nitrobenzene to form nitrobenzene anion radical ($ArNO_2\dot{-}$) occurs in a separate wave at more positive potentials than addition of the next 3 electrons [Eq. (34)].

Micelles and Microemulsions

$$ArNO_2 + e \rightleftarrows ArNO_2^{\div} \quad (ca. -0.7 \text{ V vs. SCE}) \tag{33}$$

$$ArNO_2^{\div} + 3e + 2H_2O + H^+ \rightleftarrows ArNHOH + 3OH^-$$
$$(ca. -1 \text{ V}) \tag{34}$$

This two-wave behavior is similar to that found in aprotic organic solvents.

Including surface-active inhibitors in the solutions allowed study of nitrobenzene radical anions [Eq. (33)] by slowing down following chemical reaction steps. ESR was used to estimate stabilities of the radical anions. Half-lives for nitrobenzene and *meta*-nitrophenol were several minutes. *Para-* and *ortho*-nitrophenols decomposed faster [15]. Mechanistic changes were also found when inhibitors were present at pH < 10, but only single four-electron peaks were observed in these cases.

In 1982, McIntire and co-workers reported detailed studies on the electrochemistry of nitrobenzene in unbuffered SDS, CTAB, and Brij-35 [75,96]. Cyclic voltammetry showed a separate peak for nitrobenzene anion radical formation [Eq. (33)] only in SDS solutions (Fig. 12). In aqueous CTAB and

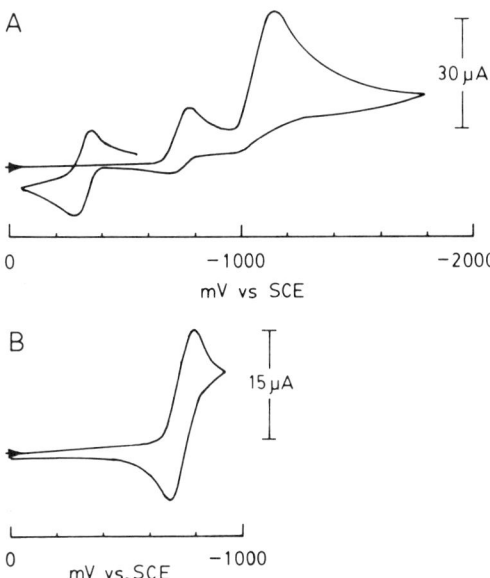

FIG. 12. Cyclic voltammograms of 0.5 mM nitrobenzene in 50 mM LiCl/50 mM SDS at 0.05 V/sec at HDME. Arrows indicate initial potential and direction of scan. (A) 0 to −2000 mV; (B) 0 to −1000 mV. (From Ref. 75.)

Brij-35, a single four-electron peak for reduction to the hydroxylamine was observed. Measurements of samples reduced in situ in an ESR cavity detected $ArNO_2$ $\dot{-}$ in micellar SDS and Brij solutions, but not in CTAB. The anion radical was more stable in SDS than in aqueous Brij-35 or surfactant-free LiCl solutions. Nitrogen hyperfine splitting constants indicated that $ArNO_2$ $\dot{-}$ in SDS was present in the polar Stern layer of the micelles. Comparisons of UV absorption maxima with those in a series of solvents of different polarities suggested that nitrobenzene was also present in the Stern layer. Electrochemical diffusion data were tested with several models for interactions of nitrobenzene with the micelles. A surface interaction between nitrobenzene and the micelles predominated for SDS and Brij-35, while CTAB and CTAC data gave better fits to a pseudophase distribution model [75].

The organic electrochemistry group at the Universite Blaise Pascal in France has studied the electrochemistry of a series of nitrobenzene derivatives. Usually they employed SDS, CTAB, and Brij-35 as surfactants. An early paper [97] explored the influence of surfactant concentration and pH on the electroreduction of nitrobenzene on glassy carbon and Hg electrodes. In CTAB solutions, a single four-electron reduction to phenylhydroxylamine was observed throughout the pH range from 2–12. In acidic solutions, a similar four-electron reduction was found in solutions of all three surfactants. Protonation followed by electron transfer to $ArNO_2H^+$ is involved in the first one-electron step in acidic solutions. In the range $6 \leq pH \leq 9$, protonation of the nitrobenzene anion radical was considered to follow electron transfer [Eq. (33)]. At $pH > 9$, a separate wave for the one-electron reduction of nitrobenzene was found in SDS and in buffer without added surfactant. Nitrobenzene anion radicals formed at pH 8.15 were studied by ESR. Stability was much greater in Brij and SDS solutions than in CTAB solutions [97], similar to the findings of McIntire et al. [75].

Substituent effects on polarographic half-wave potentials ($E_{1/2}$) of nitrobenzene derivatives were examined in acidic Brij-35 and CTAB solutions [98]. A single four-electron wave was observed in all cases. $E_{1/2}$ values gave linear correlations with Hammett/Taft σ substituent constants with slopes of 0.15 for 0.01 M CTAB, 0.20 for 0.01 M Brij, and 0.24 without surfactant. At pH 9 and above, concentrations of Brij > 1 mM caused the four-electron wave for the nitrobenzene derivatives to split into separate one-electron and three-electron waves [99]. Conversely, four-electron waves were observed in CTAB solutions at all pH values. At pH 9, $E_{1/2}$ values gave linear correlations with σ having slopes of 0.21 for 0.01 M CTAB, 0.19 for 0.01 M Brij, and 0.20 without surfactant. ESR measurements on anion radicals and UV absorbance maxima of parent compounds suggested that both species were sol-

Micelles and Microemulsions

ubilized near the surface of the micelles. Limiting currents reflected the apparent diffusion coefficients of the micelles [98].

The homologous p-nitroalkoxybenzene series C_nH_{2n+1}-Ar-NO$_2$ was studied by the Blaise Pascal group in solutions containing Brij-35. In aqueous solutions (pH 9.6) and 50% ethanol (pH 10.7) containing > 0.01 M surfactant, separate one-electron [Eq. (33)] and three-electron [Eq. (34)] waves were found [100]. The first reaction in these media was reversible in dc polarography. Spectroscopic data suggested that the nitro group resided reasonably close to the surface of the micelle for all the homologs, but the hydrocarbon chains resided in the micellar core. D' values for these homologs decreased with increasing chain length in both aqueous and ethanol/water Brij solutions. The value for the C_6 derivative, because of solubility the longest-chain derivative that could be studied in purely aqueous surfactant, had D' = 5×10^{-7} cm^2/sec. This is in reasonable agreement with a value of 3×10^{-7} cm^2/sec estimated from the Brij-35 micelle's radius in water [75]. Results were rationalized in terms of dependence of micelle-solute interactions on chain length [100].

Nitrobenzenes with charged substituents, specifically m-SO$_3^-$ and p-O(CH$_2$)$_3$N$^+$(CH$_2$)$_3$, were found to have electrochemical reactions similar to those of other derivatives with a few interesting differences. These water-soluble ions were not bound to micelles unless the surfactant head groups had an opposite charge. Large negative shifts in polarographic $E_{1/2}$ values in acidic media when Brij-35 and SDS were added to the solution were attributed to adsorption of the surfactant on the electrode and its influence on preceding protonation of the nitro group [101]. Further information about the reasons for this behavior was obtained by studies of the homologous series of p-O(CH$_2$)$_n$N$^+$(CH$_2$)$_3$-nitrobenzenes. Negative shifts of $E_{1/2}$ in CTAB and Brij-35 solutions became smaller as n increased, but diffusion currents decreased. The former was attributed to adsorption of surfactants, and the latter to mixed micelle formation [102].

The three isomers of nitroaniline gave essentially the same behavior in micellar solutions as the other nitrobenzene derivatives, except for the *meta* and *para* isomers in acidic solutions. In the absence of surfactant, these two isomers gave single six-electron polarographic waves corresponding to the formation of diaminobenzenes. In acidic buffers containing 0.01 M Brij-35, the wave splits into two waves representing successive four-electron and two-electron processes [103]. The first reaction forms the hydroxylamine. This dehydrates to quinonediamine HN=R=NH which is then reduced in the second two-electron wave [Eq. (37)].

$$H_2N\text{-Ar-NO}_2 + 4e + 4H^+ \rightleftarrows H_2N\text{-Ar-NHOH} + H_2O \qquad (35)$$

$$H_2N\text{-}Ar\text{-}NHOH \rightarrow HN=R=NH + H_2O \tag{36}$$

$$HN=R=NH + 2e + 2H^+ \rightleftarrows H_2N\text{-}Ar\text{-}NH_2 \tag{37}$$

For p-aminonitrobenzene, the pseudo–first-order rate constant for dehydration of the hydroxylamine at pH 4.3 decreased from 3.3 per second in the absence of surfactant to 0.36 per second in 0.01 M Brij-35. Addition of surfactant also allowed ESR detection of an unusual intermediate p-phenylenediammine cation radical involved in the two-electron reduction [Eq. (37)] [103].

Polarography of 1-nitronaphthalene and 9-nitroanthracene in surfactant solutions was also studied [104a]. These compounds gave four-electron reduction waves in SDS, Brij-35, and CTAB solutions buffered at pH 2.8. Voltammetric studies of the ECE reduction of p-nitrosophenylenediammine in 0.1 M NaOH showed a decrease by about half in the rate of the chemical step, hydrolysis of the hydroxylamine intermediate, upon addition of CTAB in micellar amounts [104b].

2. Reduction of Aromatic Ketones and Activated Olefins

The Blaise Pascal group also investigated electroreduction of aromatic ketones in micellar solutions. Benzophenone, fluorenone, and acetophenone were found to alter the CMC of aqueous CTAB micelles. CMCs were estimated by measuring peak potentials and currents by cyclic voltammetry on glassy carbon electrodes. Above the CMC, the above reactants were reduced by two electrons to the corresponding carbinols. However, at concentrations of acetophenone above 50 mM, one-electron dimerization to the pinacol was the predominant reaction [105].

The *para*-acetophenone derivatives $CH_3CO\text{-}Ar\text{-}O(CH_2)_xN^+(CH_2)_3$, x = 3 to 10, showed adsorption prewaves by polarography and CV in aqueous solutions. The number of electrons transferred per molecule increased from one to two as pH increased from 2 to above 6. In acidic media, a proton and an electron add to the carbonyl group to give ketyl radical $CH_3COH\cdot ArR$, which dimerizes to a pinacol. In basic solutions, two-electron reduction to the carbinol occurred. The strength of adsorption increased as x increased [106], as predicted by Traube's rule. The products of reduction are also adsorbed.

Voltammetric behavior of the above compounds was altered by the addition of surfactants. In acidic and basic micellar solutions of CTAB, and in acidic Brij-35 media, $CH_3CO\text{-}Ar\text{-}O(CH_2)_xN^+(CH_2)_3$ compounds with $x \leq 6$ were replaced by surfactant on the surface of the electrode. These ketones are thought to be reduced through a film of surfactant on the electrode. Compounds with $x \geq 8$ formed mixed micelles with CTAB and Brij-35. AC polarography suggested the presence of coadsorbed reactant and CTAB on Hg electrodes. Polarographic waves for all of the above compounds were greatly decreased in micellar solutions of SDS, with almost no wave found for the x = 10 derivative. The authors suggested formation of large aggregates of SDS and the x = 10 derivative. They proposed that the cationic acetophenone derivatives might act as counterions for SDS micelles [107]. An alternative interpretation involves an increasing distance of electron transfer as x increases.

The influence of CTAB on pinacol/carbinol ratio and the dl/meso stereoisomeric ratio of the pinacol was examined [108] for electrolysis of CH_3CO-Ar-$O(CH_2)_xN^+(CH_2)_3$, x = 3 or 10. CTAB provided some control over product ratios for the x = 3 derivative. At pH 2.7 at potentials of the first one-electron wave, 0.01 M CTAB improved the yield of the x = 3 pinacol from 73 to 100%. At pH 10.3, at the potential of the two-electron wave, a 25% yield of x = 3 carbinol was found in the absence of CTAB. Electrolysis in 0.01 M CTAB at this pH improved the yield of carbinol to 74%. The authors suggested that this result was because of stabilization of the carbanion intermediate by the cationic head groups. CTAB had only a minor influence on dl/meso ratios. Electrolysis of the x = 10 derivative was largely unaffected by CTAB. This suggested that the x = 10 derivative is always reduced within a micellelike environment, since it is itself a micelle-forming surfactant.

Reductions of α, β-unsaturated ketones take pathways similar to those of the acetophenone derivatives discussed above. For example, proton-assisted one-electron reduction of *trans*-ArCH=CHOCH$_3$ gives radicals that couple to yield dimers:

$$\begin{array}{l} ArCHCH_2C(OH)CH_3 \\ | \\ ArCHCH_2C(OH)CH_3 \end{array}$$

At a more negative potential, *trans*-ArCH=CHOCH$_3$ is reduced in a two-electron reaction to ArCH$_2$CH$_2$C(OH)CH$_3$ via a carbanion intermediate. Because of a common radical intermediate for both products, electrolysis at potentials of the two-electron reduction gives mixtures of the dimer and the ketone. Jaeger et al. [109] showed that electrolysis of *trans*-ArCH=CHOCH$_3$ in ethanol/water pH 5 gave ketone to dimer ratios of about 5, with some improvement afforded by using larger alkali cations. The use of CTAB in the reaction medium increased the ketone/dimer ratio to 94 and allowed the reaction to be done without alcohol in the solvent and at less negative potentials. As with the acetophenones, improvement in yield of the two-electron product was attributed to stabilization of the carbanion intermediate by the cationic head group. However, ketone to dimer ratios of 28 in SDS and Brij-35 suggest a micellar effect as well.

A mechanistic study of one-electron reduction of α,β-unsaturated ketone mesityl oxide, $(CH_3)_2C=COCH_3$, showed that addition of sufficient CTAB or Brij-35 to acidic solutions leads to reduction of the neutral rather than the protonated form of the reactant [110]. Electrohydrodimerization of mesityl oxide occurred in a layer of adsorbed CTAB on the electrode. The rate of this reaction was enhanced, and this was attributed to stabilization of the reacting radical anions by cationic head groups in the adsorbed CTAB layer on the electrode.

As mentioned in the introduction, electrohydrodimerizations of activated olefins are of considerable commercial importance, an example being dimerization of adiponitrile in the commercial production of its hydrodimer adiponitrile. Since the original discovery that high concentrations of tetraethylammonium p-toluenesulfonate facilitated high yields of the dimer [17], it was shown that similar selectivity could be achieved by using very low concentrations (e.g., 10^{-4}–10^{-2} M) of nonionic surfactants such as Triton X-100 [111]. The surfactant is thought to displace water molecules from the electrode, limiting the rate of proton donation to the intermediate anion radical. In this way, the anion radicals dimerize faster than when undergoing competing protonation and reduction [111a].

3. Other Organic Reductions

A study of reductions of the cyclodiene organochlorine pesticides dieldrin, endosulfan, and endosulfan sulfate showed that the best signal/noise using differential pulse polarography was obtained by using an aqueous solution of 0.2% each in Triton X-405 and Hyamine 2389 (predominantly methyldodecylbenzyltrimethylammonium chloride) at pH < 8. The authors suggest an irreversible, two-electron reduction involving cleavage of a C-Cl bond located on the bridging carbon atom [112a].

Polarography of dimethylphthalate in micellar 0.15% Hyamine 1622 solutions was consistent with an irreversible four-electron reduction [112b].

4. Micellar Effects on Organic Radicals

In preceding discussions, stabilization of ion radicals by micelles has been invoked in many interpretations of surfactant influences on electrochemical reactions. There have been a number of specific electrochemical studies of radical stability in different surfactant media. McIntire and Blount found that 10-methylphenothiazine cation radical was more stable in SDS solutions than in isotropic salt solutions, but less stable in neutral cationic and nonionic

Micelles and Microemulsions

micellar solutions [96,113a]. The stability of this cation radical in weakly acidic, nonionic Tween-20 solutions was about the same as that in weakly acidic homogeneous solutions [113b]. In unbuffered solutions containing 0.1 M NaCl, stability of the cation radical in SDS was maximum at about 0.5% by weight SDS, but decreased as concentration of cetylpyridinium chloride or Brij 35 was increased above their CMCs [113c].

Kaifer and Bard [77] reported that the equilibrium constant for formation of methyl viologen cation radical ($MV^{\cdot+}$) by the disproportionation reaction

$$MV^{2+} + MV^{o} \rightleftarrows MV^{\cdot+}$$

decreased in the order SDS micelles > water > Triton X-100 micelles > CTAB micelles. SDS micelles stabilized $MV^{\cdot+}$ against dimerization when [micelles]/[$MV^{\cdot+}$] $\geq$ 1. Similar results involving chemical reactions coupled to the one-electron reduction of MV^{2+} were found in micelles of sodium decylsulfate [114a]. However, the amount of dimerization increased when [micelles]/[$MV^{\cdot+}$] < 1, e.g., for alkylsulfate concentrations just above the CMC [114b].

One-electron reduction of tetradecylethyl viologen ($TDEV^{2+}$) was studied by spectroelectrochemistry [114c]. Below its CMC, $TDEV^{\cdot+}$ rapidly formed a dimer. This was also the case upon reduction of $TDEV^{2+}$ in micellar solutions of tetradecyltriethylammonium bromide. Above the CMC of the electroactive amphiphile, however, a mixture of dimers and monomers of $TDEV^{\cdot+}$ was found.

Several groups have reported stabilization of 1,2-dicyanobenzene anion radical in micellar CTAB [21,59,114d]. Anion radicals of 9-phenylanthracene were stable for several seconds when produced in CTAB films on Hg electrodes at -2.2 V vs. SCE, but less stable in solutions of nonionic Igepal [56]. Thus, it is apparent that surfactant aggregates with ionic head groups can stabilize radical ions of the opposite charge in micellar solutions. Stability can be improved further by hydrophobic interactions with micelles. Similar conclusions have been reached based on photochemical and photophysical studies [11].

5. *Organic Oxidations Using Hyamine 2389*

Franklin and Sidarous [115] reported in 1975 that addition of the commercial surfactant Hyamine 2389 to an emulsion of acetonitrile and 2 M NaOH increased yields in the oxidation of benzhydrol to benzophenone from < 1% (no Hyamine) to 90% in 1-hour electrolyses. Yields of 30% benzophenone were obtained in aqueous micellar solutions of Hyamine containing 2 M NaOH, but low yields were found in micellar solutions of anionic and neutral surfactants [116].

Voltammetry on Pt in 4% Hyamine/2 M NaOH showed a large anodic peak at -1.3 V on the first scan. This peak decreased greatly upon subsequent scans, and oxidation of water, representing the positive limit of the potential window, is shifted from 0.8 V (2 M NaOH alone) to 1.8 V. A peak for benzhydrol was clearly visible at -1.3 V upon addition to the Hyamine/NaOH solution, but no peak was detected in 0.1 M NaOH. Results suggested that a film formed on the Pt electrode by oxidation of Hyamine. This film is thought to exclude water from the electrode surface, thus inhibiting oxidation of water to oxygen [116].

Following these early papers, Franklin's group developed a series of applications of electrodes coated with Hyamine films. Micellar solutions and emulsions containing Hyamine 2389 were found to greatly increase yields for oxidative dimerization of diphenylacetonitrile, diethylmalonate [117], thiourea [118], and cysteine [119]. Franklin and Iwunze showed that many organic compounds not oxidizable on Pt electrodes in 0.2 M NaOH give voltammetric waves in micellar solutions of Hyamine/NaOH [120]. They also used Pt coated with Hyamine to enhance the rate of hydrolysis of ethylbenzoate [121]. The fact that this reaction was independent of potential suggested that Hyamine films formed on Pt even in the absence of oxidation.

The exact structure of the Hyamine films has not been fully elucidated. However, Hyamine films on Pt prepared by anodic oxidation are thought to form on an underlayer of Cl adsorbed on the Pt. Multilayer surfactant films were proposed to build up on top of the Cl layer with head groups of the first layer facing the electrode [4,119–123]. Pulsed potentiostatic and galvanostatic studies of the electrooxidation of benzhydrol supported this proposal, and showed that the Hyamine film decomposed at potentials of about -0.6 V vs. SCE. Increased current efficiencies when potentials were pulsed from 1.7 V to values negative of -0.6 V suggested the formation of another film on bare Pt without the adsorbed Cl [123].

A useful variation of the above films was developed by oxidizing Hyamine 2389 in an emulsion containing styrene and 0.2 M NaOH. This results in a polystyrene-Hyamine film on Pt with very low residual currents at potentials as positive as 2 V vs. SCE. Twenty-nine organic acids, aldehydes, alcohols, and nitroanilines gave well-defined oxidation waves on Pt coated with polystyrene-Hyamine [124]. The polymerized coatings gave better sensitivities for the organic compounds than Hyamine-coated Pt without polystyrene.

6. Other Organic Oxidations

Thomalla and co-workers studied several organic oxidations in aqueous micellar systems. Anodic nitration of 1,4-dimethoxybenzene in the presence of

Micelles and Microemulsions 55

NaNO$_2$ was shown to involve coupling of the cation radical of 1,4-dimethoxybenzene and oxidatively produced NO$_2$ [125]. In a study of the effects of different micellar solutions, yields of 2,5-dimethoxynitrobenzene were 70% in Brij 35 and 41% in SDS, as opposed to 28% in homogeneous media. A study of Kolbe oxidation of mixtures of sodium octanoate and octanoic acid showed that formation of tetradecane was inhibited by micelles and could be increased by addition of organic solvent [126].

Rauniyar and Thomalla [127] published an extensive study of anodic cyanation of dimethoxybenzenes in micellar media. They found that heterogeneous electron transfer to the organic reactant was slower in CTAB solutions than in SDS, Brij 35, or homogeneous acetonitrile. However, yields of cyanated product and reaction rates were highest in CTAB micelles. This is because CN$^-$ was also oxidized at the potential of oxidation of the dimethoxybenzenes (DMB). The reaction in aqueous CTAB micelles involved radical coupling between CN$\cdot$ and DMB$\cdot^+$, as opposed to addition of CN$^-$ on DMB$\cdot^+$, which predominated in the other solvent systems. The reactions in CTAB took place in a surfactant film on the electrode and were assisted by bimolecular "micellar catalysis." Because of the change in mechanism in CTAB, the distribution of the two possible cyanated products was different than in homogeneous solutions, anionic micelles, or neutral micelles. This is a good example of control of a reaction pathway by surfactant aggregates.

Voltammetric oxidation of indole in micellar solutions of Triton-405 was used as the basis for an analytical method. A detection limit of 85 ppm was achieved using differential pulse voltammetry [128a]. The detection limit was lowered to 40 ppm by using water–hexane–ethyl acetate emulsified with Triton-405. The method was used to determine indole in shrimp [128b].

As mentioned previously, the oxidation of tetrathiofulvalene (TTF) in micellar SDS proceeds in two one-electron waves of equal height [81c]. In solutions of β-cyclodextrin (β-CD), TTF is bound in the cavity of the cyclodextrin, but the product cations are not. This results in a second wave that is considerably larger than the first [128c], for the same reasons that this is observed in cationic and nonionic micelles (see Sec. IV.B) [81]. Addition of SDS to solutions of β-CD and TTF causes a release of TTF into bulk solution and its uptake by SDS micelles because of binding of SDS monomers to β-CD. Two waves of equal height are observed when the concentration of SDS sufficiently exceeds that of β-CD.

Oxidation of the coumarin drugs umbelliferone (7-hydroxycoumarin) and hymecromone (7-hydroxy-4-methylcoumarin) was investigated at glassy carbon electrodes in micellar solutions of Triton X-450 [129]. A single oxidation peak for these compounds was found at pH 4.8 at about 0.8 V vs. SCE, which was suitable for analytical determinations.

7. Films from Electrochemical Disruption of Micelles

A novel application of micellar electrochemistry has been developed by Saji and co-workers [130a]. A tetraalkylammonium surfactant with a ferrocene moiety attached on the end of its tail is used, e.g., 11-ferrocenyl(undecyl)trimethylammonium bromide. Upon oxidation of this surfactant, a dication results, which does not form micelles. Soluble dyes and insoluble dye particles or polymers can be solubilized or dispersed in solutions of such surfactant. Upon electrooxidation, the micelles are "disrupted" and their contents are released to form films on indium tin oxide electrodes. This method has been successful for making films of 1-phenylazo-2-naphthol [130a], polyvinylpyridine, a methacrylate copolymer [130b], and various metallophthalocyanines [131,132]. A nonionic surfactant, 11-ferrocenyl(undecyl)polyethyleneglycol, was used to prepare the phthalocyanine films.

8. Inorganic Ions and Organometallics

The influence of low concentrations of surfactants on inorganic cations and anions has been widely studied [3], partly because of historical use of surfactants as maximum suppressors in polarography. More recently, this topic has been reviewed in several articles with respect to analytical applications [4,20,133]. In this section, we discuss recent studies stressing the influence of micelles on electrochemical reaction pathways.

The electrochemistry of ferrocene has been extensively studied in micellar media [21,25,69b,71,76,94,125,134]. Its oxidation yields ferrocinium cation. As in organic solvents, this one-electron reaction is reasonably fast (quasireversible) in CTAB and SDS micelles (see Table 4). Ferrocene itself is rather insoluble in water and binds strongly to most micelles. It is an excellent probe for diffusion studies, and data obtained using it fits Eq. (22) for single micelle systems and Eq. (26) for two-micelle systems [21,25,71]. The water-soluble ferrocinium ion remains bound to SDS micelles, but is not bound to nonionic and cationic micelles [69b].

Kamau et al. reported on electroreductions of tris(2,2'-bipyridyl) and tris(4,4'-dimethyl-2,2'-bipyridyl) complexes of Co(II) in micellar solutions of SDS and CTAB on glassy carbon electrodes [28]. The electrode reactions are

$$Co(bpy)_3^{2+} + e \rightleftarrows Co(bpy)_3^{+} \tag{38}$$

$$Co(bpy)_3^{+} + 2e \rightleftarrows Co(bpy)_2^{-} + bpy \tag{39}$$

Heterogeneous electron transfer [Eq. (38)] for the Co(II) complexes was discussed previously (see Table 4). CV, UV absorbance, and NMR studies showed that the complexes were bound to the Stern layer of SDS micelles and

Micelles and Microemulsions

to more hydrophobic regions of CTAB micelles. Ligand dissociation from reduced forms of the complexes was more favorable in CTAB micelles than in acetonitrile. Reduced forms of the complexes were adsorbed to the electrode from CTAB solutions. The (-1)Co complexes reduced water in micellar SDS, but not in CTAB [28].

Kirchoff et al. studied a series of complexes of Re and Tc of the form $MD_2X_2^-$, where D = 1,2-bis(diethylphosphino)ethane (depe) or 1,2-bis(dimethylphosphino)ethane (dmpe), and X = Cl or Br, in micellar solutions of SDS, CTAB, and Triton X-100 [34]. The electrode reaction

$$MD_2X_2^- + e \rightleftarrows MD_2X_2^0 \tag{40}$$

could be made reversible for the dmpe complexes by addition of surfactants at micellar concentrations. Micelles solubilized the $MD_2X_2^0$ dmpe complexes and prevented their adsorption on the electrode, but the more hydrophobic depe complexes remained adsorbed.

Ouyang and Bard [135a] studied voltammetry and electrogenerated chemiluminescence (ECL) of tris(2,2'-bipyridyl)osmium(II) in micellar solutions of SDS, CTAB, and Triton X-100. Interaction of $Os(bpy)_3^{2+}$ was strongest with SDS micelles. Without salt present, no oxidation wave or ECL was observed in SDS solutions, suggesting that the complex ion was inaccessible to the electrode. Increasing salt concentration in SDS solutions increased peak current and ECL intensity. Weak interactions of $Os(bpy)_3^{2+}$ were found with CTAB, and there was almost no effect of Triton X-100.

Hoshino et al. noted a similar effect of salt on the oxidation of 11-(ferrocenyl)undecyltrimethylammonium bromide in micellar SDS [135b]. No oxidation occurred unless salt was added to the solution. Peak current increased up to about 0.5 M salt. The type of salt was not very important. These results were explained in terms of charge compensation by added salt upon oxidation of the ferrocene surfactant.

Franklin and co-workers showed that arsenite, iodide, bromide, thiosulfate, and sulfide ions gave anodic waves at Pt electrodes in micellar solutions of Hyamine 2389/2 M NaOH, but not in isotropic NaOH solutions [136]. This behavior was attributed to formation of a Hyamine film on Pt, which inhibits oxygen evolution, as discussed earlier, allowing the anion oxidations to be observed.

These researchers also reported electrooxidation of inorganic minerals as particles suspended in Hyamine 2389/styrene/NaOH solutions and in thin layer cells having Pt electrodes coated with Hyamine/polystyrene films. They demonstrated the anodic oxidation of iron pyrite (FeS_2) [137,138], galena (PbS) [139], PbO [140], and sulfides, selenides, and tellurides of Ni(II),

Mn(II), and Mo(IV) [141]. A variety of oxidation reactions were identified, especially in the thin layer cell where products of initial reactions were kept close to the electrode and could be further electrolyzed. These reactions involved oxidations of the anions as well as conversions of metals to higher oxidation states. This approach was used to develop analytical methods for determination of pyrite in ore [138] and lead in paint [140].

Franklin also found that barium peroxide was oxidized in solutions of cationic micelles to yield what is presumably barium superoxide. This oxidation product generated at anodes was used to oxidize carbon tetrachloride [142], ethylene dibromide [143], and a series of organic chlorides and fluorides [144]. The products of these reactions were halide ions, water, and carbon dioxide, trapped as barium carbonate. A study of oxidation of ethylene dibromide promoted by barium peroxide revealed that dodecyltrimethylammonium chloride solutions gave better yields than Hyamine 2389. The optimum pH was 5, and graphite electrodes gave better performance under galvanostatic conditions than Pt or Pb [143].

9. Modified Electrodes in Micellar Solutions

Octadecylsilyl-coated electrodes were studied in the hope that they would coadsorb surfactant and electroactive reactants from micellar solutions [46e]. Pt and pyrolytic graphite with stable monolayers of octadecyltrichlorosilane (ODS) bound to their surfaces were used. CTAB and SDS layers adsorbed from micellar solutions onto these ODS-coated electrodes were fragile. Analyses of these electrode surfaces by x-ray photoelectron spectroscopy (XPS) showed that moderate washing with water removed the surfactant, but not the ODS coatings. Significant concentrations of adsorbed ferrocyanide, ferrocene, or ruthenium hexammine were not achieved on surfaces of ODS electrodes in micellar solutions because of partition of solutes into the micelles. However, micellar solutions provided moderate control over heterogeneous electron transfer (ET) rates at the ODS-electrodes. For surfactant and electroactive ions of the same charge sign, ET was partly inhibited. Ferrocene and ions of charge opposite to the surfactant showed faster ET [46e].

Electrodes coated with clay films have also been examined in surfactant solutions. Natural clays are layered aluminosilicate cation exchangers. They adsorb cationic surfactants, which form bilayer or hemimicelle coatings on the clay surface. Surfactant-treated clay colloids can coadsorb nonpolar reactants [145]. Clay-modified electrodes (CMEs) were made by depositing colloidal Na-bentonite (ca. 500 nm thick) on pyrolytic graphite (PG). The tris(2,2'-bipyridyl)cobalt(II) dication was taken up by the CMEs in the absence and presence of CTAB micelles [146]. It gave separate CV reduction peaks for Co(II) (-1.2 V) and Co(I) (-1.5 V).

Micelles and Microemulsions

Shi et al. applied various characterization techniques to CMEs treated with CTAB and cationic metal complexes [147]. X-ray powder diffraction showed expansion of clay interlayers by 2–5 Å when CMEs were equilibrated with CTAB, $Ru(NH_3)_6^{3+}$, or $Co(bpy)_3^{2+}$ alone or in micellar solutions of CTAB. Results suggested independent binding sites in clay films for the metal complexes and CTAB. Co and Ru were detected by XPS on CMEs treated with $Ru(NH_3)_6^{3+}$ and $Co(bpy)_3^{2+}$. Depth profiles by Argon ion sputtering and XPS indicated distribution of Co and Ru throughout the CMEs. The amount of Co found by XPS on the surface of CMEs treated with $Co(bpy)_3^{2+}$ and CTAB was less than on CMEs treated only with $Co(bpy)_3^{2+}$.

Static secondary ion mass spectrometry (SIMS) detects ions from *only the top 2–5 Å* of the sample. A striking result from SIMS analysis of CMEs was a larger signal for cobalt on the Co/CTAB/CME surface than for a CME treated only with $Co(bpy)_3^{2+}$ [147]. Combined with the other data, this suggests that $Co(bpy)_3^{2+}$ binds to aggregates of CTAB on the outer surface of the clay [147]. Ru was not detected on the surface of the CMEs treated with $Ru(NH_3)_6^{3+}$ and CTAB. Hydrophilic $Ru(NH_3)_6^{3+}$ does not bind to CTAB aggregates and is replaced by them on the surface.

The electroactive surfactant (ferrocenylmethyl)dodecyldimethylammonium bromide was also taken up from aqueous solutions by CMEs [148]. This ion could be oxidized only when present in excess of the cation exchange capacity of the clay, suggesting that surfactant tightly bound to the clay is not oxidized. CMEs treated with CTAB were found to take up the negative ferrocyanide ion but to reject methyl viologen dications.

E. Electrochemical Catalysis in Micellar Solutions

1. Introduction

Control and enhancement of thermal and photochemical reactions with micelles is well known. Principles of rate control developed for these systems [9–11] have also been applied to electrochemical catalysis [22]. The electrode adds another interface, at which heterogeneous redox reactions occur. This electrode-solution interface can be exploited for kinetic control.

In electrochemical catalysis, a catalyst (mediator) is added to the system to shuttle electrons between electrodes and substrates that are otherwise difficult to reduce or oxidize. A major advantage is that the catalyzed reaction requires less energy (i.e., a smaller overpotential) and is faster at the formal potential of the catalyst redox couple than direct electron exchange between substrate and electrode. In some cases, the direct reaction may not be possible. Rate-determining steps in electrochemical catalyses are usually bimolecular electron transfers. As will be shown below, this means that their rates can be controlled by surfactant aggregates.

As mentioned previously, the observed rate of a chemical reaction in micellar solutions is considered to be the sum of the rates in the continuous and micellar phases [9,10]. For a bimolecular reaction between A and B in aqueous micelles:

$$k_{obs}[A][B] = k_m[A]_m[B]_m + k_w[A]_w[B]_w \tag{41}$$

where observed rate constant k_{obs} is found on the basis of the moles of A and B in the total volume of the system, V_t. Subscript m refers to the micellar phase and subscript w to the water phase. In the simplest case, both reactants are entirely bound to the micelles and the reaction takes place completely in the micellar phase. The rate of reaction in water can be neglected, so that the second term on the righthand side of Eq. (41) is negligible. This leads to the expression:

$$k_{obs} = k_m[A]_m[B]_m/[A][B] \tag{42}$$

For this special case of totally bound reactants $[A]_m = [A]/\phi_m$ and $[B]_m = [B]/\phi_m$, where ϕ_m is the volume fraction of the micellar phase. Using these relations in Eq. (42) gives:

$$k_{obs} = k_m/\phi_m^2 \tag{43}$$

Equation (43) shows that the observed rate is enhanced by compartmentalization of reactants into the reaction volume $V_t\phi_m$, producing an apparent catalysis. Rate enhancement is mainly a consequence of high reactant concentrations in the micellar volume, which contains all the reactants.

The largest rate enhancements in electrochemical catalysis have been found when the reaction occurs in surfactant aggregates on the electrode surface [22]. In this case, the volume that contains the bimolecular rds is the volume of the surface aggregates. Equations (41) and (42) can also explain kinetic control of selectivity of catalytic reactions in micelles. If the two reactants are spatially separated in micellar and water phases by virtue of their solubility properties, the rate of reaction will be decreased compared to the case where both reactants are present only in the micellar phase.

Although much more sophisticated treatments of kinetics in micelles are available [9–11,26], the simple concepts above are useful for a qualitative understanding of electrochemical catalysis in micellar solutions. Mediation of electrochemical reactions in micellar solutions seems to first have been used by Kuwana and co-workers [149]. They solubilized ferrocene in nonionic micelles and formed ferrocinium ions at electrodes for redox titrations of cytochrome c and cytochrome c oxidase.

Micelles and Microemulsions

2. Catalytic Reductions of Organohalides

These reactions (Scheme II) were among the first to be studied in micellar systems to see if increases in rates could be realized. Organohalide reductions involve two-electron cleavage of carbon-halogen bonds. These reductions can be catalyzed by organic catalysts such as anthracene derivatives and macrocyclic complexes including metal corrins and phthalocyanines. The accepted reaction pathway, where Ar is a phenyl ring, is as follows [150]:

Scheme II

$$P + e \rightleftarrows Q \qquad E^o \qquad \text{(at electrode)} \tag{44}$$

$$ArX + Q \underset{k_2}{\overset{k_1}{\rightleftarrows}} ArX^{\div} + P \tag{45}$$

$$ArX^{\div} \overset{k}{\to} Ar\cdot + X^- \tag{46}$$

$$Ar\cdot + Q \to P + Ar^- \qquad \text{(fast)} \tag{47}$$

$$Ar^- + (H+) \to ArH \qquad \text{(fast)} \tag{48}$$

Catalyst P is dissolved in the surfactant medium or immobilized on the electrode. At potentials near E^o, P is reduced to its active form Q by accepting an electron from the electrode [Eq. (44)]. Q transfers an electron to substrate ArX [Eq. (45)].

Equation 45 is the rate-determining step in many aryl halide reductions. For alkyl halides, processes in Eqs. (45) and (46) usually occur as a single concerted step. A second electron transfer and protonation [Eqs. (47) and (48)] yield RH.

The resulting advantage of electrochemical catalysis as in Scheme II is that ArX is reduced to ArH at the formal potential (E^o) of the catalyst, rather than at the more negative potential for direct, irreversible reduction of ArX at the electrode. Catalyst P is regenerated in the catalytic cycle [Eqs. (45) and (47)] and gets reduced again at the electrode. As a result, the electrochemically measured peak current for reduction of P [Eq. (44)] is larger when RX is present. This is called the "catalytic" current and can be used to obtain the chemical catalytic rate constants [150,151].

One of the first mediators evaluated for organohalide reductions in micellar solutions was 9-phenylanthracene (9-PA) in aqueous CTAB. Reduction of 9-PA at Hg at -2.2 V vs. SCE gave anion radicals [see Eq. (44)] that were stable for several seconds in a multilayer film of CTAB adsorbed on a Hg electrode [56]. CV at very low scan rates clearly showed that reactive anion radicals of 9-PA were concentrated in a thick film of CTAB on the electrode. This anion radical was less stable in nonionic surfactants and not observable in anionic SDS micelles. The positive charge of CTAB helped stabilize 9-PA anion radical in the film on the electrode.

An apparent rate constant k_1 [Eq. (45)] of about 10^7 M^{-1}/sec was estimated[56] by cyclic voltammetry for reaction of 9-PA anion radical with 4-bromobiphenyl (4-BB) in 0.1 M CTAB/0.1 M tetraethylammonium bromide (TEAB). This is much larger than k_1 in homogeneous N,N-dimethylformamide (DMF) (Table 5). The thick film of CTAB adsorbed on the electrode (Fig. 2) at -2.2 V preconcentrates large amounts of hydrophobic reactants and causes rate enhancement. In fact, the rate of electron transfer between 9-PA anion radical and 4-BB became so fast that it was no longer the sole rate-determining step. The kinetics of decomposition of ArX $\dot{-}$ in Eq. (46) also needed to be considered to explain the voltammetric results [56].

In bulk electrolysis using aqueous CTAB and 9-PA on a Hg pool electrode, 92% of 4-BB was converted to biphenyl in 2.3 hours. The yield of biphenyl based on moles of 4-BB converted was 103%, and no other products were detected [56].

Catalytic reductions of organohalides by micelle-bound catalysts at potentials more positive than -2 V in aqueous CTAB showed smaller rate enhancements (Table 5). These reactions did not occur on the electrode surface, but in diffusing micelles [21,22,151a]. Smaller apparent rates are found in such cases because the two reactants are statistically distributed among the micelles, lowering the product of their concentrations in "reactive" micelles containing both reactants. It is assumed that the largest contribution to the kinetics occurs in such "reactive" micelles containing at least

TABLE 5
Observed Catalytic Rate Constants in Micellar and Homogeneous Systems[a]

		Micellar System			Homogeneous	
Cat.[b]	$E^{o\prime c}$ (V/SCE)	Substr.	System	$k_{1,app}$ (M^{-1}/sec)	Solvent	k_1 (M^{-1}/sec)
9-PA	-2.15	4-BB	0.1 M CTAB	10^7	DMF	300
1,2-DCB	-1.55	TCB	0.1 M CTAB	3×10^3	DMF	115
Co(bpy)$_3^{2+}$	-1.1	AllylCl	0.1 M CTAB	220	MeCN	100
Co(bpy)$_3^{2+}$	-1.2	AllylCl	0.1 M SDS	90		
Co(bpy)$_2$(bpyC16)$^{2+}$	-1.6	AllylCl	0.1 M SDS	300		
Co(phen)$^{2+}$	-0.05	H$_2$O$_2$	5 mM SDS[d]	810	water	2×10^3

[a] Data summarized from Ref. 22, except last line.
[b] Abbreviations: 9-PA, 9-phenylanthracene; 1,2-DCB, 1,2-dicyanobenzene; TCB, 2,2′,5,5′-tetrachlorobiphenyl; 4-BB, 4-bromobiphenyl.
[c] In micellar solution
[d] From Ref. 152.

one of each of the reacting species. Micelles containing only a single reactant make smaller contributions to observed rates, especially if the reaction is inherently fast. The product of molar reactant concentrations at the reaction site, to which the forward rate of Eq. (45) is directly proportional, can be larger when both reactants are concentrated in a film on the electrode.

Small rate increases in CTAB were found for reduction of allyl chloride with tris(2,2'-bipyridyl)cobalt(II) [151a]. The pathway of this catalytic dimerization is shown in Scheme III.

Scheme III

$$\text{Co(bpy)}_3^{2+} + e \rightleftarrows \text{Co(bpy)}_3^{+} \quad \text{(at electrode)} \tag{49}$$

$$\text{Co(bpy)}_3^{+} + \text{RX} \xrightarrow{k_1} \text{Co(bpy)}_2(\text{RX})^{+} + \text{bpy} \tag{50}$$

$$\text{Co(bpy)}_2(\text{RX})^{+} \rightarrow \text{Co(bpy)}_2^{2+} + \text{R} \cdot + \text{X}- \tag{51}$$

$$2 \text{ R} \cdot \rightarrow \text{R-R} \tag{52}$$

$$\text{Co(bpy)}_3^{2+} \rightleftarrows \text{Co(bpy)}_2^{2+} + \text{bpy} \tag{53}$$

The Co(I) complex formed in Eq. (49) reacts with allyl chloride (RX) to give an organocobalt intermediate [Eq. (50)]. This decomposes [Eq. (51)] to yield allyl radicals (R·), which couple to form 1,5-hexadiene [R-R, Eq. (52)]. Only 73% of the catalyst is bound to CTAB micelles [151], lowering its concentration in the reactive micelles and limiting rate enhancement.

Rate constants for reduction of allyl chloride mediated in aqueous SDS micelles by tris(2,2'-bipyridyl)cobalt(II) were similar to those in acetonitrile. The long-chain derivative bis(2,2'-bipyridyl)(4,4'-dihexadecyl-2,2'-bipyridyl)cobalt(II) gave a threefold larger rate (Table 1) [151]. NMR and UV spectra showed that tris(2,2'-bipyridyl)cobalt(II) is bound to the negatively charged outer surface of SDS micelles, but to hydrophobic sites of CTAB micelles [28]. Bis(2,2'-bipyridyl)(4,4'-dihexadecyl-2,2'-bipyridyl)cobalt(II) and allyl chloride are anchored to hydrophobic regions of micelles. Comparing these observations with kinetic data (Table 5) suggests that the largest rates occur when both reactants are strongly bound in similar regions of the micelles.

Sugiyama and Aoki studied the catalytic reduction of hydrogen peroxide with Cu(II)(1,10-phenanthroline), $[\text{Cu(phen)}]^{2+}$, in SDS micelles [152]. They found that reactive species $[\text{Cu(phen)}]^{+}$ was adsorbed to the glassy carbon electrode in SDS. The rate constant for the catalytic reduction was threefold larger in water than in 5 mM SDS. This is consistent with the reaction occurring in a micellar environment in which H_2O_2 would be poorly soluble.

Catalytic efficiencies for reduction of allyl chloride with tris(2,2'-bipyridyl)cobalt(II) were about twofold larger on clay-modified electrodes (CMEs) than on bare PG in CTAB solutions. Furthermore, decomposition of the organocobalt intermediate [Eq. (51)] was catalyzed by the clay surface [147]. Reductive debromination of 4,4'-dibromobiphenyl (4,4'-DBB) was mediated by the two-electron reduction of tris(2,2'-bipyridyl)cobalt(I) [146]. Catalytic efficiency for this reaction in CTAB micelles was also enhanced up to twofold on CMEs compared to bare PG. A remarkable feature during reduction of 4,4'-DBB on CMEs in CTAB was effective protection of tris(2,2'-bipyridyl)cobalt(I), as opposed to complete decomposition of this mediator in acetonitrile in an hour (four turnovers) [146].

Electroanalytical, x-ray diffraction, and surface spectroscopy studies showed that CMEs coadsorbed CTAB, catalyst, and organohalides from CTAB solutions [147]. UV spectra of clay films treated with CTAB and either tris(bipyridyl)cobalt(II) or 4,4'-DBB demonstrated adsorption of the latter two species on the clay. Fivefold smaller UV absorbance and energy disper-

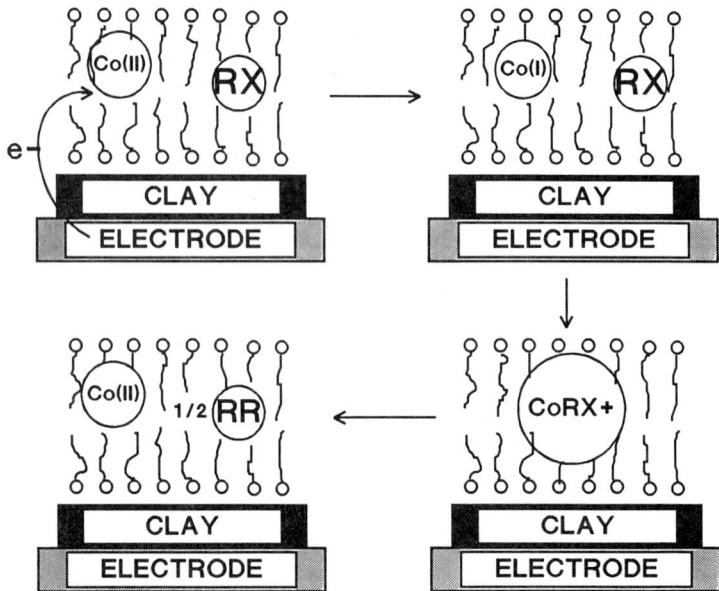

FIG. 13. Conceptual representation of electrochemical catalytic dimerization of allyl chloride (RX) to 1,5-hexadiene (RR) with tris(2,2'-bipyridyl)cobalt(II) [Co(II)] on the outer surface of a CME in CTAB solutions. The layered nature of the clay coating is not shown. (From Ref. 22.)

sive x-ray analysis signals for tris(bipyridyl)cobalt(II) when adsorbed from aqueous CTAB instead of water suggested that CTAB and catalyst compete for some similar binding sites.

The above results and those discussed earlier for CMEs in CTAB solutions suggest that reductions of organohalides by cobalt bipyridyl complexes occurred in CTAB aggregates adsorbed on the outer surfaces of the CME (Fig. 13). Adsorption of CTAB provides hydrophobic reaction sites on the CME with higher concentrations of reactants than can be achieved at bare electrodes, increasing catalytic efficiency by enhancing rates of bimolecular rate-determining steps.

V. ELECTROCHEMISTRY IN MICROEMULSIONS

A. Diffusion in Conductive Microemulsions

Mackay and co-workers pioneered the use of electroanalytical methods to characterize mass transport in microemulsions. In several early papers, they demonstrated a correlation between electrochemically measured diffusion coefficients (D') of water-soluble electroactive ions in microemulsions with the volume fraction of oil (ϕ_c) [153–155]. The equation

$$D' = D_o(1 + \phi_c)^{n+1} \tag{54}$$

where D_o is the diffusion coefficient of the probe ion in water, was found to hold for microemulsions of nonionic and ionic surfactants with a series of hydrophilic electroactive ions including Cd(II), Tl(I), $Fe(CN)_6^{3-}$, and $Fe(CN)_6^{4-}$. The value of exponent n was the same as that determined from a similar relationship between equivalent conductance and ϕ_c, where $\phi_c = 1 - W_w g$, W_w is the weight fraction of water in the microemulsion, and g is its specific gravity. Diffusion coefficients of these water-soluble ions were consistent with an obstruction effect of oil droplets in o/w microemulsions or with bicontinuous structures.

On the other hand, D' for the oil-soluble 1-dodecyl-4-cyanopyridinium ion was independent of the amount of water (35–65%) in sodium cetylsulfate (SCS)/water/pentanol/oil microemulsions and gave a D' value of 4.4×10^{-7} cm²/sec. Using the Stokes-Einstein relation, this corresponds to a droplet radius of 45 Å, in good agreement with the value of 50 Å found by x-ray scattering [155].

1-Dodecyl-4-cyanopyridinium ion was also an effective probe for diffusion of oil droplets in octane/CTAB/1-butanol/water microemulsions at high water content. Values of D' obtained by polarography were in excellent agreement with those from QELS [87]. For SCS and CTAB microemulsions with low water content, agreement of polarographic D'-values with QELS was not as good. This may be attributed to polydispersity, which is reflected in different ways by the two different methods. Polarography provides apparent self-diffusion coefficients, while QELS can give a self-diffusion coefficient, a collective diffusion coefficient, or a combination of the two. Polarographic diffusion data of a series of viologen and pyridinium ions with different alkyl chain lengths were used to support o/w or bicontinuous structural interpretations for CTAB and SCS microemulsions [87].

Mackay et al. also studied polarography of Cu(II) and Cd(II) and their complexation with quinoline in SCS/1-pentanol/oil/water and CTAB/hexadecane/1-butanol/water microemulsions [156]. Large negative shifts, e.g., 0.9 V for Cu(II), for reduction waves of these ions were observed when changing from benzene to mineral oil in the SCS microemulsions. Both ions formed only weak complexes with quinoline, with Cu(I) complexing more strongly. Polarography of Cd(II) in the CTAB microemulsion gave a D-value of 7×10^{-7} cm^2/sec that was independent of water content between 35 and 70%. This suggested that Cd(II) diffused along with the oil droplet and was bound to the CTAB-coated droplet's surface mainly as the dianion $CdBr_4^{2-}$ [156].

George and Berthod have reported a number of diffusion studies using electrochemical measurements in microemulsions. In the methylene chloride/sodium p-octylbenzenesulfonate/1-pentanol/water system, D' for oil-soluble 10-methylphenothiazine increased and D for water-soluble hydroquinone decreased as ϕ of the oil increased. Except for hydroquinone at high oil content, all D'-values were too large to reflect droplet diffusion. These data were used to infer a large bicontinuous region in the center of the phase diagram [157]. Similar conclusions were drawn from a study of reduction of oxygen and hydrogen peroxide in the same microemulsion system [158]. Diffusion data reflected a progressive change from an o/w system in the water-rich corner of the phase diagram, through a bicontinuous structure, to a w/o system in the oil-rich domains.

Electrochemically measured diffusion coefficients of water-soluble hydroquinone and oil-soluble ferrocene were obtained at a series of compositions of microemulsions of brine/SDS/dodecane with 1-pentanol or 1-heptanol as cosurfactants. D'-values were shown to be consistent with conductivity data, and microviscosity and polarity estimated by fluorescence probe studies [159]. This approach was also used to help interpret some unusual details of phase diagrams of brine/SDS/pentanol/dodecane or hexane systems [160].

Qutubuddin and co-workers [161] evaluated ferrocene as a probe to obtain self-diffusion coefficients of oil droplets in o/w microemulsions of CTAB/water + NaBr/1-butanol/n-octane. D'-values found by CV and RDV

Micelles and Microemulsions

showed good agreement with each other but poor agreement with values obtained from QELS. The authors suggested that their QELS experiments measured mainly collective diffusion. They also report that D' was independent of ferrocene concentration, and considered the measured values to give good estimates of droplet diffusion coefficients.

The same group investigated the electrochemistry of methyl viologen (MV^{2+}) in o/w microemulsions of CTAB, SDS, and Triton X-100 [80]. The microemulsions blocked adsorption of reduction products found in isotropic aqueous media. The two redox couples $MV^{2+}/MV^{+}\cdot$ and $MV^{+}\cdot/MV^{0}$ were reversible in these microemulsions. Measured D'-values for MV^{2+} indicated its binding to oil droplets in SDS systems, but no binding in cationic and nonionic microemulsions. After obtaining the droplet D'-value using ferrocene, a slow droplet-probe equilibrium model similar to Eq. (14) was used to estimate the fraction of MV^{2+} bound in various SDS systems. The validity of this approach was not verified by the authors. As discussed previously, the slow equilibrium model is not expected to hold for most microemulsions.

Qutubuddin et al. [162] also attempted to determine distribution coefficients of ferrocyanide ion, 4-amino-3-methyl-N,N-diethylaniline (PPDS), and 4-amino-3-methyl-N-ethyl-N-(β-sulfoethyl)aniline (PPD1) in o/w microemulsions of CTAB and octadecyldimethyl betaine (ODMB). The authors again used Eq. (14) (CV data) and its analog for RDV to convert the measured diffusion data to distribution coefficients (P). These P-values were shown to be a lower bound on the actual distribution coefficients as a consequence of neglecting probe-droplet kinetics near the electrode. Using this approach, ferrocyanide ion was reported to be very weakly bound to CTAB-coated oil droplets. The order of binding was ferrocyanide << PPD1 << PPD2, with the same binding order of PPD1 and PPD2 in the ODMB microemulsion.

Mackay et al. [163] recently used a series of probes with different solubility properties to examine mass transport in a CTAB/1-butanol/hexadecane/water microemulsion. The probes were ferricyanide, ferrocene (Fc), diferrocenylethane (Fc-Fc), acetyl ferrocene (AcFc), and methyl viologen. For ferricyanide, D' was on the order of 10^{-6} cm^2/sec between 19 and 60% water, suggesting that ferricyanide traveled with the surfactant. D' for Fc decreased from 4 to 0.6×10^{-6} cm^2/sec when water content was varied from 20 to 90%, suggesting that the system converts from a bicontinuous to an o/w structure. D'-values for Fc-Fc suggested that this probe was strongly bound and perhaps not fully accessible to the electrode. Values for water-soluble MV^{2+} and AcFc were larger, suggested that these species resided mainly in the continuous water phase. A qualitative attempt was made to use Eq. [31] to explain $E_{1/2}$ values of the probe.

Ferrocene was found [163] to give a concentration dependent D' as reported for several micellar systems [25]. However, concentration-independent D'-values for this probe were indicated in similar CTAB microemulsions [161]. Also, ferrocyanide was reported to be strongly bound to CTAB structures [163] but was also reported to be only weakly bound in a similar CTAB microemulsion [162]. Clearly additional work, preferably within a framework of reliable models for the concentration dependence of D', is needed to resolve these inconsistencies.

B. Diffusion Studies with Microelectrodes

With oil as the continuous phase, w/o microemulsions are quite resistive. Voltammetry in these systems is plagued by a large iR drop, and meaningful D'-values are difficult or impossible to obtain with conventional-sized electrodes. However, by virtue of their size, microelectrodes with dimensions in the μm range are nearly unaffected by iR drop and can be used to study diffusion in highly resistive systems.

Chen and Georges [164] first used steady-state microelectrode voltammetry to study diffusion in w/o microemulsions. They used ferrocene to probe diffusion in the oil phase of an SDS/dodecane/1-heptanol/water system. The diffusion coefficient of Fc reflected the microviscosity of the oil phase, rather than the bulk viscosity of the microemulsion.

Owlia et al. [23] reported using water-soluble probes and microelectrodes to study diffusion of water droplets in w/o microemulsions. The resistivity of the Aerosol OT (AOT, bis(2-ethylhexyl)sulfosuccinate) w/o microemulsions was 40-fold greater than that of neat acetonitrile without added electrolyte. Using the water-soluble cobalt(II) corrin complex, vitamin B_{12r}, in a microemulsion of 0.2 M AOT, 4 M water buffered at pH 3, and isooctane, they found a dependence of D' on C_X that was readily fit with Eq. (22) (Fig. 14) by using n = 3. A droplet diffusion coefficient of 0.63×10^{-6} cm^2/sec indicated a hydrodynamic water pool radius of 75 Å by using the Stokes-Einstein relation [23]. This was larger than the value of 50 Å found for solute-free water droplets, but in agreement with an earlier report that vitamin B_{12} caused solute-induced size increases in water pools of dodecylammonium propionate/benzene/water microemulsions [165].

Wang et al. [82] showed that adsorption of MV^{2+} on carbon microdisk electrodes led to incorrect diffusion coefficients in AOT/isooctane/water microemulsions. Although Eq. (22) gave a good fit to the data, values of D_0 and D_1 were unreasonably large. This was caused mainly by adsorption of MV^{2+} on the carbon microdisk electrodes from AOT microemulsions, which was demonstrated by square-wave voltammetry (SWV) following medium exchange. The influence of electrical migration was also addressed in this paper.

Micelles and Microemulsions

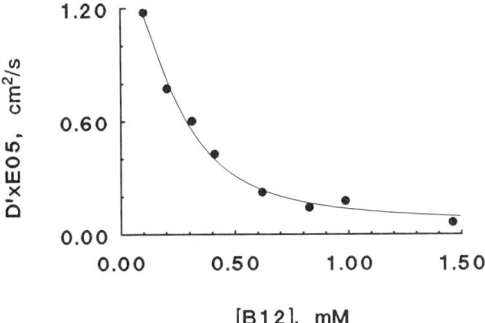

FIG. 14. Influence of concentration of vitamin B_{12r} on apparent diffusion coefficient in microemulsion of 0.2 M AOT, 4 M water pH 3, and isooctane. Squares are experimental data; line is best fit by nonlinear regression onto eq 22 for n = 3. (From Ref. 23.)

C. Electrochemical Reactions in Microemulsions

Owlia et al. [23] studied reductions of the central Co(III) atom of vitamin B_{12a} in w/o microemulsions of AOT/isooctane by using microelectrode voltammetry. Water pools were buffered with phosphate at a nominal pH 3.0, although the formal potential of the pH-sensitive B_{12r} Co(II)/Co(I) redox couple in the microemulsions suggested an apparent pH of about 3.7. The electrochemistry of vitamin B_{12} depends strongly on pH because of its basic benzimidazole side chain that can act as an axial ligand. Protonation of the benzimidazole side chain (L) gives "base off" forms, which are reduced more rapidly at electrodes. Comparisons of formal potentials and heterogeneous rate constants of the Co(III)/Co(II) and Co(II)/Co(I) with those in buffered homogeneous solutions led to identification of the following reactions at carbon microdisks in microemulsion water pools:

$$[LL'Co(III)] + e \rightarrow [L'Co(II)] + L \quad E^{o'} = -0.03 \text{ V/SCE} \quad (55)$$

$$H^+ + [L'Co(II)] + e \rightarrow [Co(I)] \sim LH+ \quad E^{o'} = -0.80 \text{ V/SCE} \quad (56)$$

Both reactions are quasireversible in the AOT microemulsion water pools and in weakly acidic aqueous solutions.

Iwunze et al. showed that quasireversible electrode reactions of several oil and water-soluble redox couples in *bicontinuous* microemulsions of DDAB, oil, and water could be described by voltammetric theory developed for homogeneous solutions [166]. Cyclic voltammetry was used to study the electrochemistry of water-soluble ruthenium(III) hexammine, ferrocyanide, and vitamin B_{12}, as well as oil-soluble ferrocene and polycyclic aromatic hydrocarbons (PAHs) in microemulsions of DDAB/dodecane/water. Results were in good agreement with simulated voltammograms assuming that the bicontinuous medium was homogeneous. This is because both water and oil are continuous. Each microemulsion phase behaves as a homogeneous medium. This is in sharp contrast to micelles, w/o, and o/w microemulsions, for which coupled diffusion and dynamic binding equilibria of reactants with surfactant aggregates must be considered for a full interpretation of voltammetric results.

Standard heterogeneous rate constants (k^o) for the above species in DDAB microemulsions were similar to those found in isotropic solutions (Table 6). Diffusion of solutes did not reflect the high bulk viscosities (19–38 cp) of the DDAB microemulsions. Hydrophilic ions diffused at rates characteristic of the water phase; nonpolar molecules diffused at rates similar to the self-diffusion of oil in the oil phase [166].

Surfactant aggregates in microemulsions might be expected to influence stability of ion radicals in a similar way to micelles. However, George and Berthod [113c] reported a significant decrease in half-life ($\tau = 0.6$ hr) of electrochemically produced 10-methylphenothiazine cation radical in methylene chloride/sodium p-octylbenzenesulfonate/1-pentanol/water microemul-

TABLE 6

Apparent Electrochemical Parameters at Glassy Carbon Electrodes in Bicontinuous Microemulsions

Species	%DDAB[a]	$10^6 D$ (cm^2/sec)	k^o (cm/sec)	$E^{o\prime}$ (V vs. SCE)
Ferrocyanide	21	1.0	0.027	0.027
	13	0.79	0.017	0.022
Ru(II)hexammine	21	0.68	0.016	−0.213
Cob(II)alamine	21	0.3	0.0002	−0.87
Ferrocene	21	6.3	0.010	0.34
	13	5.6	0.004	0.38
Perylene	21	5	0.2	−1.64
Pyrene	21	6	0.1	−2.06
9-PA	21	8	0.2	−1.95

[a] Compositions of microemulsions 21% DDAB/40 dodecane/39% water and 13% DDAB/59% dodecane/28% water.
Source: Ref. 166.

sions, compared to 0.1 M NaCl (τ = 7 hr) and micellar SDS solutions (τ = 3–9 hr). Instability in the microemulsions was attributed to the alcohol, not the surfactant, since addition of pentanol to SDS micelles gave similar decreases in lifetime.

Polyaromatic hydrocarbons (PAHs) were found to undergo ECE reactions at low scan rates in bicontinuous microemulsions of DDAB [166,167]. The ECE process was similar to that found for PAHs in organic solvents in the presence of proton donors. It consists of electron transfer to the PAH to form an anion radical, protonation of the anion radical to give a neutral radical, and a second electron transfer to this neutral radical. This process was studied in some detail for perylene in DDAB microemulsions [167]. Good quality cyclic voltammograms were obtained in bicontinuous systems with as little as 1% water, made by using cyclohexane as the oil. Voltammetric theory for the ECE reaction in a homogeneous medium gave a good description of the data. A self-consistent mean second-order rate constant of 3.7 ± 0.7 M^{-1}/sec was found for reaction of perylene anion radical with water in microemulsions containing 1, 26, and 39% water. The stability of the anion radical increased with decreasing water content as it would in a homogeneous mixture of water and organic solvent.

Garcia et al. recently reported a novel application of w/o microemulsions to polymerization of acrylamide [168]. To overcome the ohmic resistance of an AOT/toluene/water microemulsion, they used a Pt/Nafion solid polymer electrode (SPE) to separate the microemulsion from an aqueous electrolyte in the counter electrode compartment. Oxidation of persulfate ion in the microemulsion at 1.2 V was used to initiate polymerization. Latex particles and solid polyacrylamide were obtained as products. Efficient stirring was important for good yields of latex particles. The microemulsions used for polymerization were characterized by microelectrode voltammetry on a variety of electroactive probes.

D. Electrochemical Catalysis

Owlia et al. [23] were the first to study the kinetics of an electrochemical catalytic reduction in w/o microemulsions. They investigated reductions of several alkyl vicinal dibromides catalyzed by vitamin B_{12}. The rate-determining step in this reaction (Scheme IV) is an *inner sphere* electron transfer between $B_{12}Co(I)$ and the alkyl dibromide [Eq. (58)]. This can occur by a radical mechanism (Scheme IV) or by a concerted E_2 elimination. These two pathways are kinetically indistinguishable. Both give an alkene as the product.

Scheme IV

$$Co(II) + e \rightleftarrows Co(I) \text{ (at electrode)} \quad (57)$$

$$Co(I) + RX_2 \xrightarrow{k_1} RX\cdot + X^- + Co(II) \text{ (rds)} \quad (58)$$

$$RX\cdot + Co(I) \rightarrow \text{alkene} + X^- + Co(II) \quad (59)$$

In w/o microemulsions of AOT/water/isooctane, the highly water-soluble vitamin B_{12a} resides entirely in water pools. Substrates ethylene dibromide (EDB), 1,2-dibromobutane (DBB), and *trans*-1,2-dibromocylohexane (t-DBCH) are present mainly in the continuous isooctane phase. Thus, catalyst and substrate are spatially separated in the two phases of the microemulsion.

Because of the high resistance of w/o microemulsions, ultramicroelectrode voltammetry was required. Addition of vicinal dihalides to the microemulsion containing vitamin B_{12} caused an increase in limiting current (Fig. 15), from which k_1 was obtained. The reaction occurred at diffusing water microdroplets, rather than on the electrode surface. Observed apparent

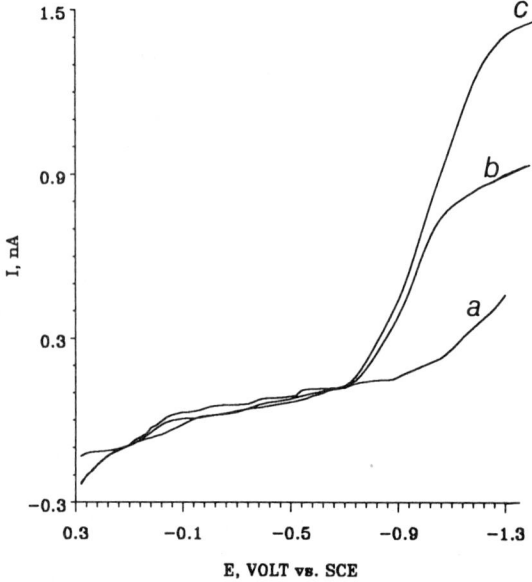

FIG. 15. Steady state voltammograms at 10 mV s^{-1} at a carbon microdisk electrode (r = 6 μm) in microemulsions of 0.2 M AOT/4 M water (pH 3)/isooctane: (a) background for microemulsion; (b) 0.5 mM vitamin B_{12} alone; (c) 2 mM 1,2-dibromobutane and 0.5 mM vitamin B_{12}. (From Ref. 71.)

k_1s were three orders of magnitude smaller in the w/o microemulsion than in acetonitrile/water (Table 7). Attenuation of rates occurs because of the spatial segregation of the reactants in the microemulsion. Relative k_1-values for DBB:EDB:t-DBCH were 1:2:4 in water/MeCN and 1:4:20 in the microemulsion [23]. Thus, the microemulsion alters the relative reactivities of the alkyl dibromides. The increase in reactivity of t-DBCH in the microemulsion may be caused by selective interaction of the cyclic t-DBCH with the AOT hydrocarbon tails at the outer surface of the water pools.

The above catalytic reduction was used as the basis for determination of EDB in leaded and unleaded gasolines. Square wave voltammetric analysis was done by standard addition after adding vitamin B_{12} to an emulsion prepared from the gasoline sample, AOT, and water [169a].

DDAB microemulsions and dispersions gave 10-fold better rates of bulk electrolysis of 4,4′-dichlorobiphenyl (4,4′-DCB) than aqueous CTAB micelles. In general, DDAB was found to be much more efficient than CTAB at coadsorbing reactants at carbon felt cathodes and enhancing rates [22,169b].

Recent work in our laboratory investigated bicontinuous DDAB/water/dodecane (21/57/22, weight%) microemulsions on carbon felt and Pb cathodes for the catalytic dechlorination of 4,4′-DCB [170]. In 5-hour electrolyses using zinc phthalocyanine as catalyst, 76% of 4,4′-DCB was fully dechlorinated on Pb as opposed to 66% on carbon felt. Also, the ratio of biphenyl to reduced BPs is about 4 for Pb and 1 for carbon felt, implying that

TABLE 7

Apparent Rate Constants[a] for Catalytic Reduction of Alkyl Vicinal Dibromides by Vitamin B_{12s}

Substrate	0.2 M AOT/4 M water/isooctane[b]		pH 2.3 acetonitrile/water[c]	
	$10^{-3} k_{obs}$ $(M^{-1}s^{-1})$	rel k	$10^{-6} k_{obs}$ $(M^{-1}s^{-1})$	rel k
DBB	0.31 ± 0.19	1	1.3 ± 0.2	1
EDB	1.34 ± 0.17	4.3	2.6 ± 0.5	2
t-DBCH	6.2 + 1.4	20	5.1 ± 1.3	3.9

[a] Data obtained at 10 mV/sec at carbon microdisk (r = 6 μm) electrodes.
[b] 17 mM phosphate (pH3) in water pools, total concentrations: 0.5 mM vitamin B_{12r}, 2 mM substrate.
[c] 17 mM phosphate buffer, 0.5mM vitamin B_{12}, 5 mM substrate.
DBB, 1,2-dibromobutane; EDB, ethylene dibromide; 1,2-DCBH, trans-1,2-dibromocyclohexane
Source: Ref. 23

Pb is more efficient for dechlorination since the system does not use excess energy in further reducing biphenyl. Preliminary experiments showed that it may be possible to improve results by optimizing surface pretreatment of the Pb cathode.

Preliminary work by Kamau and co-workers [171] explored catalytic dehalogenation of trichloroacetic acid and alkyl vicinal dibromides in bicontinuous microemulsions of DDAB/dodecane/water (21/39/40). Mediators used were nickel(II) phthalocyaninetetrasulfonate (NiPCTS) and copper(II) phthalocyaninetetrasulfonate (CuPcTS). These catalysts are water soluble, as is trichloracetic acid (TCA), but the alkyl vicinal dihalides 1,2-dibromobutane (DBB) and *trans*-1,2-dibromocyclohexane (t-DBCH) are expected to reside predominantly in the oil phase. The reactions are similar to that in Scheme IV, giving acetic acid from TCA and alkenes from the vicinal dibromides. CV and SWV showed much larger catalytic efficiencies at glassy carbon electrodes for DBB and t-DBCH in the microemulsions than in homogeneous acetonitrile/water. However, catalytic efficiency for TCA was larger in the homogeneous solution than in the microemulsion. A tentative explanation of these data may involve rate enhancement in the microemulsion by coadsorption of catalysts, DDAB, and the nonpolar substrates DBB and t-DBCH at the electrode, as found previously for a number of the bimolecular reactions in micellar solutions [22].

VI. ELECTROCHEMISTRY IN LAMELLAR AND VESICLE DISPERSIONS

A. Lamellar Dispersions

Dispersions of DDAB with tetraethylammonium bromide added to decrease viscosity were explored as media for catalytic reductions of polychlorinated biphenyls (PCBs) [169,170]. These dehalogenation reactions employed catalyst zinc phthalocyanine (ZnPc) and carbon felt cathodes at -2.3 V vs. SCE with ultrasonic mixing. Ultrasound increased reaction rates about threefold by enhancing the rate of mass transport. An apparent diffusion coefficient for ferrocene, presumably bound to the lamellar aggregates, was constant in the dispersion at $0.68 \pm 0.1 \times 10^{-6}$ cm^2/sec both before and after electrolysis, suggesting that ultrasound does not influence aggregate size in this system.

The commercial PCB mixture Aroclor 1016 (42% Cl by weight), 12 mg in 25 ml of DDAB dispersion, was completely dehalogenated in 25 hours in this system. In contrast, less than 10% dehalogenation was found for analogous reaction conditions in aqueous CTAB [169]. DDAB facilitated coadsorption of ZnPc and PCBs onto the negative cathode. In acidified DDAB dispersions, the reaction takes place predominately in a DDAB film on the electrode surface. After electrolysis, the major fraction of products and reactants was recovered by washing the carbon felt cathode with DMF. This and other results suggested that dehalogenation rates are enhanced by coadsorbing reactants with surfactant on the electrodes. Coadsorption of nonpolar reactants with DDAB on the electrode improves rates of dehalogenation compared to micellar CTAB, which is less strongly adsorbed.

Improvements in electrolysis rates at carbon felt cathodes using ZnPc as catalyst were achieved by buffering the DDAB/TEAB dispersions at pH 3.5 with acetate. As seen by SWV (Fig. 16), catalytic increases in current occur at the third and fourth reduction peaks of ZnPc in dispersions containing 4,4′-DCB. Thus, catalytic reductions of PCB mixtures probably involve addition of three and/or four electrons to the neutral ZnPc.

Studies of the above system on Pb, carbon felt, and Hg cathodes showed that the best conversion rates are obtained with Hg. Electrolysis of 0.1 g of Aroclor 1016 using 2 mM ZnPc in the pH 3.5-buffered DDAB dispersion at a Hg pool cathode gave nearly quantitative yields of the fully dechlorinated

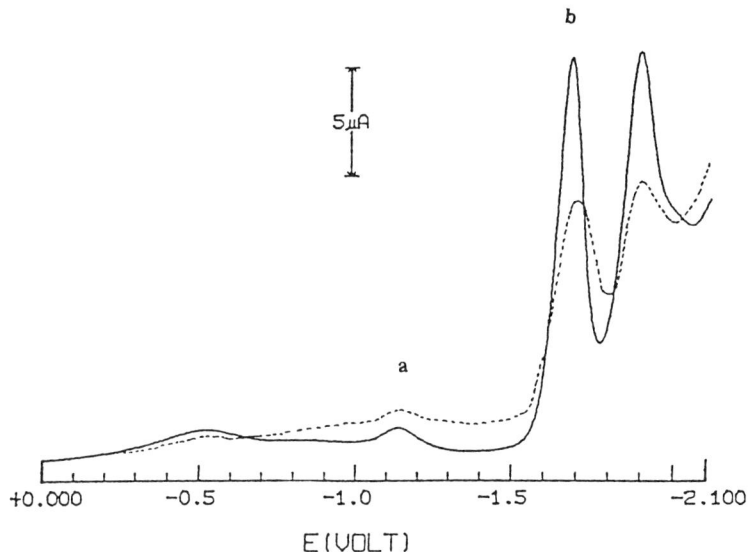

FIG. 16. Square wave voltammograms on glassy carbon disk (A = 0.07 cm^2) in dispersion of 0.08 M DDAB/0.05 M TEAB buffered with acetate at pH 3.5: (a) 1 mM ZnPc; (b) 1 mM ZnPc and 0.76 mM 4,4′-dichlorobiphenyl. SWV at 15 Hz, 25 mV pulse amplitude, 4 mV step.

biphenyl (35%) and biphenyl reduction products (60%) after 18 hours of electrolysis [170]. This represents a 10-fold increase in the amount of Aroclor 1016 dechlorinated in 18 hours compared to previous 24-hour electrolyses with acidified DDAB dispersions on carbon felt. Inclusion of acetate in the DDAB dispersion is likely to result in a mixture of lamellar and micellar aggregates [170]. This may help solubilize reaction products and move them away from the electrode more efficiently.

B. Vesicle Dispersions

Kaifer was the first to report voltammetric studies of electroactive species bound to vesicles [172]. He compared the voltammetry of diheptylviologen (HV^{2+}) in vesicle suspensions (also called liposomes) of the zwitterionic phosphatidylcholine (PC) and in CTAC micelles. Two pairs of redox peaks were found, corresponding to the well-known alkylviologen reductions:

$$HV^{2+} + e \rightleftarrows HV^{+} \cdot \qquad (60)$$

$$HV^{+} \cdot + e \rightleftarrows HV^{o} \qquad (61)$$

Both micelles and liposomes solubilized $HV^{+}\cdot$ and inhibited its precipitation on the electrode. However, HV^o precipitated on the electrode in CTAB micelles, but not in PC vesicle preparations. The apparent diffusion coefficient of HV^{2+} found by CV was 2.3 times larger in CTAC micelles than in PC vesicles, in accord with the larger size of the vesicles.

Lu and Cotton examined voltammetry, SERS and resonance Raman spectroscopy of HV^{2+} and its reduction products in PC vesicles using Pt and Ag electrodes [173]. The vesicle suspensions provided better reversibility than aqueous solutions and inhibited film formation during both electrode reactions. The vesicles also inhibited the disproportionation reaction:

$$2\ HV^{+}\cdot \rightleftarrows HV^o + HV^{2+} \qquad (62)$$

Lu et al. [174] studied voltammetry of asymmetric viologens with one methylated nitrogen and the other bearing a long hydrocarbon chain, denoted C_8MV^{2+}, $C_{12}MV^{2+}$, and $C_{16}MV^{2+}$. In aqueous solution, the presence of sharp anodic CV peaks indicated a greater tendency of the reduction products to precipitate on the electrode as chain length increased. CVs tended toward more reversibility in PC vesicle suspensions. MV^{2+} gave reversible peaks, but some precipitation of products was found for C_xMV^{2+}, with $x \geq 8$. In dispersions of anionic dihexadecylphosphate (DHP), very strong binding of all the viologens to the vesicles resulted in quite small voltammetric peaks. This is because electrostatic and hydrophobic interactions are operative be-

Micelles and Microemulsions

tween the cationic viologens and negative DHP vesicles, whereas interactions with PC vesicles are mainly hydrophobic.

Lei and Hurst measured formal potentials ($E^{o\prime}$) for the first one-electron reduction of homologous asymmetric C_xMV^{2+} viologens [see Eq. (60)] in vesicle suspensions by thin layer spectroelectrochemistry and compared them to values obtained by CV [175]. $E^{o\prime}$-values showed no clear trends with increasing chain length except for a distinct 100 mV difference between $C_{10}MV^{2+}$ and $C_{12}MV^{2+}$ in the DHP vesicles. Similar discontinuity in measured properties had also been seen in a number of other types of experiments. The authors attributed this to an abrupt change in binding interactions as chain length increases [175]. For $x < 12$, the viologens are probably adsorbed on the surface of the DHP vesicles, but at $x \geq 12$ the viologen is intercalated into the vesicle bilayer. Hydrophobic and electrostatic interactions of C_xMV^{2+} with the vesicles was reflected in the order of $E^{o\prime}$ values, which was DHP < PC < DODAB. Thus, the most negative potentials are found in DHP vesicles where both electrostatic and hydrophobic interactions stabilize the viologens. The zwitterionic vesicles bind the viologens mainly through hydrophobic interactions. For DODAB vesicles, the hydrophobic effect needs to be strong enough to overcome electrostatic repulsion between the dicationic viologen and the positively charged vesicle surface. The influence of vesicles on CV behavior increased with chain length because of the presence of hydrophobic interactions for both the PC and DHP vesicles.

The latter study also showed that the extent of dimerization and further aggregation of C_xMV^{2+} increased as chain length increased. The free energies of viologen disproportionation [see Eq. (62)] were similar to those in water and indicated equilibrium constants of the order of 10^{-4}–10^{-5} [175].

C. Detecting Photochemistry at CdS Particles in Vesicles

Electrochemical measurements can be used to study photochemical reactions of semiconductor particles in vesicles. Chang and Fendler [176] used electrochemical oxidation of a mediator at a Pt anode to study electron transfer reactions of small CdS particles on the surface of DHP vesicles. The use of a Pt electrode with 3 mm^2 area allowed the experiments to be done without added electrolyte. The aqueous suspensions used contained CdS/DHP vesicles, glucose (RCH_2OH) as an electron donor, and MV^{2+} as electron acceptor. Upon illumination at the dark rest potential, the sequence of reactions in Scheme V caused current to flow in the cell. The photocurrent was mediated by MV^{2+} above pH 10.1, where the conduction band potential of CdS becomes positive of the formal potential of MV^{2+}.

Scheme V

$$CdS + h\nu \rightarrow CdS(e + h) \quad (63)$$

$$CdS(e + h) + RCH_2OH \rightarrow CdS(e) + RCHO \quad (64)$$

$$CdS(e) + MV^{2+} \rightarrow CdS + MV^{+\cdot} \quad (65)$$

$$MV^{+\cdot} \rightarrow MV^{2+} + e \text{ (at electrode)} \quad (66)$$

Photocurrents reached a steady state in about 120 seconds, and decayed for about 100 seconds after the lamp was turned off. Particles of 5 nm diameter gave about twice the photocurrent as 10 nm particles, in accord with the higher energy of conduction band electrons in the smaller particles [176]. This electrochemical method was shown to be an excellent tool for characterizing redox properties of colloidal semiconductors on vesicles.

Tricot and Manassen combined electrochemical measurements and optical absorbance to study similar CdS/DHP systems [177]. They used 20 cm^2 area Pt mesh working electrodes, so that bulk electrolysis of $MV^{+\cdot}$ could be done at -0.3 V vs. Ag/AgCl. The reaction pathway was similar to Scheme V, but benzyl alcohol was used as electron donor and electrolyte ions were added to the solution. This combined analytical approach gave sensitivity to events inside and outside the vesicles. Electrochemical measurements monitor $MV^{+\cdot}$ outside the vesicles only, while optical absorbance monitored total $MV^{+\cdot}$ in the system. The authors attempted to measure photoinduced transmembrane electron transfer with MV^{2+} bound on one side of the membrane and CdS on the other. This was achieved only when MV^{2+} was inside the vesicles and a relatively high concentration of CdS existed on the outer surfaces. $MV^{+\cdot}$ formed in this reaction [Eq. (65)] leaked across the membrane and also dimerized. These side reactions eventually shut down transmembrane electron transfer.

D. Immunological Detection: Marker Release from Liposomes

The insensitivity of electrochemical measurements to substances inside vesicles (liposomes) had been employed previously by D'Orazio and Rechnitz to detect antibodies on liposome surfaces, or to measure protein complements that bind to these antibodies and destroy the membrane [178]. They used reconstituted blood cell liposomes with trimethylphenylammonium chloride (TMPACl) isolated inside. The liposomes had antibodies on their surfaces. When the active protein complement was added, the liposomes were broken (lysed), and TMPA$^+$ marker cations released into solution were detected potentiometrically with a TMPA$^+$ ion–selective electrode. In this way they

Micelles and Microemulsions

detected rabbit anti-sheep hemolysin [178] and later extended the method to serum antibodies to bovine serum albumin [179].

The same approach was used with amperometry by Haga et al. [180] to detect antitheopyllin antibodies using liposomes loaded with the enzyme horseradish peroxidase, which reduces oxygen in the presence of NADH. The decrease in oxygen content was followed with a Clark oxygen electrode. A similar approach was used to detect glucose released from immune lysis of liposomes loaded with glucose with an amperometric glucose enzyme electrode [181].

Durst and co-workers [182] reported a detailed study on the liposome marker release method. They used liposomes made from a combination of PC, cholesterol, and diacetylphosphate. Although the more hydrophobic markers leaked out of these vesicles, ferrocyanide ions showed no leakage over several weeks. At high volume fraction, liposomes passivated Pt electrodes. At low volume fraction the ferrocyanide released by lysis of the liposomes with Triton X-100 was difficult to detect. Good sensitivity was achieved by using an N-methylpyridinium polysiloxane polymer coated on the electrode, which preconcentrated ferrocyanide ions on its surface for the measurement [182].

VII. SUMMARY AND CONCLUSIONS

This review illustrates the wide variety of specific applications of electrochemistry in microheterogeneous fluids organized by surfactants. Applications are of two general types: (1) control of reaction pathways and kinetics by surfactant aggregates and (2) use of electrochemical methods for fundamental characterization of organized fluids.

In the first category, we saw that surfactant aggregates on electrode surfaces and suspended in solutions can control reaction pathways, as in the synthesis of adiponitrile and reductions of aromatic nitro compounds, in enhancing and controlling rates of bimolecular catalytic reactions, and in inhibiting or facilitating formation of solid films. The most effective kinetic control is afforded by aggregates on electrode surfaces. Thus, development of a fundamental understanding of supramolecular structures and their dynamics on electrode surfaces will be an important factor for future progress in this field. Reaction control with surfactants shows great promise for design of synthetic and analytical methods. An interesting example of the latter is the detection of biomacromolecules that cause immunological lysis by using vesicles filled with an electroactive marker.

In the second category, electrochemical probe methods can be used to obtain self-diffusion coefficients of micelles and microemulsion droplets and to characterize diffusion in continuous phases. Conditional binding constants for probe-micelle and probe-droplet equilibria can be obtained from both diffusion and formal potential measurements, but a more accurate and general theoretical framework is needed before true equilibrium constants can be obtained.

Electrochemical diffusion studies can be done even in nonconducting w/o microemulsions by using μm-sized microelectrodes. As demonstrated in a few cases, kinetics of solute-micelle and solute-droplet interactions are also amenable to measurement by electroanalytical methods. Perhaps because of the rather fast reaction rates involved in many such systems, electroanalytical kinetic measurements in microheterogeneous fluids have not yet reached their full potential. However, the faster kinetic regimes should fall into the range of the smallest ultramicroelectrodes. Electrochemical methods are also useful for characterization of dark and photoexcited electron transfer processes of molecules and small semiconductor particles on vesicles.

ACKNOWLEDGMENTS

The author is grateful for financial support for the preparation of this article and for his unpublished work described herein to U.S. PHS for NIH Grant ES03154 awarded by the National Institute of Environmental Health Sciences, Connecticut Department of Higher Education, and the National Science Foundation (INT-9002223). He is also grateful to students and colleagues named in references to joint publications whose hard work, insight, and encouragement were essential to much of the research described herein. Thanks also to Dr. Inam Ul Haque for helpful suggestions concerning the preparation of this chapter.

REFERENCES

1. D. Atwood, and A. T. Florence, *Surfactant Systems*, Chapman and Hall, London, 1983.
2. P. C. Hiemenz, *Principles of Surface and Colloid Chemistry*, Marcel Dekker, New York, 1986.
3. J. Heyrovsky, and J. Kuta, *Principles of Polarography*, Academic Press, New York, 1966.
4. T. C. Franklin, and S. Mathew, in *Surfactants in Solution*, Vol. 10 (K. L. Mittall, Ed.), Plenum, New York, 1989, p. 267.

5. A. W. Adamson, *Physical Chemistry of Surfaces*, 5th Ed., Wiley, New York, 1990.
6. P. G. Grimes, in *Surfactants in Emerging Technologies* (M. J. Rosen, Ed.), Marcel Dekker, New York, 1987, p. 101.
7. L. Meites, *Polarographic Techniques*, 2nd Ed., Wiley, New York, 1965.
8. H. H. Bauer, Electroanalyt. Chem. *8*:169 (1975).
9. J. H. Fendler, E. Fendler, *Catalysis in Micellar and Macromolecular Systems*, Academic Press, New York, 1975.
10. J. H. Fendler, *Membrane Mimetic Chemistry*, Wiley, New York, 1982.
11. M. Gratzel, *Heterogeneous Photochemical Electron Transfer*, CRC Press, Boca Raton, FL, 1989.
12. E. L. Colichman, J. Am. Chem. Soc. *72*:4036 (1950).
13. L. Meites, and T. Meites, J. Am. Chem. Soc. *73*:177 (1951).
14. L. Holleck, and H. J. Exner, Z. Elektrochim. *56*:46 (1952).
15. W. Kemula, and T. M. Krygowski, in *Encyclopedia of Electrochemistry of the Elements*, Vol. XIII (A. J. Bard and H. Lund, Eds.), Marcel Dekker, New York, 1979, p. 77.
16. D. E. Danly, in *Organic Electrochemistry*, 2nd Ed. (M. M. Baizer and H. Lund, Eds.), Marcel Dekker, New York, 1983, p. 959.
17. M. M. Baizer, J. Electrochem. Soc. *111*:215 (1964).
18. G. E. O. Proske, Anal. Chem. *24*:1834 (1952).
19. For examples of such work see (a) S. Hayano, and N. Shinozuka, Bull. Chem. Soc. Japan *42*:1469 (1969); (b) S. Hayano, and N. Shinozuka, Bull. Chem. Soc. Japan *43*:2083 (1970); (c) P. G. Westmoreland, R. A. Day, and A. L. Underwood, Anal. Chem. *44*:737 (1972); (d) T. Erabi, H. Hiura, and M. Tanaka, Bull. Chem. Soc. Japan *48*:1354 (1975).
20. (a) G. McIntire, CRC Crit. Rev. Anal. Chem. *21*:257 (1990); (b) E. Pelizzetti, and E. Pramauro, Anal. Chim. Acta, *169*:1 (1985).
21. J. F. Rusling, C.-N. Shi, E. C. Couture, and T. F. Kumosinski, in *Redox Chemistry and Interfacial Behavior of Biological Molecules* (G. Dryhurst and K. Niki, Eds.), Plenum, New York, 1988, p. 565.
22. J. F. Rusling, Accts. Chem. Res. *24*:75 (1991).
23. A. Owlia, Z. Wang, and J. F. Rusling, J. Am. Chem. Soc. *111*:5901 (1989).
24. C. Tanford, *The Hydrophobic Effect*, 2nd Ed., Wiley, New York, 1980.
25. J. F. Rusling, C.-N. Shi, and T. F. Kumosinski, Anal. Chem. *60*:1260 (1988).
26. J. H. Fendler, *J. Phys. Chem. 89*:2730 (1985).
27. (a) R. Zana, and J. Lang, in *Solution Behavior of Surfactants* Vol. 2 (K. L. Mittal and E. J. Fendler, Eds.), Plenum, New York, 1980, p. 1195; (b) R. Zana, in *Surfactants in Solution* Vol. 4 (K. L. Mittal and P. Bothorel, Eds.), Plenum, New York, 1984, p. 115.
28. G. N. Kamau, T. Leipert, S. S. Shukla, and J. F. Rusling, J. Electroanal. Chem. *233*:173 (1987).
29. K. Kalyanasundarum, *Photochemistry in Microheterogeneous Systems*, Academic, New York, 1987.

30. P. L. Luisi, and L. J. Magid, CRC Crit. Rev. Biochem. *20*:409 (1987).
31. (a) M. Kahlweet, Science *240*:617 (1988); (b) C. A. Miller, and S. Qutubuddin, in *Interfacial Phenomena in Apolar Media* (H. F. Eicke and G. D. Parfitt, Eds.), Marcel Dekker, New York, 1987, p. 117; (c) J. H. Fendler, and K. Kurihara, in *Metal Containing Polymeric Systems* (J. E. Sheats, C. E. Carraher, and C. U. Pittman, Eds.), Plenum, New York, 1985, p. 341.
32. D. F. Evans, D. J. Mitchell, and B. W. Ninahm, J. Phys. Chem. *90*:2817 (1986).
33. K. Shinoda, and B. Lindman, Langmuir *3*:135 (1987).
34. A. J. Bard, and L. R. Faulkner, *Electrochemical Methods*, Wiley, New York, 1980.
35. B. Aveyard, in *Surfactants* (Th. F. Tadros, Ed.), Academic, London, 1984, p. 153.
36. (a) J. H. Harwell, J. C. Hoskins, R. S. Schecter, and W. H. Wade, Langmuir *1*:251 (1985); (b) M. A. Yeskie, and J. H. Harwell, J. Phys. Chem. *92*:2346 (1988).
37. P. Chandar, P. Somasundaran, and N. Turro, J. Coll. Interface Sci. *117*:31 (1987).
38. J. F. Scamehorn, R. S. Schechter, and W. H. Wade, J. Coll. Interface Sci. *85*:463 (1982).
39. B. H. Bijsterbosch, J. Coll. Interface Sci. *47*:186 (1974).
40. A. R. Rennie, E. M. Lee, E. A. Simister, and R. K. Thomas, Langmuir *6*:1031 (1990).
41. N. Shinozuka, and S. Hayano, in *Solution Chemistry of Surfactants*, Vol. 2 (K. L. Mitall, Ed.), Plenum, New York, 1979, p. 599.
42. R. Guidelli, and M. L. Foresti, J. Electroanal. Chem. *77*:73 (1977).
43. R. A. Marcus, Ann. Rev. Phys. Chem. *15*:155 (1964).
44. G. L. Closs, and J. R. Miller, Science *240*:440 (1988).
45. For examples of such work see: (a) B. N. Zaba, M. C. Wilkinson, D. M. Taylor, T. J. Lewis, and D. L. Laidman, FEBS Lett. *213*:49 (1987); (b) M. D. Porter, T. B. Bright, D. L. Allara, and C. E. D. Chidsey, J. Am. Chem. Soc. *109*:3559 (1987); (c) H. O. Finklea, S. Avery, M. Lynch, and T. Furtsch, Langmuir *3*:409 (1987); (d) K. A. Bunding Lee, R. Mowry, G. McLennan, and H. O. Finklea, J. Electroanal. Chem. *246*:217 (1988); (e) C. E. D. Chidsey, Science *251*:919 (1991); (f) C. Miller; P. Cuendet, and M. Gratzel, *J. Phys. Chem.* *95*:877 (1991); (g) Y. S. Obeng, and A. J. Bard, Langmuir *7*:195 (1991).
46. Examples of films bound through silicon-oxygen linkages are discussed in (a) L. Netzer, and J. Sagiv, J. Am. Chem. Soc. *105*:674 (1983); (b) H. O. Finklea, L. R. Robinson, A. Blackburn, B. Richter, D. Allara, and T. Bright, Langmuir *2*:239 (1986); (c) E. Sabatani, I. Rubinstein, R. Moaz, and J. Sagiv, J. Electroanal. Chem. *219*:365 (1987); (d) E. Sabatani, and I. Rubinstein, J. Phys. Chem. *91*:6663 (1987); (e) J. F. Rusling, H. Zhang, and W. Willis, Anal. Chim. Acta *235*:307 (1990).

47. D. Schuhman, P. Vanel, E. Tronel-Peyroz, and H. Raous, in *Surfactants in Solution*, Vol. 2 (K. L. Mitall and B. Lindman, Eds.), Plenum, New York, 1984, p. 1233.
48. D. C. Graham, Chem. Rev. *41*:441 (1947).
49. P. Delahey, *New Instrumental Methods in Electrochemistry*, Interscience, New York, 1954.
50. G. L. Besio, R. K. Prud'homme, and J. B. Benzinger, Langmuir *4*:140 (1988).
51. M. Kaisheva, V. Kaishev, and M. Matsumoto, J. Electroanal. Chem. *171*, 111 (1984).
52. J. B. Hayter, and R. J. Hunter, J. Electroanal. Chem. *37*:71, 81 (1972).
53. J. B. Hayter, M. W. Humphries, R. J. Hunter, and R. Parsons, J. Electroanal. Chem. *56*:160 (1974).
54. M. Zembala, J. Electroanal. Chem. *66*:45 (1975).
55. M. W. Humphries, Ph.D. thesis, University of Bristol, 1975.
56. J. F. Rusling, C.-N. Shi, D. K. Gosser, and S. S. Shukla, J. Electroanal. Chem. *240*:201 (1988).
57. J. F. Rusling, Trends in Anal. Chem. *7*:266 (1988).
58. J. L. Stickney, M. P. Soriaga, A. T. Hubbard, and S. E. Anderson, J. Electroanal. Chem. *125*:73 (1981).
59. J. F. Rusling, and E. C. Couture, Langmuir *6*:425 (1990).
60. S. Sun, R. L. Birke, J. R. Lombardi, J. Phys. Chem. *94*:2005 (1990).
61. J. F. Rusling, and M. F. Ahmadi, Langmuir *7*:1529 (1991).
62. S. Dong, Y. Zhu, and G. Cheng, Langmuir *7*:389 (1991).
63. J. S. Facci, Langmuir *3*:525 (1987).
64. (a) J. J. Donohue, and D. A. Buttry, Langmuir *5*:671 (1989); (b) H. C. DeLong, J. J. Donohue, and D. A. Buttry, Langmuir *7*:2196 (1991).
65. L. L. Nordyke, and D. A. Buttry, Langmuir *7*:380 (1991).
66. A. Diaz, and A. E. Kaifer, J. Electroanal. Chem. *249*:333 (1988).
67. C. A. Widrig, and M. Majda, Langmuir *5*:689 (1989).
68. (a) H. W. Hoyer, and J. Novodoff, J. Colloid Interf. Sci. *26*:490 (1968); (b) J. Novodoff, H. L. Rosano, and H. W. Hoyer, J. Colloid Interf. Sci. *38*:424 (1972).
69. (a) S. Hayano, M. Fujihira, in *Proc. of Int. Conf. on Colloid and Surf. Sci.* (E. Wolfram, Ed.), Elsevier, New York, 1975, p. 609; (b) J. Georges, and S. Desmettre, Electrochim. Acta *29*:521 (1984).
70. R. Zana, and R. A. Mackay, Langmuir *2*:109 (1986).
71. J. F. Rusling, Z. Wang, and A. Owlia, Colloids and Surfaces, *48*:173 (1990).
72. M. Fleischmann, S. Pons, D. R. Rolinson, and P. P. Schmidt, *Ultramicroelectrodes*, Datatech Systems, North Carolina, 1987.
73. J. Osteryoung, and J. J. O'Dea, in *Electroanalytical Chemistry*, Vol. 14 (A. J. Bard, Ed.), Marcel Dekker, New York, 1986, p. 209.
74. J. F. Rusling, and M. Y. Brooks, Anal. Chem. *56*:2147 (1984).
75. G. L. McIntire, D. M. Chiappardi, R. L. Casselberry, and H. N. Blount, Phys. Chem. *86*:2632 (1982).

76. Y. Ohsawa, and S. Aoyagui, J. Electroanal. Chem. *136*:353 (1982).
77. (a) A. E. Kaifer, and A. J. Bard, J. Phys. Chem. *89*:4876 (1985); *91*:2007 (1987); (b) J. W. Park, and J. H. Paik, J. Phys. Chem. *91*:2005 (1987).
78. D. H. Evans, J. Electroanal. Chem. *258*:451 (1989).
79. A. B. Mandal, B. U. Nair, and D. Ramaswamy, Langmuir *4*:736 (1988).
80. E. Dayalan, S. Qutubuddin, and A. Hussam, Langmuir *6*:715 (1990).
81. (a) M. J. Eddowes, and M. Gratzel, J. Electroanal. Chem. *163*:31 (1984); (b) M. J. Eddowes, and M. Gratzel, J. Electroanal. Chem. *152*:143 (1983); (c) J. Georges, and S. Desmettre, Electrochim. Acta *31*:1519 (1986).
82. Z. Wang, A. Owlia, and J. F. Rusling, J. Electroanal. Chem. *270*:407 (1989).
83. (a) D. F. Evans, S. Mukherjee, D. J. Mitchell, and B. W. Ninham, J. Colloid Interf. Sci. *93*:184 (1983); (b) R. M. Weinheimer, D. F. Evans, and E. L. Cussler, J. Colloid Interf. Sci. *80*:357 (1981).
84. H. Fabre, N. Kamenka, A. Khan, G. Lindblom, B. Lindman, and G. J. T. Tiddy, J. Phys. Chem. *84*:3428 (1980).
85. (a) R. B. Dorshow, J. Briggs, C. A. Bunton, and D. F. Nicoli, J. Phys. Chem. *86*:2388 (1982); (b) J. Briggs, R. B. Dorshow, C. A. Bunton, and D. F. Nicoli, J. Chem. Phys. *76*:775 (1982).
86. B. Lindman, M.-C. Puyal, N. Kamenka, R. Rymden, and P. Stilbs, J. Phys. Chem. *88*:5048 (1984).
87. R. A. Mackay, N. S. Dixit, R. Agarwal, and R. P. Seiders, J. Disp. Sci. Technol. *4*:397 (1983).
88. (a) R. E. Verrall, S. Milioto, A. Giraudeau, and R. Zana, Langmuir *5*:1242 (1989); (b) J. Texter, F. R. Horch, S. Qutubuddin, and E. Dayalan, J. Colloid Interf. Sci. *135*:263 (1990).
89. J. P. Kratohvil, and T. M. Aminabhavi, J. Phys. Chem. *86*:1254 (1982).
90. S. J. Candau, E. Hirsch, and R. Zana, J. Phys. (Les Ulis, Fr.) *45*:1263 (1984).
91. R. M. Wightman, Science *240*:415 (1988).
92. (a) R. F. Noftle, and D. Pletcher, J. Electroanal. Chem. *293*:273 (1990). (b) J. F. Rusling, H. Zhang, and W. S. Willis, Anal. Chim. Acta *235*:307 (1990).
93. (a) Y. Ohsawa, Y. Shimazaki, and S. Aoyagui, J. Electroanal. Chem. *114*:235 (1980); (b) Y. Ohsawa, Y. Shimazaki, K. Suga, and S. Aoyagui, J. Electroanal. Chem. *123*:409 (1981).
94. A. P. Abbott, C. L. Miaw, and J. F. Rusling, J. Electroanal. Chem. *327*:31 (1992).
95. Y. Ohsawa, and S. Aoyagui, J. Electroanal. Chem. *145*:109 (1983).
96. G. L. McIntire, and H. N. Blount, in *Solution Behavior of Surfactants*, Vol. 2 (K. L. Mitall and E. Fendler, Eds.), Plenum, New York, 1982, p. 1101.
97. P. Pouillen, A.-M. Martre, and P. Martinet, Electrochim. Acta *27*:853 (1982).
98. S. Bencheikh-Sayarh, P. Pouillen, A.-M. Martre, and P. Martinet, Electrochim. Acta *28*:627 (1983).
99. S. Bencheikh-Sayarh, A.-M. Martre, G. Mousset, and P. Pouillen, Bull. Soc. Chim. Fr. I No. 11–12:329 (1984).
100. S. Bencheikh-Sayarh, A.-M. Martre, G. Mousset, and P. Pouillen, Electrochim. Acta *28*:1105 (1983).

101. C. Mousty, B. Devaux, G. Mousset, P. Pouillen, and P. Martinet, Electrochim. Acta 30:1733 (1985).
102. C. Mousty, P. Pouillen, A.-M. Martre, and G. Mousset, J. Colloid Interf. Sci. 113:521 (1986).
103. S. Bencheikh-Sayarh, B. Cheminat, G. Mousset, and P. Pouillen, Electrochim. Acta 29:1225 (1984).
104. (a) M. Verniette, P. Pouillen, and P. Martinet, Bull. Soc. Chim. Fr. I No. 5–6:141 (1984); (b) A. Davidovic, I. Tabakovic, D. Davidovic, and L. Duic, J. Electroanal. Chem. 280:371 (1990).
105. P. Pouillen, A.-M. Martre, and P. Martinet, Electrochim. Acta 26:1035 (1981).
106. C. Mousty, P. Pouillen, and G. Mousset, J. Electroanal. Chem. 236:253 (1987).
107. C. Mousty, and G. Mousset, J. Colloid Interf. Sci. 128:427 (1989).
108. C. Mousty, B. Cheminat, and G. Mousset, J. Org. Chem. 54:5377 (1989).
109. D. A. Jaeger, D. Bolikal, and B. Nath, J. Org. Chem. 52:276 (1987).
110. K. Bennis, and P. Martinet, Bull. Soc. Chim. Fr. I 799 (1988).
111. (a) M. R. Moncelli, F. Pergola, G. Aloisi, and R. Guidelli, J. Electroanal. Chem. 143:233 (1983); (b) M. R. Moncelli, L. Nucci, P. Mariani, and R. Guidelli, J. Electroanal. Chem. 172:83 (1984).
112. (a) A. J. Reviejo Garcia, A. Ruiz Barrio, J. M. Pingarron Carrazon, and L. M. Polo Diez, Anal. Chem. Acta 246:293 (1991); (b) A. Gonzalez Cortes, J. M. Pingarron Carrazon, and L. M. Polo Diez, Electrochim. Acta 36:1573 (1991).
113. (a) G. L. McIntire, and H. N. Blount, J. Am. Chem. Soc. 101:7720 (1979); (b) M. Genies, and M. Thomalla, Electrochim. Acta 26:829 (1981); (c) J. Georges, and A. Berthod, Electrochim. Acta 28:735 (1983).
114. (a) P. A. Quintela, and A. E. Kaifer, Langmuir 3:769 (1987); (b) P. A. Quintela, A. Diaz, and A. E. Kaifer, Langmuir 4:663 (1988); (c) M. Lapkowski, and W. Szulbinski, J. Electroanal. Chem. 300:159 (1991); (d) G. Meyer, L. Nadjo, and J. M. Saveant, J. Electroanal. Chem. 119:417 (1981).
115. T. C. Franklin, and L. Sidarous, J. Chem. Soc. Chem. Commun. 741 (1975).
116. T. C. Franklin, and L. Sidarous, J. Electrochem. Soc. 124:65 (1977).
117. T. C. Franklin, and T. Honda, in *Micellization, Solubilization, and Microemulsions*, Vol. 2 (K. Mitall, Ed.), Plenum, New York, 1976, p. 617.
118. T. C. Franklin, and M. Iwunze, J. Electroanal. Chem. 108:97 (1980).
119. T. C. Franklin, M. Iwunze, and S. Gipson, in *Inorganic Reactions in Organized Media* (S. L. Holt, Ed.), ACS Symposium Series No. 177, Am. Chem. Soc., Washington, D.C., 1982, p. 139.
120. T. C. Franklin, and M. Iwunze, Anal. Chem. 52:973 (1980).
121. T. C. Franklin, and M. Iwunze, J. Am. Chem. Soc. 103:5937 (1981).
122. T. C. Franklin, D. Ball, R. Rodriguez, and M. Iwunze, Surf. Technol. 21:223 (1984).
123. T. C. Franklin, and T. Jimbo, J. Electrochem. Soc. 134:2169 (1987).
124. T. C. Franklin, and M. Ohta, Surf. Technol. 18:63 (1983).
125. E. Laurent, G. Rauniyar, and M. Thomalla, Bull. Soc. Chim. Fr. :I78 (1984).

126. E. Laurent, E. Moraes-Kraus, and M. Thomalla, C. R. Acad. Sci. Paris *t.306*:1073 (1988).
127. G. Rauniyar, and M. Thomalla, Bull. Soc. Chim. Fr. :156 (1989).
128. (a) J. M. Pingarron Carrazon, A. J. Reviejo Garcia, and L. M. Polo Diez, J. Electroanal. Chem. *234*:175 (1987); (b) J. M. Pingarron Carrazon, A. J. Reviejo Garcia, and L. M. Polo Diez, Analyst *115*:869 (1990); (c) J. Georges, and S. Desmettre, J. Colloid Interf. Sci. *118*:192 (1987).
129. J. M. Pingarron Carrazon, A. Gordon Vergara, A. J. Reviejo Garcia, and L. M. Polo Diez, Anal. Chim. Acta *216*:231 (1989).
130. (a) K. Hoshino, and T. Saji, J. Am. Chem. Soc. *109*:5881 (1987); (b) K. Hoshino, M. Goto, and T. Saji, Chem. Lett. :547 (1988).
131. (a) T. Saji, Chem. Lett. :693 (1988); (b) T. Saji, K. Hoshino, Y. Ishii, and M. Goto, J. Am. Chem. Soc. *113*:450 (1991).
132. Y. Harima, and K. Yamashita, J. Phys. Chem. *93*:4184 (1989).
133. G. L. McIntire, Am. Lab. *18*:173 (1986).
134. J. R. Kirchoff, W. R. Heineman, and E. Deutsch, Inorg. Chem. *27*:3608 (1988).
135. (a) J. Ouyang, and A. J. Bard, Bull. Chem. Soc. Japan *61*:17 (1988); (b) K. Hoshino, K. Suga, and T. Sagi, Chem. Lett. :979 (1986).
136. T. C. Franklin, and S. Gipson, S. Surf Technol. *15*:345 (1982).
137. T. C. Franklin, R. Nnodimele, and W. K. Adeniyi, *J. Electrochem. Soc. 134*:2150 (1987).
138. T. C. Franklin, and W. K. Adeniyi, Anal. Chim Acta *207*:311 (1988).
139. T. C. Franklin, R. Nnodimele, W. K. Adeniyi, and D. Hunt, J. Electrochem. Soc. *135*:1944 (1988).
140. T. C. Franklin, and R. Nnodimele, Anal. Chim. Acta *229*:291 (1990).
141. T. C. Franklin, W. K. Adeniyi, and R. Nnodimele, J. Electrochem. Soc. *137*:480 (1990).
142. T. C. Franklin, J. Darlington, and W. K. Adeniyi, J. Electrochem. Soc. *137*:2124 (1990).
143. T. C. Franklin, J. Darlington, and T. Solouki, J. Electrochem. Soc. *138*:747 (1991).
144. T. C. Franklin, J. Darlington, T. Solouki, and N. Tran, J. Electrochem. Soc. *138*:2285 (1991).
145. (a) J. K. Thomas, J. Phys. Chem. *91*:267 (1987); (b) T. Nakamura and J. K. Thomas, Langmuir *3*:234 (1987).
146. J. F. Rusling, C. Shi, and S. L. Suib, J. Electroanal. Chem. *245*:331 (1988).
147. C. Shi, J. F. Rusling, Z. Wang, W. S. Willis, A. M. Winiecki, and S. L. Suib, Langmuir *5*:650 (1989).
148. B. Brahimi, P. Labbe, and G. Reverdy, J. Electroanal. Chem. *267*:343 (1989).
149. (a) Y. Fujihira, T. Kuwana, and C. R. Hartzell, Biochem. Biophys. Res. Comm. *61*:488 (1974); (b) P. Yeh, and T. Kuwana, J. Electrochem. Soc. *123*:1334 (1976).
150. (a) C. P. Andrieux, C. Blocman, J.-M. Dumas-Bouchiat, and J. M. Saveant, J.

Am. Chem. Soc. *101*:3431 (1979); (b) T. F. Connors, J. F. Rusling, and A. Owlia, Anal. Chem. *57*:170 (1985); (c) J. V. Arena, and J. F. Rusling, J. Phys. Chem. *91*:3368 (1987).
151. G. N. Kamau, and J. F. Rusling, J. Electroanal. Chem. *240*:217 (1988).
152. K. Sugiyama, and K. Aoki, J. Electroanal. Chem. *271*:249 (1989).
153. R. A. Mackay, C. Hermansky, and R. Agarwal, in *Colloid and Interface Science*, Vol. 2 (M. Kerker, Ed.), Academic, New York, 1976, p. 289.
154. R. A. Mackay, and R. Agarwal, J. Coll. Interf. Sci. *65*:225 (1978).
155. R. A. Mackay, in *Microemulsions* (I. D. Robb, Ed.), Plenum Press, New York, 1982, p. 207.
156. R. A. Mackay, N. S. Dixit, and R. Agarwal, ACS Symposium Ser., *177*:179 (1982).
157. A. Berthod, and J. Georges, J. Chim. Phys. (Fr.) *80*:245 (1983).
158. A. Berthod, and J. Georges, J. Colloid Interf. Sci. *106*:194 (1985).
159. J. Georges, and J.-W. Chen, Coll. and Polymer Sci. *264*:896 (1986).
160. J. Georges, J.-W. Chen, and N. Arnaud, Coll. and Polymer Sci. *265*:45 (1987).
161. K. Chokshi, S. Qutubuddin, and A. Hussam, J. Colloid Interf. Sci. *129*:315 (1989).
162. E. Dayalan, S. Qutubuddin, and J. Texter, J. Colloid Interf. Sci. *143*:423 (1991).
163. R. A. Mackay, S. A. Myers, L. Bobalbhai, and A. Bratjer-Toth, Anal. Chem. *62*:1084 (1990).
164. J.-W. Chen, and J. Georges, J. Electroanal. Chem. *210*:205 (1986).
165. J. H. Fendler, F. Nome, and H. C. Van Woert, J. Am. Chem. Soc. *96*:6745 (1974).
166. M. O. Iwunze, A. Sucheta, and J. F. Rusling, Anal. Chem. *62*:644 (1990).
167. M. O. Iwunze, and J. F. Rusling, J. Electroanal. Chem. *303*:267 (1991).
168. E. Garcia, L. E. Oppenheimer, and J. Texter, in *Electrochemistry in Colloids and Dispersions*, (R. A. Mackay and J. Texter, Eds.), VCH Publishers, New York, p. 257.
169. (a) J. F. Rusling, T. F. Connors, and A. Owlia, Anal. Chem. *59*:2123 (1987); (b) M. O. Iwunze, and J. F. Rusling, J. Electroanal. Chem. *266*:197 (1989).
170. E. C. Couture, J. F. Rusling, and S. Zhang, Inst. Chem. Eng. Sympos. Series (U.K.) No. 127:177 (1992).
171. G. N. Kamau, N. Hu, and J. F. Rusling, Langmuir *8*:1042 (1992).
172. A. E. Kaifer, J. Am. Chem. Soc. *108*:6837 (1986).
173. T. Lu, and T. M. Cotton, J. Phys. Chem. *91*:5978 (1987).
174. T. Lu, T. M. Cotton, J. K. Hurst, and D. H. P. Thompson, J. Electroanal. Chem. *246*:337 (1988).
175. Y. Lei, and J. K. Hurst, J. Phys. Chem. *95*:7918 (1991).
176. A.-C. Chang, and J. H. Fendler, J. Phys. Chem. *93*:2538 (1989).
177. Y.-M. Tricot, and J. Manassen, J. Phys. Chem. *92*:5239 (1988).
178. P. D'Orazio, and G. A. Rechnitz, Anal. Chem. *49*:2083 (1977).

179. P. D'Orazio, and G. A. Rechnitz, Anal. Chim. Acta *109*:25 (1979).
180. M. Haga, H. Itagaki, S. Sugawara, and T. Okano, Biochem. Biophys. Res. Commun. *95*:187 (1980).
181. Y. Umezawa, S. Sofue, and Y. Takamoto, Anal. Lett. *15*:135 (1982).
182. R. M. Kannuck, J. M. Bellama, and R. A. Durst, Anal. Chem. *60*:142 (1988).

MECHANISM OF CHARGE TRANSPORT IN POLYMER-MODIFIED ELECTRODES

György Inzelt

Eötvös University
Budapest, Hungary

I. Introduction   90
II. General Remarks and Scope   91
    A. Definition and classification   91
    B. Comparison of traditional and polymer film electrodes   92
III. Theories of Electron Transport in Polymer Film Electrodes   95
    A. Early models of charge propagation   95
    B. Theory of the electron exchange reaction   97
    C. New theories predicting nonlinear $D(c)$ function   103
    D. Potential dependence of diffusion coefficient   118
    E. Models of charge transport in electronically conducting polymer films   120
IV. Results on Charge Transport in Polymer Films   124
    A. Ion-exchange polymers containing electrostatically bound redox centers   124
    B. Fixed-site redox polymers   138
    C. Conducting polymer films   184
V. Effect of Film Morphology on Charge Transport in Polymers   218
    A. Theories of de Gennes on the adsorption, diffusion, and charge transport of polymers   219
    B. Polymer self-diffusion   221
    C. Mechanical and electrochemical equilibria in polymer layers   225
VI. Conclusion   228
    References   231

I. INTRODUCTION

In the past 10–15 years much effort has been made concerning the development and characterization of electrodes modified with electroactive polymeric materials. Undoubtedly, the wide range of promising applications in the field of electroanalysis, organic and bioelectrochemistry, electrocatalysis, photoelectrochemistry, corrosion protection of metals and semiconductors, energy storage, electronic devices, and electrochromic displays has given an impetus to find out more about these systems. Another motivation of the electrochemists is to gain a better understanding of the processes occurring in these surface films.

The dual motivation of this research, i.e., the pursuit of a deeper understanding of the mechanism and kinetics, as well as the fabrication of new systems and a search for their application, can be perceived as we follow the history of polymer-modified electrodes from their inception [1–4]. The most important results on chemically modified electrodes, including their preparation, characterization, and application, were reviewed in this series by Murray [5] in 1984.

Since 1984, significant developments have come about in the field of polymer film electrodes. A variety of systems have been studied, and concomitant theories and applications have been put forward and continue to mushroom.

The elucidation of the nature of charge transfer and charge transport processes in electrochemically active polymer films may be the most interesting theoretical problem of the field. It is also a question of great practical importance, because in the majority of their applications fast charge propagation through the film is needed. It has become clear that the elucidation of their electrochemical behavior proves to be a very difficult task due to the complex nature of these systems. The application of new and powerful techniques, such as electrochemical impedance spectroscopy (EIS) [6–36], quartz crystal microbalance (EQCM) [37–62], radiotracer method [63–72], fast scan rate cyclic voltammetry on ultramicroelectrodes [73–75], sandwich systems [76–78], and scanning electrochemical microscopy (SECM) [79–83], has opened up new vistas concerning these issues. In addition, recent studies in which the chemical environment and the temperature were varied have provided a better insight into the nature of charge transfer and charge transport processes occurring in redox and conducting polymer films.

This chapter is not an exhaustive review on polymer-modified electrodes, inasmuch as there exist several monographs on the subject that emphasize various aspects of the field, mostly covering the relevant literature up to about 1986–87 [84–90]. It rather focuses on the problem of charge transport in

Charge Transport in Polymer-Modified Electrodes

polymer film electrodes, which clearly appears to be a key question in this field, and on other closely related phenomena. It is intended here to summarize and systematize the knowledge accumulated in this respect.

The topics discussed here are those of the greatest interest at present. Therefore, selected examples are presented instead of a detailed survey of the field. This may result in some incompleteness and subjectivity. On the other hand, the emphasis on topics of current interest may offer some helpful guidelines for the study of the vast literature of polymer-modified electrodes. (According to our compilation, the publications on modified electrodes and on closely related topics between 1978 and 1991 number more than 2500.) Throughout this chapter, attention is given to those thermodynamic and kinetic properties of polymer films that may lead to the establishment of a general concept of the behavior of these systems.

II. GENERAL REMARKS AND SCOPE
A. Definition and Classification

A polymer film electrode can be defined as an electrochemical system in which at least three phases are contacted successively in such a way that between a first-order conductor (usually a metal) and a second-order conductor (usually an electrolyte solution) is an electrochemically active polymer layer: this polymer is, in general, a mixed (electronic and ionic) conductor. A transfer of electrons to solution species may occur at the two interfaces (phase boundaries) and as a mediated reaction inside the film. In a narrow sense, the first-order conductor (metal) may be called an electrode, following common usage in the literature. The use of this term is especially reasonable when the electrochemical transformation and the processes occurring in the polymer film are the subject of the study. (Murray [91] has given a somewhat different, but very useful and descriptive definition as well. He thinks of a redox polymer film electrode as any conductor covered with a molecular layer that contains more than one monolayer—equivalent of electroactive centers. The essential characteristic is that a mechanism for charge transport, e.g., electron hopping and ion migration, must exist for the film to be electroactive. A dielectric polymer film electrode is then a different form of polymer film electrode, since it acts as a barrier.)

In view of this interpretation, electrodes covered with monolayers of polymeric material do not belong to the category of polymer film electrodes, because in this case there is no separate polymer phase. The behavior of monolayer films can be described by the theories elaborated for strongly adsorbed species [92]. Thick films are frequently referred to as multilayer films, which is misleading, because there are no actual layers in these films; true multilayer films (e.g., bilayer films) exist, however. The films, which are several dozen nanometers to several micrometers thick and are essentially homogeneous, were recently called monolithic films by Faulkner [93].

It is useful to classify polymer film electrodes into several categories, according to their chemical structure, which is loosely connected with the possible modes of charge propagation through the surface layer. In this respect, we can distinguish two main classes of polymers, namely, redox polymers and conducting polymers (e.g., poly(aniline), poly(pyrrole)). The redox polymers contain electrostatically and spatially localized redox sites, which can be oxidized or reduced, while the electrochemical transformation of conducting polymers usually leads to a reorganization of the bonds of the macromolecule. In a partially oxidized state, the latter systems show high electronic conductivity.

The class of the redox polymers can be divided into subclasses:

1. Polymers that contain covalently attached redox sites, either built in the chain or as pendant groups. The redox centers may be organic molecules (e.g., tetrathiafulvalene, tetracyanoquinodimethane), organometallic molecules (ferrocene), or coordinatively attached redox couples (polymerized metal bipyridine (bpy) complexes).
2. Ion-exchange polymeric systems, where the redox-active ions are held by electrostatic binding (e.g., $Fe(CN)_6^{3-/4-}$ in protonated poly(vinylpyridine) or $Ru(bpy)_3^{3+/2+}$ in Nafion).

B. Comparison of Traditional and Polymer Film Electrodes

In the case of traditional electrodes, the electrode reaction involves the mass transfer of the electroactive species from the bulk solution to the electrode surface and the electron transfer step at the electrode surface. These processes may be accompanied, and thus complicated by, other processes such as homogeneous chemical reactions (protonations, dimerizations) and surface processes (adsorption, crystallization). When an electric current flows in an electrochemical cell, the current in solution is carried by the movement of ions. These ions are not necessarily those that react at the metal surface. The mass transfer in solution occurs because of a gradient of electrochemical potential ($\tilde{\mu}$), i.e., by diffusion and/or migration, and by convection. It can be described with the help of the well-known Nernst-Planck equation [94]

$$J_i = -D_i \nabla c_i - \frac{z_i F}{RT} D_i c_i \nabla \Phi + c_i v \tag{1}$$

Charge Transport in Polymer-Modified Electrodes

or for one-dimensional mass transfer along the x-axis

$$J_i(x) = -D_i \frac{\partial c_i(x)}{\partial x} - \frac{z_i F}{RT} D_i c_i \frac{\partial \Phi(x)}{\partial x} + c_i v(x) \qquad (2)$$

where J_i is the flux, D_i is the diffusion coefficient, c_i is the concentration, and z_i is the charge of the species i, respectively, $v(x)$ is the velocity with which a volume element in solution moves in the solution (or along the x-axis), and Φ is the electrostatic potential. The three terms on the right-hand side represent the contributions of diffusion, migration, and convection to the flux, respectively.

In the case of polymer-modified electrodes, the situation is somewhat different from that of traditional electrodes. In the absence of mediated reaction, the convection term in Eqs. (1) and (2) can be omitted, because any stirring of solution has no effect inside the film, as shown by the experimental results. However, when attempting to interpret diffusion and migration of the electrochemically active species, electrochemists working in the field encounter special problems. Especially in the case of redox polymers belonging to subclass 1 where the redox sites are covalently bound to the polymer chain, no free diffusion of the sites occurs; thus the explanation of results obtained by transient electrochemical methods (cyclic voltammetry, chronoamperometry, or chronocoulometry) requires some additional assumptions.

One also has to bear in mind that the polymer chains are trapped in a tangled network, and the layer is more or less stably attached to the metal, mainly by adsorption (adhesion). The free energy of sticking (relative surface energy) is small for a monomer, which is why stable films cannot be made from small molecules, but the energy per chain is large [95].

The fundamental observation that should be brought into harmony with the theory is that even rather thick films, in which most of the sites are as far from the surface as 100–10,000 nm, i.e., the surface concentration, $\Gamma = 10^{-8}$–10^{-6} mol/cm^2, may be oxidized or reduced in slow sweep rate cyclic voltammetric experiments. However, in many cases, not all the redox sites undergo electrochemical transformations.

Consequently, an explanation should be provided as to how the electrons traverse the film, since it was reasonably assumed that the chains cannot diffuse to the metal surface during the time scale of the experiment. We should not, however, exclude the possibility that polymer diffusion may play a role in carrying charges when the film is held together by physical forces. Electrochemical transformations of redox sites were observed when the polymer chains were connected by chemical cross-linkages, and thus only segmental motions were possible.

Therefore, the transport of electrons can be assumed to occur via an electron exchange reaction (or electron hopping) between neighboring redox sites if the segmental motions make it possible. Electron exchange reactions coupled to diffusion have been detected in solutions of redox species [96–101], which we will discuss later. However, the main difference is that in solution it may enhance—usually to a very small extent—the rate of diffusion, whereas in our case, this is the main mechanism of the electron transport, though the overall rate may be determined by the rate of the segmental motion needed to bring the sites together. However, in the case of polymer-modified electrodes belonging to subclass 2, where the incorporated redox ions can move within the polymer network only their motion is hindered due to the interactions with the functional groups of the polymer and the high viscosity, a situation similar to that in solution arises and the contribution of the electron exchange reaction to the transport rate, depending on the relative rate of these processes, may be substantial. In special circumstances, e.g., at high redox site concentrations, we may envisage a situation where electrons move among the fixed redox sites by mixed-valence conduction mechanism, by extended electron transfer through space, or through intervening bonds.

In the case of conducting polymers, the delocalized electrons can move through the conjugated systems (intrachain conduction), but an electron-hopping mechanism is likely to be operative between the chains (interchain conduction) and defects. This amounts to the fact that in contrast to the usual electrode reaction, we should assume long-range electron transport, which presumes an electronic conduction and/or a chemical reaction.

In almost every case, the charge is also carried by the motion of electroinactive ions during electrolysis, because it is necessarily coupled to the transfer of electrons in order to preserve electroneutrality within the film. The motion of counterions may also be the rate-determining step. This situation arises especially when a neutral film is converted into a polyelectrolyte as a function of the potential and the incorporation of a large amount of counterions into the film from the bulk solution phase is necessary. Therefore, the electrochemical response depends on the cell arrangement and on the amplitude of the perturbation (voltage, current), as the dominating process in the charge transport may change.

The thermodynamic equilibrium between the polymer phase and the contacting solutions requires $\tilde{\mu}_i(\text{film}) = \tilde{\mu}_i(\text{solution})$ for all mobile species. In fact, we may regard our film as a membrane or a swollen polyelectrolyte gel, i.e., the charged film contains solvent molecules and, depending on the conditions, co-ions, in addition to the counterions. As a consequence of the

Charge Transport in Polymer-Modified Electrodes

incorporation of ions and solvent molecules into the film, swelling or shrinkage of the polymer matrix takes place. Depending on the nature and the extent of cross-links, reversible elastic deformation or irreversible changes (e.g., dissolution) may occur. As with the usual electrode reactions, other effects, such as dimerization, ion-pair formation, etc., should also be considered. The chemical reaction may lead to cross-linking, which alters the morphology of the film and restricts the chain and segmental motions.

We have already mentioned several effects that are connected with the polymeric nature of the layer. It is evident that all the charge transport processes listed are affected by the physicochemical properties of the polymer. Therefore, we also must deal with the properties of the polymer layer if we wish to understand the electrochemical behavior of these systems.

Because of the interrelated nature of the phenomena, it is hard to separate the effects of the different processes and to avoid some overlapping between the topics. Therefore, we first recapitulate the models and theories elaborated with respect to the charge transport processes, including the experimental tests of their predictions. Second, we deal with the properties of the polymer layer, which involves an investigation of the morphological changes induced by the variation of the experimental conditions (potential, electrolyte composition, temperature, etc.) in the light of theory and the experimental evidence. It is also our aim to illustrate the effect of different factors on the mechanism and kinetics of charge transport while providing an abundant range of experimental examples.

III. THEORIES OF ELECTRON TRANSPORT IN POLYMER FILM ELECTRODES

A. Early Models of Charge Propagation

The first models and their mathematical formulations appeared in a closely spaced series of publications by Andrieux and Savéant [102], Laviron [103], and Peerce and Bard [104]. They used the idea introduced by Kaufman and Engler [105] that the electron transfer between neighboring redox sites may take place via an electron-hopping mechanism. They relied on the early results obtained by potential sweep and potential step methods, which showed that the charge transport process can be described within the framework of diffusion kinetics.

It is worthwhile to give a brief account of these theories, because they serve as a starting point for the subsequent models. The three models share several common features. First, the polymer film was treated as a macroscopic assembly of planar layers containing equal concentrations of evenly distributed redox centers, although in the model of Peerce and Bard [104] the consequences of the uneven distribution of nonequivalent, interacting sites were also emphasized. Second, it was presumed that direct electron transfer between the metal and the film involves only those redox sites situated in the layer immediately adjacent to the metal surface, and third, the film is electronically nonconductive.

Andrieux and Savéant [102] assumed that the electron propagation through the film occurs as a sequence of successive steps between redox centers located in sublayers, and rationalized the sweep rate dependence of the cyclic voltammetric curves in terms of a linear diffusion process occurring within a finite space.

It was pointed out that as far as the charge transport under a concentration gradient is concerned, the electron hopping between adjacent redox sites of different oxidation states can be described by Fick's law, i.e., as a diffusion process of two immobile redox sites with the same diffusion coefficient,

$$D_e = k_e \delta^2 c \tag{3}$$

where k_e is the bimolecular electron-transfer rate constant, c is the total redox site concentration, and δ is the mean distance between two adjacent redox sites, which is equal to the thickness of a layer (Δx) in the film.

Relying on this theory, Andrieux and Savéant calculated the values of D and k_e for poly(p-nitrostyrene) films. It was found that $D = 2 \times 10^{-12}$ cm^2/s and $k_e = 10^4$ dm^3/mol/s taking $\Delta x = 0.8$ nm and using the heterogeneous rate constant, $k_s = 5 \times 10^{-5}$ cm/s. When these values were compared with the respective values obtained in solution, it was revealed that both the charge transport and charge transfer processes are much slower in polymer films, which was assigned to the different collisional and environmental situations, as well as to ohmic drop effects.

The model elaborated by Laviron [103] is basically similar to the model described above. Laviron emphasized that the model is valid when the transport of counterions and solvent molecules do not appear as a limiting factor. In his model, the thickness of the first layer, Δx_1, may differ from Δx.

In the model proposed by Peerce and Bard [104], it was assumed that the electron transport within the layer is driven by the difference between the ratio of oxidized and reduced sites in the film. The rate at which equilibrium is attained depends on the rate of transport of counterions into or out of the film and the rate of electron transfer between the electroactive groups. They also considered that the appropriate orientation of the redox groups is related to the self-diffusion of the polymer chains. The combined effect of electron hopping and ion movement was treated within the model as a diffusional process. On the basis of this model a digital simulation was used to estimate the effect of D and k_s on the shape of the cyclic voltammograms in different cases. They analyzed the effect of site-site interactions, the distribution of nonequivalent redox sites, and their interconversion as a function of potential. The treatment was applied to simulate the results obtained for two ferrocene-containing polymers in which the spacing between the electroactive groups was different.

For poly(vinylferrocene) (PVF) film in 0.1 mol/dm^3 tetrabutylammonium perchlorate (TBAP)/acetonitrile D = 2 × 10^{-10} cm^2/s and k_s = 100–250 s^{-1} were obtained, and it was found that the values of the diffusion coefficient depend on the film thickness (dry thickness was used for the simulation), the nature of counterions (explained on the basis of ion-pairing effect and the structure modifying ability of the counterions), and the spacing between the redox groups. The application of this model fits the experimental results well, which attests that, in the first approximation, the cyclic voltammetric behavior of these films can be described in terms of a model including a heterogeneous electron transfer step and a diffusion process, even when interactions between the sites and interconversion of the different sites can be assumed.

Because of the phenomenological nature of these theories and the applied simplifications, several questions have remained unanswered concerning the actual mechanism of charge transport. We have not obtained an answer for the fundamental question: What does the measured diffusion coefficient represent? This is identical with the problem: What is the rate-determining step? Among the possible explanations are the diffusion of the redox sites, the local segmental or long-range chain motion, and diffusion or migration of counterions. One must confront the problem of coupling between the electron transport and the counterion displacement.

We will address these problems in the following sections. First the theory of the electron exchange reaction will be inspected, then the more sophisticated theories of charge transport in redox polymers will be surveyed and tested by investigating the concentration dependence of D.

B. Theory of the Electron Exchange Reaction

For the fast electron exchange processes coupled to isothermal diffusion in solution, the theoretical description and its experimental verification were given by Dahms [97] and by Ruff and co-workers [98–101]. Ruff and co-workers studied the displacement of the centers of mass of particles, which is brought about by both common migrational motion and chemical exchange reaction of the type

$$A + AX = AX + A \tag{4}$$

It was concluded that the exchange reaction (electron hopping) can be described in terms of second-order kinetics, and a relationship was given between the measured diffusion coefficient (D) and the diffusion coefficient (D_0) that would be measured in the absence of any electron exchange reaction:

$$D = D_0 + \frac{k_e \delta^2 \Pi}{4} c = D_0 + D_e \tag{5}$$

where k_e is the second-order rate constant of the exchange reaction and c and δ are the concentration and the distance between the centers of the chemically equivalent species involved in the electron exchange, respectively. In fact, δ is the distance with which the hydrodynamic displacement is shortened for the species and can be determined or at least estimated on the basis of independent data, e.g., the results of neutron scattering.

Later, Ruff and Botár [106] corrected the original thermodynamic derivation, which was affected by some errors. According to the corrected thermodynamical treatment, the constant factor in Eq. (5) is 1/6, 1/4, or 1/2 (instead of $\Pi/4$) for three, two-, and one-dimensional diffusion, respectively, i.e.,

$$D = D_0 + k_e \delta^2 c/6 \tag{6}$$

They also used a random walk treatment to describe the electron-hopping process coupled to physical diffusion [107]. In this model, two real components A and AX are replaced by two formal components, A and X. The motion of X is restricted only to those lattice points that are occupied by an A, and an X is allowed to jump over to the neighboring A only if its orientation is favorable, i.e., if it is on the edge, pointing towards the "receiving" A. The motion of A and AX, i.e., their hydrodynamic diffusion along some lattice points, is unrestricted. The random walk of AX concurs with the events, when an A happens to be at one of the neighboring lattice points and X jumps over to the other carrier A. This would model the transfer diffusion step. Thus, the resultant motion of X is the superposition of two random walks.

On the basis of the random walk treatment, Ruff and Botár [107] derived two equations for the diffusion coefficients A and AX:

$$D_{AX} = \frac{1}{6} \left\{ \lambda_{AX} \delta^2 [1 - \Phi_A(j)] + \lambda_X \delta^2 \Phi_A(i) \right\} \tag{7}$$

$$D_A = \frac{1}{6} \left[\lambda_{AX} \delta^2 - \lambda_X \delta^2 \right] \Phi_{AX}(i) \tag{8}$$

where λ_{AX} and λ_X are the frequency of jumps, δ is the distance between the lattice points, $\Phi_A(i)$ and $\phi_{AX}(i)$ are the one-position distribution functions, which suggest the probability that position i is occupied by particle A or AX,

Charge Transport in Polymer-Modified Electrodes

respectively. The $\Phi(i)$ functions can be regarded as concentrations. Having applied the suitable simplification, i.e., neglecting the cross-terms, we may give the following equation for the concentration dependence of D:

$$D = D_0 (1 - \alpha c) + k_e \delta^2 c/6 \qquad (9)$$

where α is a factor converting $\Phi_A(i)$ into concentration.

One can see that the diffusion coefficients are concentration dependent for two reasons. First, the concentration of species A appears in the hopping diffusion term of AX—as in the original formula—since A is needed for the exchange reaction to take place and vice versa. Second, there is a so-called blocking factor of $[1 - \Phi_A(i)]$ in the new equations, which expresses the hindering effect of the occupied position in the physical diffusion.

Buck [108] criticized the original treatment applied by Ruff et al. [98–101]. In his opinion, two anomalous steps were made in the derivation. The first one occurs in the application of the time-dependent Fick's second law, the second appears in the differential formulation of the chemical rate expression. The details of the mathematical derivations are available in the original paper [108], but some of the results will be used later, when the effect of the local electric field will be discussed. Buck found that the spherical diffusion coefficient ($\Pi/4$) in Eq. (5) was not necessary [109].

Neither Ruff's own correction nor Buck's critique have had any kind of impact on the researches carried out in this field, as practically every researcher has used the original Dahms-Ruff relationship, as given in Eq. (5), since it was applied first by Buttry and Anson [110] to describe the behavior of ion-exchange polymer films containing two pairs of electrostatically bound isomorphous redox ions. (The papers of Ruff and Botár [106,107] were first cited in the literature of polymer film electrodes in 1989 [111,112], and the use of the corrected numerical prefactor gained ground in the papers published in 1991–1992 by the top groups in the field [78,113–118].) They pointed out that while the condition $D_e = D_0$ is seldom met in solution of small molecules, because $D_0 >> D_e$, in polymer environments, e.g., in concentrated polymer solutions or gels, the influence of electron hopping is expected to be more pronounced, because the D_0 values may be many orders of magnitude lower than they are in solution. It was also realized that as follows from the theory, D should increase linearly with c, whenever the contribution from electron exchange reaction is important (mobile redox centers), or if the electron hopping is rate-determining (fixed redox sites), i.e., the concentration dependence of D may be the test of theories based on the electron exchange reaction mechanism.

Despite the fact that considerable efforts have been made to find the linear concentration dependence of D predicted by the Dahms-Ruff theory, it has not been observed yet; perhaps the $Co(bpy)_3^{3+/2+}$/Nafion system (bpy = 2,2′-bipyridine) [119] is the only exception. (The behavior anticipated from Eq. (6) was found by Murray et al. [113] for LiTCNQ in network poly(ethylene oxide) polymer solvent, a system closely related to the polymer films on electrodes.)

Therefore, we are inclined to come to the conclusion that either this simple model is not suitable for the adequate explanation of the observations or our experimental techniques and data evaluation do not supply correct information. There may be several reasons why this model has not fulfilled our expectations:

1. The uncertainty in the determination of D by potential-step, impedance, or other techniques is too substantial to obtain reliable data, due to problems such as the extraction of D from a $D^{1/2}c$ product (this combination appears in all the methods), the difficulty arising from the in situ thickness estimation, nonuniform thickness [30,32,120,121], incomplete electroactivity [111,122,123], film inhomogeneity [12,18,22], and ohmic drop effect [124–129]. It may be forecast, for example, that the film thickness increases, thus c decreases, due to the solvent swelling of the film, however, D_0 increases because the size of the solvent-filled cavities in the film increases, making the physical diffusion of ions and the segmental motions less hindered simultaneously. In addition, the solvent swelling changes with the potential, and it is sensitive to the composition of the supporting electrolyte. Because of the interactions between the redox centers or between the redox species and the film functional groups, the morphology of the film will also change with the concentration of the redox groups. In fact, in many cases, the lack of the concentration dependence or even the decrease of D with c have been explained with these effects. Because for every system, these problems give rise to special difficulties, we will deal with them in the section reviewing the accumulated experimental material for different polymer-modified electrodes.

2. The model is accurate, but the values of the electron jump distance δ used for the calculation are not correct, or even may change with c. The electron jump distance is the redox site separation at the moment of the reaction, which is, in the first approximation, equal to the sum of the radii of Ox and Red. (Sometimes δ is called simply the site-site distance, which is incorrect.)

The jump distance was calculated for different cases by Levich [96]. The jump distances of maximum probability are similar in aqueous solutions and close to twice the radius (r_0) of the ion including the dielectric-saturated solvent layer. However, according to the theory of the extended electron transfer, δ may be higher than $2r_0$.

The implications of extended heterogeneous electron transfer coupled to semi-infinite linear diffusion were analyzed by Feldberg [130]. Feldberg started with the results of Miller et al. [131–133]. They investigated the distance dependence of electron transfer rates for homogeneous bimolecular reactions, which gives

$$k = k_0 \exp \frac{-(r - r_0)}{s} \qquad (10)$$

where r is the separation of the reactant pair, r_0 is the distance of the closest approach, and s is a characteristic distance ($\sim 10^{-8}$ cm) for the decay of interaction.

Feldberg pointed out that the effect of extended electron transfer might be measured in the case of polymer films by potential step methods, because the diffusion of redox species is hindered in the polymer network, consequently D_0 becomes low in comparison with the solution value. The most important implication of this work, however, is that k_e may increase as the site concentration is increased, since r decreases in the film, because

$$k_e = A \exp \frac{-(r_{NN} - r_0)}{s} \qquad (11)$$

It should also be taken into account that r_{NN} may be smaller than the average site-site distance calculated from the concentration and density of the film, because sites are free to oscillate over significant fractions of the site-site distance.

It is worth mentioning that long-distance electronic couplings for electron-transfer reactions are also calculated on the basis of the super-exchange-pathway model [134].

3. The rate of electron exchange reaction is not activation controlled, but diffusion controlled. This should be considered, in the case of fixed-site redox polymers, because the redox species, which are capable of exchanging electrons, are bound to a polymer chain or network; consequently their motion is governed by the chain or segmental motions. In this case, however, the segmental motion is the physical motion characterized by D_0, and at the same time, it makes the encounter of the sites possible. A similar situation arises in the case of ion-exchange layers when $D_0 << D_e$.

This means that only a k_{app} can be determined by measuring the concentration dependence of D, which is equal to the rate constant of the electron exchange reaction only in that case, when the rate of the encounter is much higher than k_e. The rate of the collision, k_d, can be estimated by using Smoluchowski's equation [135]:

$$k_d = \frac{4\Pi\delta(2D_0)N_A}{10^3} \tag{12}$$

where N_A is the Avogadro number. Therefore,

$$\frac{1}{k_{app}} = \frac{1}{k_e} + \frac{1}{k_d} \tag{13}$$

and consequently

$$D = D_0 + k_{app}\,\delta^2 c/6 \tag{14}$$

or

$$D = D_0 + \frac{\delta^2 c}{6}\left(\frac{1}{k_e} + \frac{1}{8\Pi\delta D_0 N_A}\right)^{-1} \tag{15}$$

Thus, in this case the model is correct, i.e., both the physical diffusion of ions and electron hopping occur in the system, but the latter is not the rate-determining step. However, in the present thinking [113–118], the combinations of Eqs. (6) and (12) is erroneous, and Eq. (12) does not represent the "upper measurable limit" of the reaction rate constant measurement. The physical reason for this is that Eq. (12) gives the rate at which reactants diffuse together and collide but contains no information about the reaction's spatial displacement of charge that follows a reaction-producing collision. The result is that the analysis of electrochemical charge transport by Eq. (6) actually allows measurement of a rate constant larger than that given by Eq. (12). Formally it means that D should be used instead of D_0 in Eq. (12) and in Eq. (15) in the brackets.

4. Rubinstein [136] considered a situation when k_e is activation-controlled, but the dielectric permittivity of the medium, ϵ changes with c. Because k_e depends on ϵ in the form

$$\ln k_e = \ln k_e^0 - \frac{Nz_A z_B e^2}{\epsilon RTr} \tag{16}$$

where z_A and z_B are the unit charges, e is the electronic charge, and r is the distance between the ions in the activated complex, and if z_A and z_B have the same sign, k_e will become smaller when ϵ decreases.

5. The electron hopping makes no contribution to the diffusion as in solution when $D_0 >> D_e$.

6. The most hindered process is the counterion diffusion, coupled to electron transport.

We will see later that the majority of the observations may be explained in terms of these assumptions, with the help of either one or a combination of several of them, and also consider the possibility that the rate-determining step may change with the concentration.

In order to explain the experimental results and also on a theoretical basis, new models have been developed in recent years. We will give a brief account of these theories later.

C. New Theories Predicting Nonlinear D(c) Function

1. He-Chen Model

He and Chen [137] proposed a modified version of the original Dahms-Ruff theory. They assumed that in systems where the electron exchange rate is high, the rate of the electron transport is determined by the physical diffusion of the redox species incorporated in an ion-exchange membrane, when the concentration of the redox couples is low. As the concentration increases in the film, and consequently some of the redox centers become immediately adjacent, several electron hops may become possible, i.e., the charge donated to a given redox ion through diffusional encounter may propagate over more than one site in the direction of the concentration gradient. This enhances the total electron flux. Formally, this is equivalent to an increase of the electron-hopping distance by a certain factor, f, thus D can be expressed as follows:

$$D = D_0 + \frac{\Pi k_e c (\delta f)^2}{4} \tag{17}$$

Assuming a Poisson distribution of the electroactive species, the enhancement factor can be expressed as a power series of a probability function, which is related to the concentration. At low concentrations the probability of finding more than one molecule in the hemisphere with the radius of molecular collision distance is nearly zero and $f = 1$. The factor, and therefore D_e, increases noticeably at higher concentrations, as shown in Fig. 1.

He and Chen have found such concentration dependence of the electron diffusion coefficient for $Ru(bpy)_3^{3+/2+}$ and $Os(bpy)_3^{3+/2+}$ in Nafion films in contact with 0.05 mol/dm^3 H_2SO_4 solutions containing no mediators as shown in Fig. 2. In Fig. 2 the values of relative electron diffusion coefficient obtained by chronocoulometry and impedance spectroscopy were plotted as a function of surface concentration (loading value), Γ, because the exact film thickness was not known. The authors assumed, however, that the film thickness may change only to a small degree. It is of interest that the data obtained from chronocoulometric experiments match reasonably well with

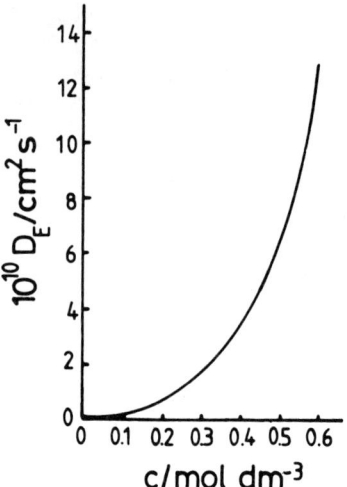

FIG. 1. Concentration dependence of the electron diffusion coefficient calculated according to Eq. (17) assuming $k_e = 10^4$ dm^3/mol s and $\delta = 1.36 \times 10^{-7}$ cm. (From Ref. 137, with permission of Elsevier Scientific Publishing Company.)

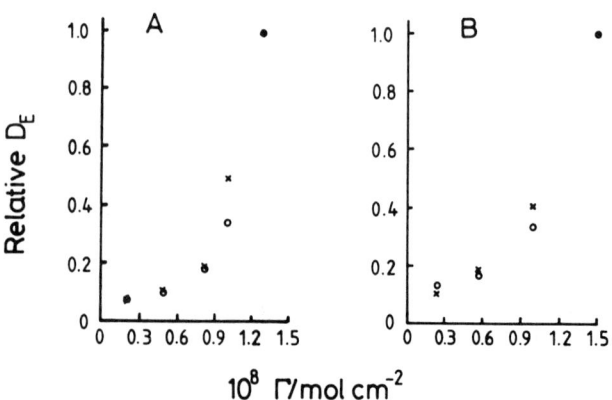

FIG. 2. Concentration dependence of the relative electron diffusion coefficient measured by chronocoulometry (o) and by ac impedance (x). (A) Ru(bpy)$_3^{2+}$ and (B) Os(bpy)$_3^{2+}$ loaded Nafion film electrodes. (From Ref. 137, with permission of Elsevier Scientific Publishing Company.)

Charge Transport in Polymer-Modified Electrodes

those determined by impedance technique. It may indicate the fact that the counterion transport plays no role in the transportation of charge through these films at the experimental conditions used. However, it should be mentioned that the determination of D_e by the impedance technique was not very accurate, because only an approximate formula was used for the calculation.

For Nafion–methyl viologen and poly(vinylpyridine) (PVP) crosslinked with 1,2-dibromoethane containing $Fe(CN)_6^{3-/4-}$ redox couple—no enhanced electron transport was observed [138]. This was explained by the effect of the slow electron exchange rate and/or by the small radius of collision (r), because electron hopping may not take place immediately after the molecules are brought together and/or the probability of two or more molecules falling in a sphere volume would be low ($k_e = 1 \times 10^9$ and 1×10^3 dm^3/mol s, $r = 1.36 \times 10^{-7}$ and 6.7×10^{-8} cm for $Ru(bpy)_3^{2+/3+}$ and $Fe(CN)_6^{3-/4-}$, respectively).

Nafion/$Os(bpy)_3^{3+/2+}$ system in 0.1 mol dm^{-3} Na_2SO_4 solution has also been investigated by Sharp et al. [139]. The dependence of D on the concentration of the $Os(bpy)_3^{3+/2+}$ in the film observed experimentally and simulated by using the He-Chen model is shown in Fig. 3. The values of D were derived from chronocoulometric and impedance data at different film thicknesses. At low c values, D increases approximately linearly with c in accordance with the prediction of the Dahms-Ruff equation. At c > ca. 0.35 mol/dm^3, a rapid rise in D can be observed. One can see, however, that the curve calculated by the treatment of He and Chen does not fit the points determined experimentally. (For the calculation of the curve, the values of film thickness, in dry and wet states, measured by ellipsometry as well as reasonable values for other quantities such as $\delta = 1.4$ nm, $D_0 = 2 \times 10^{-11}$ cm^2/s, and $k_{app} = 4.2 \times 10^4$ dm^3/mol s have been used.) Sharp et al. have concluded that at concentrations < 0.35 mol/dm^3, the variation of D with c cannot be described in terms of the model, while at high concentrations, where the redox sites are closely spaced, the mechanism of electron transfer proposed by He and Chen may be operative. It should be mentioned that an attempt was made to explain the sharp increase of the D vs. c curve at high concentrations in terms of parallel conduction pathways, where the internal structure of hydrated perfluorosulfonic acid membrane was taken into account [140]. This model provided a satisfactory explanation for the data obtained at low c values, but gave overestimated values at high concentrations. Sharp et al. have also tried to explain the result by the migrational enhancement, based on the model of Andrieux and Savéant [141]. Thus, we will return to the analysis of the results mentioned above when we describe and discuss the model of Andrieux and Savéant.

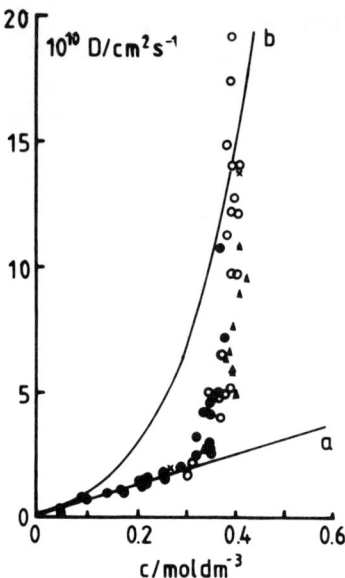

FIG. 3. The dependence of D, for charge propagation through $Os(bpy)_3^{+/2+}$/Nafion films, on the concentration. The experimental points correspond with the following film thicknesses: (●) 525 nm, (○) 210 nm, (△) 70 nm, (◗) 360 nm, and (▲) 145 nm. Points marked with stars were derived from impedance spectroscopy. Lines a and b reflect the D vs. c dependencies predicted by Eqs. (5) and (17), respectively. (From Ref. 139, with permission of Elsevier Scientific Publishing Company.)

2. Fritsch-Faules–Faulkner Model

Fritsch-Faules and Faulkner [142] developed a microscopic model to describe electron (or hole) diffusion in a rigid three-dimensional network. Their model takes excluded volume into consideration, and it is based on simple probability distribution arguments and on a random walk. The following conditions are assumed: (1) there is a homogeneous random distribution of hard-sphere redox centers of diameter r_0 (the hard-sphere approximation simplifies the geometry and eliminates the orientational effects); (2) all redox centers are identical, noninteractive, and immobilized (i.e., there is no molecular motion); (3) each step corresponds to an electron transfer across the average nearest neighbor distance between redox species, r_{NN}; (4) the electron transfer rate constant, k_e, decreases exponentially with the edge-to-edge separation between reacting pairs, $r_{NN} - r_0$; and (5) charge compensation occurs suffi-

Charge Transport in Polymer-Modified Electrodes

ciently rapidly on a local basis so that it does not limit the random hopping, except for its effect on the reorganization energies governing extended electron transfer.

The conception of Fritsch-Faules and Faulkner is closely related to that of the earlier work of Feldberg [130], who investigated the problem of the coupling of semi-infinite linear diffusion and heterogeneous electron transfer with an exponentially decaying distance dependence, and it has some features in common with the distance dependence of electron transfer rates for homogeneous bimolecular reactions [131–133]. The diffusion coefficient for the succession of electron transfer events in the network was derived from the relationship originated from the three-dimensional random walk model:

$$D_e = \frac{f \langle l^2 \rangle}{6} \qquad (18)$$

where f is the number of displacement per unit time ($1/\tau$) and $\langle l^2 \rangle$ is the mean square displacement distance. The rate of the electron exchange was expressed by Eq. (11).

The mean nearest neighbor distance was derived as a function of concentration for a system of hard spheres having finite volume, hence it became possible to predict the concentration dependence of D_e. The model suggests that D_e should first have an exponential-like rise with increasing c, then flatten at high concentrations. The exponential rise occurs, because as the concentration increases, r_{NN} becomes smaller, which promotes intersite electron transfer, according to Eq. (11). As the minimum center-to-center separation is approached, when each redox center has a nearest neighbor practically in contact, D_e asymptotically nears it theoretical maximum value. The calculated curves for two different r_0 values are shown in Fig. 4. Contact radii $r_0 = 1.3$ and 0.6 nm are the approximate values of Ru(bpy)$_3^{3+/2+}$ bound in poly(styrenesulfonate) and ferrocene moieties immobilized in films of PVF, respectively.

The shortcomings of this model are inevitable. First of all, this model is valid only for systems where redox species are truly immobilized in a rigid network. Polymer film electrodes, in contact with supporting electrolytes, do not usually fall into this category. In real systems, the physical diffusion of the redox centers, the segmental and chain motions of the polymer, probably take place, i.e., models that predict sharp rise of D with c at low concentrations but do not preclude bimolecular collisions seem to be more realistic. At high concentrations, there are surely interactions between the redox sites, and between the functional groups and redox species causing a change of the film morphology D may depend on the oxidation state of the redox sites, etc.

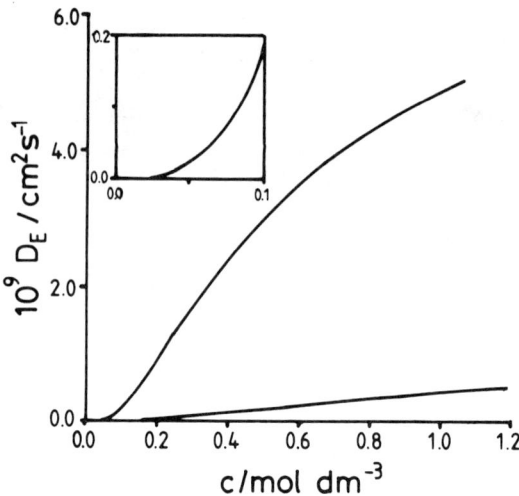

FIG. 4. Concentration dependence of the electron diffusion coefficient calculated by Fritsch-Faules and Faulkner using $A = 3.06 \times 10^6$ and $\delta = 0.105$ nm, respectively. Upper curve and inset, $r_0 = 1.3$ nm, lower curve, $r_0 = 0.6$ nm (see text). (From Ref. 142, with permission of Elsevier Scientific Publishing Company.)

It should be mentioned that in some cases concentration dependence of D similar to that shown in Fig. 4 has been observed, e.g., for quarternized poly(4-vinylpyridine) (QPVP) films containing $Ru(bpy)_2Cl^{2+/1+}$ [143], as illustrated in Fig. 5. However, it was rendered probable that the segmental motion of the polymer, leading to the local diffusion of the pendant group, is the rate-controlling element [93,143]. Nevertheless, this model should be considered a useful one, because a situation can be envisaged where an extended electron transfer is efficient and the chain motion has the effect of producing a larger-than-actual molecular diameter, which makes a long-distance electron transfer possible. Consequently, electrons are transported over long distances by exchanges between sites without the rate limitation of the segmental motion. Another extreme case is when the intersite coupling of electrons between centers is very strong, and the high electron conductivities of conducting polymers may be described by the model elaborated by Fritsch-Faules and Faulkner [93].

3. Percolation Theory

Leiva et al. [144] described the electron transport that occurs via electron hopping between localized sites in films, and the mediation reaction in terms of percolation theory. They used a model that is applied in solid state physics

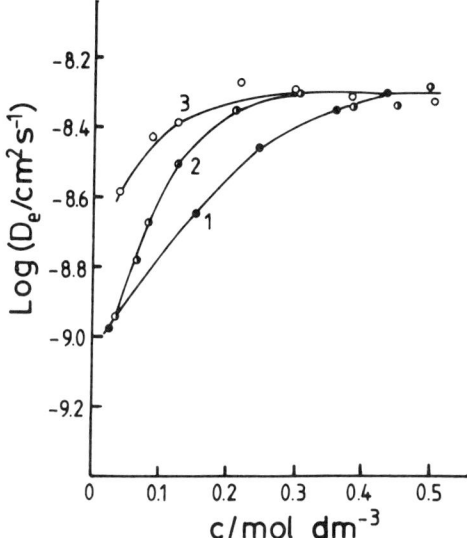

FIG. 5. Electron-diffusion coefficients measured by chronocoulometry vs. the concentration of $Ru(bpy)_2Cl^{2+/1+}$ in QPVP film. (○) 4% cross-linked; (◐) 8% cross-linked. (●) 12% cross-linked. (From Ref. 143, with permission of the American Chemical Society.)

for electron hopping through insulating film. According to this theory, the probability of electronic transition between two localized states depends on their spatial separation, r_{ij}, and the difference $E_i - E_j$ in energy:

$$P_{ij} = A \exp [-\chi r_{ij} - | E_i - E_j | /kT] \qquad (19)$$

where χ characterizes the localization of the electronic wavefunction at a particular state. The preexponential factor, A, depends on the properties of the electronic state and the surrounding medium.

The simulated current-potential relationship for a mediated reaction occurring at the film/solution interface was similar to the experimentally observed ones. It is of interest that the current is not proportional to the concentration of localized states, but follows a power law of the type $j \sim c^m$, with $m \sim 2$–3.

4. Theories of Coupled Electron Hopping–Ion Displacement

As has been mentioned earlier, when the oxidation state of the film is changed, in the simplest case counterions enter the film in order to maintain the bulk electroneutrality in the surface layer of finite thickness. This phenomenon is independent of how the electrons traverse the film. Andrieux and Savéant [141] considered the consequences of the electroneutrality coupling of electron hopping between localized sites with electroinactive counterion displacement for the steady-state [145] and transient current responses [141]. They recognized that the electron-hopping process cannot be described by the usual combination of the classical Fick and Nernst-Planck laws, when the effect of electric field is considered, but rather a second-order law should be derived from the bimolecular character of electron hopping, as opposed to the unimolecular character of ion displacement. They investigated systems containing fixed redox ions, mobile electroinactive counterions, and fixed electroinactive counterions, under semi-infinite diffusion-migration conditions.

The variation of the diffusion and migration potential-step plateau current was analyzed in terms of the ratio of the mobilities of electrons and electroinactive counterions characterized by the respective diffusion coefficients, D_e and D_i. It was found that migration results in an enhancement of the current over the pure diffusion current. Migration diminishes in all cases as the relative concentration of electroactive fixed counterions is increased, i.e., the fixed counterions play a role similar to that of the supporting electrolyte in solution studies. This is especially so when the diffusion coefficient of the mobile counterions is small compared to the diffusion coefficient for electron hopping.

An interesting feature of this theory is that it predicts an enhancement of the electron propagation, thus the enhancement of current, independent of whether the electroactive ions are anions or cations, for both the oxidation and reduction processes. This is a consequence of the electron-hopping mechanism, because where electron transport occurs by means of physical displacement of the redox ions, an enhancement may come about only in the cases where, for example, the mobile electroinactive ions are anions and the electroactive ions are cations in the course of reduction. When anions are reduced in the presence of mobile countercations, the potential gradient diminishes the current.

It was found in all cases that the Cottrell behavior is retained, i.e., the current is proportional to the inverse of the square root of time. Thus, a D value can be determined in the usual way, from the Cottrell slope of the chronoamperometric experiment. However, this apparent diffusion coefficient (D) may be substantially higher than the true diffusion coefficient of electron hopping (D_e) depending on the ratio $\sigma = D_i/D_e$. The relationship between D and D_e was given for two extreme cases as follows:

$$D = (\Pi\Psi^2)D_e \qquad \text{for } \sigma \to \infty \tag{20a}$$

and

$$D = (\sigma\Pi\Psi^2)D_e^2/D_i \qquad \text{for } \sigma \to 0 \tag{20b}$$

where Ψ is a current function, which involves the ratio of the diffusion currents perturbed (i) or unperturbed (i_o) by migration effects:

$$\Psi = \frac{1}{\Pi^{1/2}} \frac{i}{i_0} \tag{21}$$

According to the theory of Laviron [103] and Andrieux-Savéant [102] D_e is proportional to the redox site concentration, as described by Eq. (3). Therefore, when $\sigma \to \infty$, i.e., when the mobility of the electroinactive counterions is much greater than the mobility of electrons, D_e is independent of the diffusion coefficient of the mobile counterions. The contribution of migration is very small, and D increases almost proportionally to c—only a modest excess variation can be observed. The situation is quite different when $\sigma \to 0$. D is inversely proportional to D_i, which is small, and the migration may lead, depending on the concentration of the fixed electronactive ions, to a substantial increase of D. Since D_e is proportional to c, D would vary proportionately to the square of electroactive species concentration. In fact, a variation as large as c^3 is possible due to the decrease of the dimensionless parameter γ, which results in an increase of $(i/i_0)\sigma^{1/2}$. The γ parameter decreases with c, since it depends on the charge number of the initial reactant (z_A), the number of electrons transferred (n), as well as on the charge number (z_F) and the relative concentration (c_F) of the fixed electroinactive sites as follows:

$$\gamma = \frac{z_A - n + \frac{z_F c_F}{c}}{n} \tag{22}$$

The authors also came to an interesting conclusion concerning the ratio between the potential-step reduction current of a fully oxidized film and the oxidation current of the same fully reduced film. This current ratio is slightly larger than the one for a large value of σ, and decreases with σ until reaching a limiting value between 1 and 0.5 (the smaller γ, the smaller the limiting value of the ratio).

The authors themselves tested their theory using the data of Facci et al. [146], obtained for a redox copolymer film prepared from monomers of osmium (II) and ruthenium (II) bis (2,2'-bipyridyl) bis(N-4-pyridylcinnamamide) hexafluorophosphate, respectively. In this case, Os sites were diluted with Ru sites, which are approximately identical in all structural and electrostatic respects, but not electroactive at the same potential. In this way Facci et al. attempted to avoid extraneous effects leading to structural changes of the polymer film. The copolymer film was investigated in contact with 0.1 mol/dm^3 tetraethylammonium perchlorate/acetonitrile solution, and D values were measured by chronoamperometry. Andrieux and Savéant assumed that

D_i (ClO$_4^-$) is small, thus the limiting situation $\sigma \rightarrow 0$ is allowed to be used for the simulation of the D vs. c function. The measured data and the simulated curve can be seen in Fig. 6. Andrieux and Savéant claim that the simulated curve satisfactorily fits the experimental data. They criticized the original explanation given by Facci et al. [146], which was based on a mechanism in which electron hopping is coupled with the concentration-dependent physical motion of redox sites and with polymer self-diffusive motion. According to this interpretation, at low c values osmium sites are separated at distances greater than the amplitude of the average self-diffusive motion of the polymer, and the increase of c results in an increase in D as the distance decreases. At medium and high concentrations the rate-controlling steps are the physical motion (polymer self-diffusion) of the osmium site and the electron exchange reaction between the redox sites, respectively. However, Facci et al. distinguished three distinct regions, and as their paper shows, D is even independent of c in the concentration region of 0.2–0.6 mol/dm^3. According to Andrieux and Savéant, in this case, a linear D vs. c function should have been observed, while they could simulate the curve with a $c^{2.89}$ dependence. The simulated curve fits the experimental points only at high concentrations, which may be the consequence of the current enhancement of migration. It is

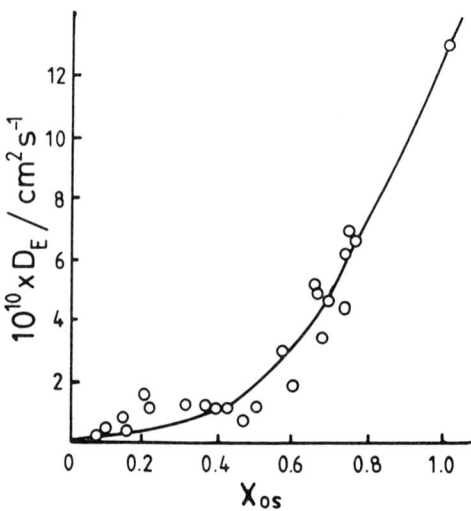

FIG. 6. Simulation of the Murray et al. Os(II)-Ru(II) copolymer data [146] with an electron-hopping model taking into account the effect of migration with the mobility of the electroninactive counterions, ClO$_4^-$, being much slower than the mobility of electrons. (From Ref. 141, with permission of the American Chemical Society.)

clear, however, that the theory of Andrieux and Savéant provides an approximate description of the concentration dependence observed.

Within the framework of this theory, no explanation can be given either for the changes of the activation energy (E_a) in the three regions mentioned above. However, this variation of E_a clearly indicates that the rate-determining step changes with temperature and at least three different coupled processes should be taken into consideration. Murray et al. [77] were of the opinion that while this theoretical analysis was correct, the assumption of small D_i was not. Murray et al. measured the permeability and diffusion coefficient of chloride ions in these poly($[Os(bpy)_2(p\text{-cinn})_2]^{2+}$) films. Assuming that D_i for Cl^- ions is a good approximation of that of ClO_4^- ions, and that the polymer exhibits no strong partitioning preference for Cl^- vs. ClO_4^-, it was calculated that $D_i = 100 D_e$. Andrieux and Savéant also emphasized that other factors (activity, ion pairing) may interfere, thus modifying the estimated increase caused by the migration effect. Clearly, the most important point that Andrieux and Savéant have made is that the transient response depends upon the mobility of the electroinactive counterions: the smaller the mobility, the larger the current due to the corresponding increase of the contribution of migration to the current. This is at variance with the idea generally used, i.e., that the overall rate is determined by the slower, or more precisely by the most hindered, process when the electrons are transported by hopping between sites and that this process is accompanied by the motion of counterions.

Sharp et al. [139] used the theory of Andrieux and Savéant to describe the concentration dependence of D observed for Nafion films containing Os $(bpy)_3^{3+/2+}$ redox species. (See also the description of the He-Chen model.) The values of D were derived from chronocoulometric and impedance data at several film thicknesses. In this case, the situation differs from that discussed previously, because in the perfluorosulfonic acid membrane the physical diffusion of the $Os(bpy)_3^{3+/2+}$ species should be taken into account, considering that they are bound only electrostatically. The mobile counterions were sodium ions, because the experiments were carried out in 0.1 mol/dm^3 Na_2SO_4 solutions, and their mobility may be high. The simulated D vs. c curves at three different σ values are shown in Fig. 7. One can see that $\sigma \to \infty$ appears to fit the data best at low concentrations, whereas that for $\sigma \to 0$ provides the best correspondence at high c values. According to the authors this effect is due to the diminishing mobility of sodium counterions, caused by the morphological change of the film as the concentration of the redox ions is increased.

Buck also dealt with the problem of the coupling of electron hopping and counterion transport in several papers [108,109,147–151]. He analyzed the

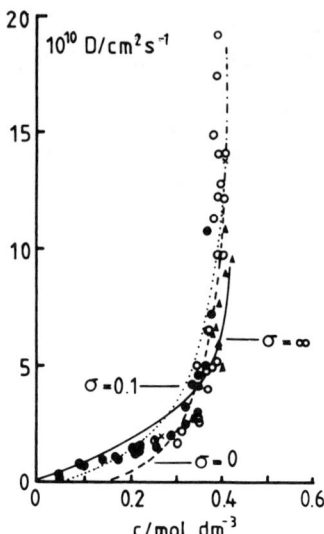

FIG. 7. Comparison of experimental D vs. c data obtained for Os(bpy)$_3^{+/2+}$/Nafion films described in Fig. 3 with predictions of the Andrieux-Savéant theory of migrational enhancement. The solid line for $\sigma = \infty$ was calculated from $D_e = k_e \delta^2 c$, where $k_e = 4.2 \times 10^4$ dm^3/mol s and $\delta = 1.4$ nm. (From Ref. 139, with permission of Elsevier Scientific Publishing Company.)

behavior of membranes of fixed redox sites coated with a porous, inert metal on both sides and submerged in electrolyte solutions containing electroinactive counterions. Buck also described the charge transport in an ion-exchange membrane with mobile redox counterions. In his model, the equation of motion for second-order electron hopping shows diffusion and migration terms, but—similarly to the conception of Andrieux and Savéant—it differs from the usual form of the Nernst-Planck equation. For uniform distribution of sites the equation turns out to be a combination of a numerically corrected Dahms-Ruff diffusion term and a Levich migration term:

$$J_e = - \frac{k\delta^2}{6} \left(\frac{c_{Ox} \partial c_{Red}}{\partial x} - \frac{c_{Red} \partial c_{Ox}}{\partial x} \right) - \frac{k\delta^2}{6} \frac{zF}{RT} \frac{c_{Ox} c_{Red} \partial \Phi}{\partial x} \tag{23}$$

or

$$J_e = \frac{-\tilde{D}_e \partial c_{Red}}{\partial x} - \frac{u_e zF c_{Red} \partial \Phi}{\partial x} \tag{24}$$

When the electron transport manifests itself as the motion of one of the redox species (e.g., Red), the new diffusion coefficient of this species at constant c_{Ox} is

$$\tilde{D}_{Red} = \frac{D_{Red}^2 + D_e^2}{D_{Red} + D_e} \qquad (25)$$

For mobile redox species the maximum value of D_e should not exceed the ion encounter value, provided that the jump distance is the sum of radii Ox + Red. The collisional rate constant (k_d) can be estimated on the basis of the Smoluchowski's equation [Eq. (12)]. The maximum value D_e (max) = $k\delta^2 c/6$ must be near $D_e(max) = 4\Pi\delta^3 cD_{Red}N_A/6 \times 10^3$, and this value should be the same order of magnitude as D_{Red}. Assuming that all forces influencing the energy of electrons are electrostatic, the electron diffusion coefficient can be expressed as follows:

$$\tilde{D}_e = \frac{k_e\delta^2 c}{6}\left(1 + \frac{c_{Ox}}{c}\frac{d \ln \gamma_{Red}}{d \ln c_{Red}} + \frac{c_{Red}}{c}\frac{d \ln \gamma_{Ox}}{d \ln c_{Ox}}\right) \qquad (26)$$

i.e., D_e depends upon the activity coefficients (γ) of the redox species.

Buck [109] analyzed the effect of the local electrostatic field arising from the difference in mobilities of electroinactive counterions and electrons. In his opinion, the system is expected to show a single coupled diffusion coefficient that controls the electron and ion fluxes, and the slower-moving species dominates the fluxes:

$$\tilde{D} = \frac{D_e D_i[(c_{Red} + c_i)/c - (c_{Red}/c)^2]}{D_i c_i/c + D_e[(1 - c_{Red}/c)(c_{Red}/c)]} \qquad (27)$$

In a recent paper [150] Buck discussed two extreme cases. When counterions are very mobile ($D_i >> D_e$) it was found that $1.0627 D_e < D < 1.1 D_e$. When $D_e >> D_i$ (slow counterion, fast electron transport) electrostatic coupling limits the size of D to values comparable with D_i, but in excess. Although the sandwich-type membrane arrangement investigated by Buck differs from the usual one, it provides us an opportunity to study and calculate D_e and D_i for redox polymer films and to estimate the effect of the local electric field arising from the different mobilities of the charge-carrying species.

Baldy et al. [152] derived a general equation that accounts for both diffusive and electric field–assisted electron hopping in immobilized polyvalent and/or multicouple redox systems. They generalized the equation derived separately by Buck [148] and by Savéant [141] for the electronic flux [see Eq. (23)], when the films contain a single redox couple. They studied only those cases where the explicit consideration of the transport of counterions was unnecessary, i.e., when mobile counterions are present in large excess or when there are no mobile counterions.

5. Ion Association and Electric Field Effects

Anson et al. [117] analyzed a case where, beside the electric field effects, the ion association within the polymer film plays an important role in the dynamics of electron hopping within the films. They emphasized that due to the high ionic content and the low dielectric constant that prevail in the interior of many redox polymers, an extensive ion association might be expected. The strikingly nonlinear increase in the apparent diffusion coefficient of the incorporated redox ions as their concentration increases, e.g., the case of the $Os(bpy)_3^{3+/2+}$/Nafion test system, is related to the common effects of electric fields present within the redox polymer and the ion association. According to their model, the sharp rise in the apparent diffusion coefficient as the concentration of the redox couple in the film nears saturation is an expected consequence of the shifting of the ionic association equilibrium to produce larger concentrations of the oxidized half of the redox couple, which is well matched for rapid electron acceptance from the reduced half of the couple. Electric fields present in films also contribute to the sharp increase of D with c, but the calculated effects of the electric fields by themselves are too small to account for the observed behavior. It should be mentioned that in this study particular attention was devoted to measuring the molar fraction of the electroactive material in the coatings and to controlling the film preparation in order to avoid the problems of previous measurements, when large fractions of the incorporated complexes were found to be electroinactive and data evaluation became uncertain.

6. Diffusion-Migration Model for Long-Distance Electron Hopping

Srinivasa Mohan and Sangaranarayanan [153] introduced a generalized diffusion-migration equation for long-distance electron hopping between redox centers. In their model, the potential and the concentration fluctuations were also considered. They demonstrated the advantages of employing a nondiscrete approach for the derivation of transport equations incorporating distance-dependent electron hopping. The explicit form for a typical rate constant depicting its dependence on the potential difference and distance was given as follows:

$$k_1(x,y) = k_0 \exp(-\beta y) \exp[(\alpha F/RT)(\Phi(x) - \Phi(x+y))] \quad (28)$$

where the variables x and y denote, respectively, the spatial coordinate of species A [see Eq. (4)] and the hopping distance. In Eq. (28) $\exp(-\beta y)$

Charge Transport in Polymer-Modified Electrodes   117

depicts the distance-dependent electron hopping. In their derivations they used the approximations that the potential difference term occurring in the exponent can be linearized and $\Phi(x + y) - \Phi(x)$ can be taken as $(d\Phi/dx)_y$. Assuming a second-order process for the electron transport and uniform discrete jumps to a distance, Δx, the concentration dependence of D_e was given as follows:

$$D_e = k_e \, c(\Delta x)^2 \, \exp(-\beta \Delta x) \tag{29}$$

which expression indicates the point of extremum in D_e with respect to Δx. They identified the mean square displacement $<x^2>$ as

$$\langle x^2 \rangle = \int_0^\infty y^2 \exp(-\beta y)\omega(y) dy \tag{30}$$

where $\omega(y)$ a probability distribution of hopping distances. By the help of this function the diverse concentration dependence of D_e can be analyzed. The interactions between sites can also be taken into consideration.

7. Blauch-Savéant Theory: Transition Between Percolation and Diffusion Behaviors

Blauch and Savéant [118] systematically investigated the interdependence between physical displacement and electron hopping in propagating charge through supramolecular redox systems. They concluded that when physical motion is either nonexistent or much slower than electron hopping, charge propagation is fundamentally a percolation process, because the microscopic distribution of redox centers plays a critical role in dictating the rate of charge transport. Any self-similarity of the molecular clusters between successive electron hops imparts a memory effect, making the exact adjacent-site connectivity between the molecules important. In the opposite extreme, rapid molecular motion thoroughly rearranges the molecular distribution between successive electron hops, thus leading to mean-field behavior. Monte Carlo simulations were employed to study the transition between static percolation and mean-field behaviors as a function of the relative rates of electron hopping and physical motion.

It was pointed out that the mean-field approximation presupposes that the diffusion-controlled rate constant k_d, is higher than the activation-controlled rate constant for the electron exchange, k_{act}. For this reason, the Dahms-Ruff approach does not accurately describe charge propagation occurring by means of electron hopping in systems, where the redox centers are irreversibly attached to the supramolecular structures, and they are thereby unable to contribute directly to charge transport through their physical motion. In their opinion, the mean-field approximation leads to Dahms-Ruff–type behavior for freely diffusing redox centers, but the following corrected equation should be applied:

$$D = D_0 (1 - x) f_c + D_e x \tag{31}$$

where x is the fractional loading, which is the ratio of the total number of molecules to the total number of lattice sites. The factor $(1 - x)$ in the first term accounts for the blocking of physical diffusion [see Eq. (9)], and f_c, introduced by Blauch and Savéant, is a correlation factor, which depends on x. In the framework of this model, however, the fulfilment of mean-field conditions implies a descending (or horizontal) variation of the apparent diffusion coefficient with x and not the commonly expected ascending variation. When D_0 becomes less than D_e, percolation effects appear. If $D_e >> D_0$, a characteristic static percolation behavior (D = 0 below the percolation threshold and an abrupt onset of conduction at the critical fractional loading) should be observed. They emphasized that when the redox centers are irreversibly attached to the supramolecular structures, but the rate of the bounded diffusion—a concept introduced by them—is large enough to allow interactions between neighboring redox centers, the rate constant in Eq. (3) is the activation-controlled rate constant for electron exchange and not a combination of diffusion- and activation-controlled rate constants, as was thought earlier [see Eq. (13)].

Moreover, the characteristic mean-squared displacement, Δx^2, is not equal to the square of the electron hopping distance, δ^2, but rather to $\delta^2 + \nu \lambda^2$, where λ characterizes the range of physical displacement permitted by the irreversible attachment of redox centers to the polymer structure and ν is the dimensionality of the system, 2 or 3.

D. Potential Dependence of Diffusion Coefficient

Chidsey and Murray [154] developed a microscopic model including the effects of both the counterion activity and the weak electron-electron interaction. The dc electron conductivity, σ_e, is parameterized as a product of the electron diffusion coefficient, D_e, and of the redox capacity of the material, ρ. D_e is independent of the potential in the presence of large excess of supporting electrolyte in the film, i.e., there is no change in the electrochemical potential of counterions as the redox state of the film is changed; also, all electrons go into equivalent sites and do not interact. However, in real systems there is interaction between the electrons, and the counterion population is also limited. The electron diffusion coefficient then depends on the potential and is expressed in the following equation:

$$D_e = k_e \delta^2 \{1 + [z_i^{-1}(x_e - z_s)^{-1} + (g/kT)] x_e (1 - x_e)\} \tag{32}$$

where x_e is the fraction of sites occupied by electrons, z_s and z_i are the charge of the sites and counterions, respectively, and g is the occupied site interaction energy.

One can observe that in the case of noninteracting sites (g = 0) and in the presence of large excess of supporting electrolyte ($z_s = \infty$), $D_e = k_e \delta^2$ and is a diffusion coefficient. In general, D_e is not constant as the potential, i.e., the film redox composition is changed, as illustrated in Figs. 8 and 9.

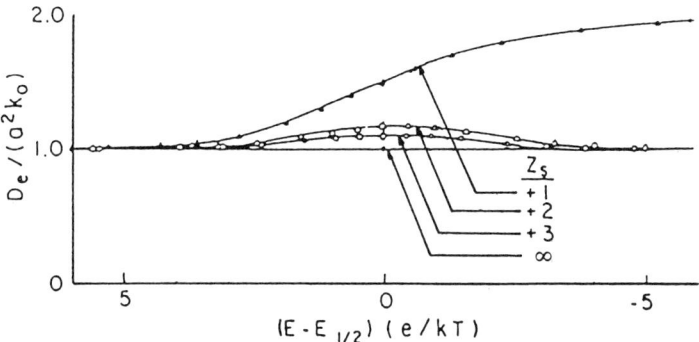

FIG. 8. Electron diffusion coefficients calculated from Eq. (32) for various values of z_s with g = 0 and $z_i = -1$. (From Ref. 154, with permission of the American Chemical Society.)

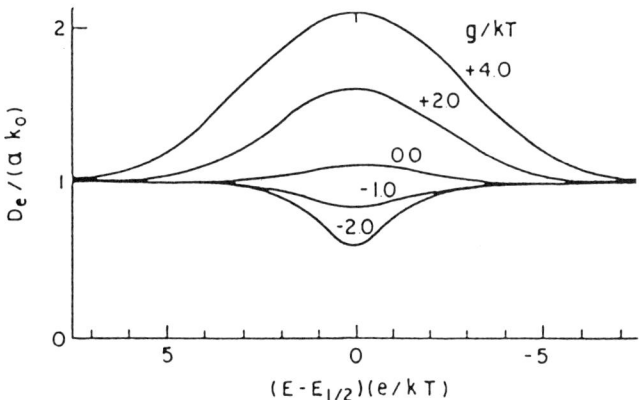

FIG. 9. Electron diffusion coefficients calculated from Eq. (32) for various values of g/kT with $z_s = +3$ and $z_i = -1$. (From Ref. 154, with permission of the American Chemical Society.)

E. Models of Charge Transport in Electronically Conducting Polymer Films

The so-called electronically conducting polymers consist of polyconjugated, polyaromatic, or polyheterocyclic macromolecules and have the unusual property of possessing high electrical conductivity in the "doped" state. This is the main reason that considerable attention has been devoted to these materials not only by chemists (electrochemists), but also by physicists during the last few years.

Because of the different approaches of chemists and physicists, two separate theoretical interpretations have been developed to describe the mechanism of charge transport in these systems. While physicists are interested in the properties (mainly in the conductivity) of doped/undoped conducting polymers in their dry state, electrochemists want to understand their behavior in contact with electrolyte solutions. Polymer film electrodes based on conducting polymers are especially interesting, as we may expect electron transport through the surface layer by a delocalized band structure as distinct from the electron-hopping mechanism in redox polymer films. However, one should bear in mind that interactions between the polymer chains and the solution species (counterions, solvent molecules) may influence the electron transport properties of the polymer. In addition, even when assuming intrachain electron transport, it is very likely electron hopping that ensures the interchain electron transport. Electron hopping may also contribute to the intrachain conduction between different conjugated segments on the same chain, because it is unlikely that a chain is fully conjugated and contains no imperfections. In the overall charge transport process during charging/discharging the film, counterions also participate, thus the sorption/desorption of those may also be rate-determining, as in the case of redox polymers.

In fact, there are many similarities between the two classes of polymeric films. First of all, conducting polymers also exhibit well-defined redox properties. Even the term "doping" as it is applied to the charging process of the film in many papers is misleading. In semiconductor physics, doping describes a process when dopant species of small quantities occupy positions within the lattice of the host material, resulting in a large-scale change in the conductivity of the doped material, as compared to the undoped one. The "doping" process in conjugated polymers is, however, essentially a charge transfer reaction, resulting in the partial oxidation (or less frequently reduction) of the polymer. Conducting polymer films have several peculiarities when compared to the redox polymers. In a half-oxidized state, the film conductivity is much higher than that in a reduced state or that observed in redox polymer films [87,89]. After the completion of the redox process, there

Charge Transport in Polymer-Modified Electrodes 121

is an unusually high capacitive current [13,15,19,34,155–160], and there is no saturation of the counterion concentration in the film [66–72]. Protonation of the film in some cases (e.g., poly(aniline)) causes an increase in the film conductivity. Both the charge transport and charge transfer processes are very fast [20,73–75,161], and the energies of activation are small as compared to the redox polymer films [162].

Various models have been developed to explain the mechanism of charge transport in conducting polymer film electrodes. Two extreme approaches exist. Some researchers [14,23,40,163–168] argue that a delocalized band model is operative, i.e., that charges and unpaired electrons are delocalized over a large number of monomer units. Others prefer a chemical model, in which the charge is localized in the polymer chain, or at most only some monomer units are involved [24]. We won't discuss the semiconductor model in detail, as it lies outside the scope of this chapter and a vast literature is available on it [169]. However, we consider it useful to present some of its fundamental ideas in order to help introduce terminology, used also in the studies discussing polymer films.

It should be stated in advance that the precise nature of charge carriers in conjugated systems varies from material to material, and the mechanism of charge transport is not yet fully known. Even when taking only one material, several structures may be considered, and consequently the nature of the charge carriers and mechanism may differ.

In general, charge carriers may be solitons (neutral defect state), polarons (a neutral and a charged soliton in the same chain, which are essentially radical cations for chemists), and bipolarons (two charged defects form a pair, a doubly oxidized dication form).

For instance, in trans-poly(acetylene), solitons and polarons are mobile along the polymer chain, giving rise to the inherent conductivity of the material. According to the theoretical calculations two polarons will repel each other, which leads to two isolated charged defects. In contrast to the poly(acetylene), two charged defects tend to form a pair (bipolaron) in poly(p-phenylene), which may contain a nondegenerate benzenoid and a higher energy quinoid configuration. Both polaron and bipolaron defects are delocalized over five rings. At high doping levels, bipolarons form bipolaron bands within the gap of the valence and conduction bands [84,169,170].

The situation is further complicated, for example, in the case of poly(aniline). While the terms polaron and bipolaron are used to describe the states resulting from $1e^-$ and $2e^-$ oxidation processes, respectively, this nomenclature is questionable, because none of these species need to be charged due to the protonation of the polymer [163,171–173]. For instance, on the basis of simultaneous cyclic voltammetric/esr results, Genies and

Lapkowski [171,172] have assumed that the electron-transfer processes result in the formation of both polarons and bipolarons, but their characteristics differ because of the structural change of the polymer in the course of oxidation. They have also emphasized that in high molecular mass poly (aniline), the cation radicals are not stabilized via resonance, but that stabilization occurs via deprotonation. Glarum and Marshall [14,163] used a one-dimensional band model to explain the results of the electrochemical esr experiments, but they also recognized the effect of the structural inhomogeneity of the polymer, as well as the role of protonation equilibria. There is disagreement regarding the explanation of esr data for oxidized poly(aniline) in terms of localized Curie spins, as against Pauli susceptibility of metallic islands [166–175].

Some esr results supported the suggestion that the fundamental charged species responsible for the charge transport in poly(pyrrole) is the bipolaron [165]. Other results showed a good correlation between the conductivity and spin (polaron) concentration, and the decrease of conductivity at higher doping level was attributed to the formation of bipolaron [23,176,177], which seems to be more reasonable.

The electrochemical esr data both for poly(aniline) and poly(pyrrole) are also interpreted in terms of a localized redox site model, assuming a phase transition, as the number of oxidized sites increases [24–26,178].

The measured diffusion coefficients are attributed to spin diffusion [163], to interchain electron hopping [26], to intrachain and interchain electron transport, coupled to counterion motion [24,179] or to ion diffusion [19,21, 22,33,129,161,180,181]. In the latter case, not only a rate-determining counterion diffusion was assumed, but also co-ion motion [28,40,46,50], which explains the very rapid switching rate observed for poly(aniline), prototropic proton migration [73]. In general, the experimental findings attest that the rate of charge transport very much depends on the morphology of the polymer, including its change with the potential (phase transition). As for redox systems, the morphology of conducting polymer films is influenced by the nature of the supporting electrolyte and the conditions of film preparation.

Another conclusion that can be drawn from the results is that the effect of the uncompensated ohmic potential drop is more substantial in the electrochemical switching of conducting polymers as compared to the redox systems, due to the large-scale change in the film conductivity during charging [73,129,161,182]. When the ohmic drop was successfully eliminated (or at least substantially reduced), a very high switching rate was detected [73–75,161], which may call into question the data extracted previously.

The RC behavior is certainly relevant in large-amplitude potential-step experiments. The D values derived from the impedance data are model-dependent. Also, because of the absence of the diffusion-controlled region in the impedance spectra of the conducting polymers, in many cases D values were derived from the low-frequency data, which may prove to be seriously in error. This serves as an explanation for why D values, ranging from 10^{-3} to 10^{-11} cm^2/s, were extracted from the experimental results.

For example, for poly(aniline) in acid media: from fast sweep rate voltammetry, $D = 2.5 \times 10^{-6}$ cm^2/s [161]; from impedance measurements, 10^{-7} or 10^{-10} cm^2/s (potential dependent) [13,19,33]; from esr measurements, 10^{-6} cm^2/s [163]; from chronocoulometry 1.4–5.7×10^{-10} cm^2/s (pH dependent between pH 0.3 and 3) [182]; and from normal pulse voltammetry $D = 9.4 \times 10^{-9}$ cm^2/s [183] were derived. Considering the high conductivity of oxidized poly(aniline) one can calculate an even much higher D value for the electron transport.

By using the relation introduced by Chidsey and Murray [154]:

$$\sigma = D_E \rho \qquad (33)$$

where ρ is the redox capacitance per unit volume and σ is the conductivity, and taking $\sigma = 1\ \Omega^{-1}$/cm for poly(aniline) in the range of potentials and pHs, where it is conductive, $D = 3 \times 10^{-3}$ cm^2/s can be obtained, which gives a time constant for the diffusion process $< 10^{-6}$ s at film thickness < 500 nm [20]. If this value is correct, one may easily understand why the diffusion-controlled region is inaccessible when using the usual techniques, e.g., impedance spectroscopy.

The role of iR drop was also emphasized in other works. For instance, Gottesfeld et al. [184] were able to explain the origin of the quasi-reversible behavior and the appearance of prepeaks or shoulders in the cyclic voltammograms with a simple model. According to this model the film resistivity is inversely related to the amount of oxidizing charge by a simple relationship. The occurrence of sharp current peaks, however, can also be explained within the framework of the percolation theory. Aoki [185] investigated the connection between the electrode potential and the distribution of conductive species in the film by using the percolation theory. When the ratio of the number of conductive elements to that of all the lattice elements in the cubic lattice model reaches the threshold value of the percolation (0.31), the percolation theory predicts that the maximum radius of a cluster rapidly increases up to the film thickness. The rapid growth of clusters is associated with a second-order phase transition. Conductive sites for $p < 0.31$ are localized to the electrode, whereas those for $p > 0.33$ are uniform over the film. This model explains the appearance of a sharp current peak near to the formal potential of the system in the voltammogram.

Although much effort has been spent [13,15,34,66–70,155] to clarify the origin of the capacitive charging current in the cyclic voltammograms of conducting polymers, no complete understanding of the problem has been achieved yet. Our lack of comprehension can be attributed to our inability to distinguish between faradaic pseudo-capacitance and double-layer capacitance on the basis of electrochemical measurements alone.

The combined application of electrochemical and nonelectrochemical techniques (e.g., radiotracer [66–70], ac modulated optical [34]) may offer an opportunity to gain a better insight into the mechanism. At present, it seems to be a reasonable assumption that the double-layer capacitance is associated with the metallic behavior, i.e., arising from an excess of surface density of mobile electronic charge carriers and an ionic double layer, which was proved by radiotracer measurements [66,67]. Feldberg [155] suggested a model according to which the double-layer capacitance linearly increases with the fraction of the polymer converted to its oxidized (conducting) form.

IV. RESULTS ON CHARGE TRANSPORT IN POLYMER FILMS

A. Ion-Exchange Polymers Containing Electrostatically Bound Redox Centers

It was Buttry and Anson [110] who carried out the first detailed investigation for Nafion-coated pyrolytic graphite electrode containing either $Co(bpy)_3^{2+}$ and $Ru(bpy)_3^{2+}$ (bpy = 2,2'-bipyridine) or $Co(NH_3)_6^{3+}$ and $Ru(NH_3)_6^{3+}$ complexes in the presence of 0.2 mol/dm^3 CF_3COONa (pH3) supporting electrolyte. A remarkably large difference in the D values was observed for the different species, despite their very similar structures and essentially equal diffusion coefficients in aqueous solution ($D = 6 \times 10^{-6}$ cm^2/s). They indicated that $D = 1.5 \times 10^{-9}$ and 1.2×10^{-8} cm^2/s for the oxidation of $Co(bpy)_3^{2+}$ and $Ru(bpy)_3^{2+}$, respectively. It was concluded that the key difference lies in the relative rates of electron exchange reaction exhibited by the two complexes: $k_e = 20$ (or 2 [119]) and 10^9 dm^3/mol s for the Co and Ru complexes, respectively. By using $\delta = 1.4 \times 10^{-7}$ cm and 1.36×10^{-7} cm, $D_e = 5 \times 10^{-18}$ and 7×10^{-7} cm^2/s were calculated for the Co and Ru complexes, respectively. Because $D = 1.5 \times 10^{-9}$ cm^2/s was measured, it was concluded that electron hopping is of importance only in the case of $Ru(bpy)_3^{2+}$.

Following a similar argument, the contribution of electron hopping was rejected in the case of $Ru(NH_3)_6^{3+}$, because $k_e = 4 \times 10^3$ dm^3/mol s, and thus $D_e = 10^{-12}$ cm^2/s, while $D_0 = 2 \times 10^{-8}$ cm^2/s. Another criterion that distinguishes between electron hopping and molecular diffusion mechanism

Charge Transport in Polymer-Modified Electrodes 125

was suggested, i.e., equal diffusion coefficients for the oxidized and reduced halves of the redox couple point to electron hopping, while a larger D for the less highly charged ions may signal the dominance of physical transport of the incorporated ions. The participation of the electron exchange reaction in charge transport was also calculated by Buttry and Anson at several k_e/D_0 ratios.

The linear D vs. c function, which is the main diagnostic criterion for the electron hopping mechanism, was not found for the systems investigated. Even in the case of $Ru(bpy)_3^{2+}$ no increase with c in the range of 0.01–0.5 mol/dm^3 was detected. It has since become clear that, similar to the earlier results of Buttry and Anson [110,119], the theory of electron exchange reaction, coupled to physical diffusion, cannot be applied without further considerations, because in most cases the predicted concentration dependence of D has not been observed.

Buttry and Anson [119] continued the investigation of $Co(bpy)_3^{3+/2+/+}$/Nafion system (using 0.5 mol/dm^3 Na$_2$SO$_4$ supporting electrolyte), and they revealed that D is much higher for $Co(bpy)_3^{2+/+}$ redox couple than for $Co(bpy)_3^{3+/2+}$, even though the same species $Co(bpy)_3^{2+}$ is diffusing in both cases. In addition, in the former case, a linear concentration dependence, predicted by Eq. (5), was found in the concentration interval of 0.15–1.2 mol/dm^3. The D vs. c plots obtained by potential-step for the reduction and oxidation of $Co(bpy)_3^{2+}$ are shown in Fig. 10. The diffusion coefficient measured by oxidizing $Co(bpy)_3^{2+}$ to $Co(bpy)_3^{3+}$, $D_{2/3}$, was attributed to the physical diffusion of $Co(bpy)_3^{2+}$ species. The decrease in $D_{2/3}$ was explained in terms of a membrane model, where the diffusing species move between charged sites, and their motion is limited by the decreasing availability of sites as the concentration increases. By correcting the increase of $D_{2/1}$ with the decrease in $D_{2/3}$, a linear concentration dependence of D was obtained, as shown in Fig. 11. However, it was left to be explained why the enhancement of diffusion was observed in the case of $Co(bpy)_3^{2+}$ and no c dependence was found when $Ru(bpy)_3^{2+}$ and $Ru(NH_3)_6^{2+}$ were investigated, especially because it was recognized that $D_{2/3} = 1 \times 10^{-12}$, 5×10^{-10}, and 2×10^{-9} cm^2/s for these systems, respectively. (The D value for the oxidation of $Co(bpy)_3^{2+}$ determined in this study [119] was considerably different from that obtained earlier [110]. No explanation was given by the authors. This difference may be attributed to the nature of supporting electrolyte or to the problems of the thickness measurements and incomplete electroactivity.) Considering D_0 values obtained for Na$^+$ and Cs$^+$ ions in Nafion, which are 9×10^{-7} and 5×10^{-8} cm^2/s, respectively, it was concluded that the partition of cations between the hydrophobic and hydrophilic phases in Nafion should be taken into account. The partitioning of the ions between the two phases depends on

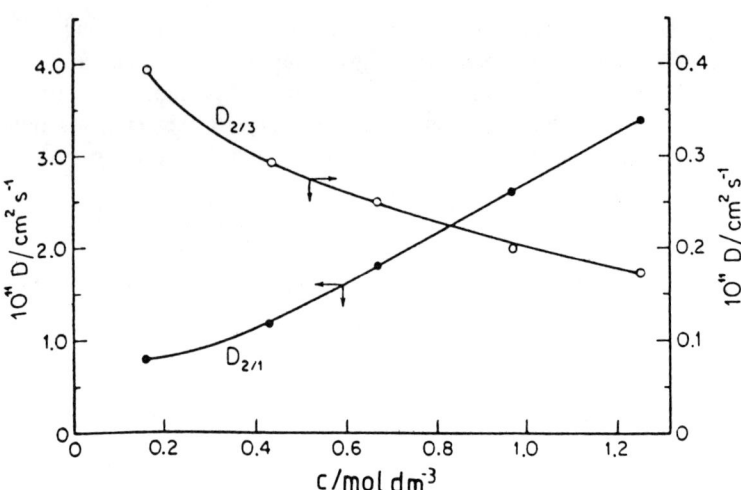

FIG. 10. Concentration dependence of diffusion coefficients for $Co(bpy)_3^{2+}$ in 0.6 μm Nafion coating: (●) $D_{2/1}$ measured by reducing $Co(bpy)_3^{2+}$ to $Co(bpy)_3^{+}$; (○) $D_{2/3}$ measured by oxidizing $Co(bpy)_3^{2+}$ to $Co(bpy)_3^{3+}$. (From Ref. 119, with permission of the American Chemical Society.)

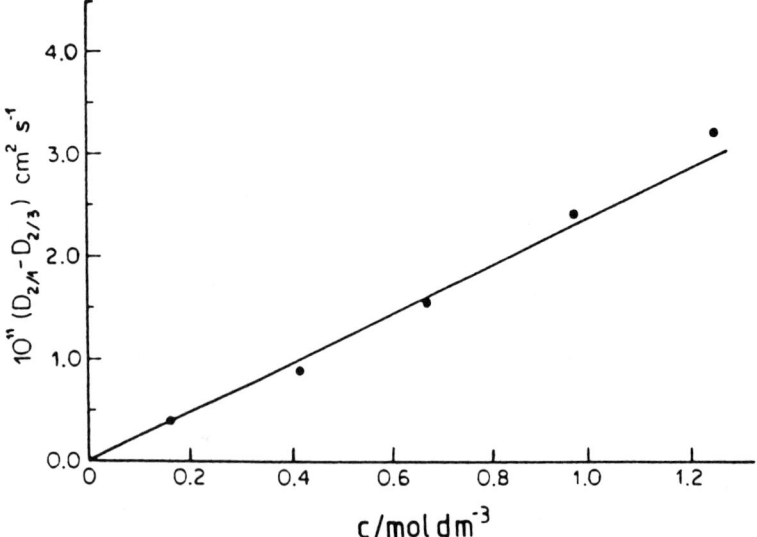

FIG. 11. Concentration dependence of the difference between the values of $D_{2/1}$ and $D_{2/3}$ in Fig. 10. (From Ref. 119, with permission of the American Chemical Society.)

Charge Transport in Polymer-Modified Electrodes 127

the nature of the species, and thus more hydrophobic $Co(bpy)_3^{2+}$ ions partition primarily in the less aqueous phase inside the coating, where much smaller diffusion coefficients prevail, e.g., $1–5 \times 10^{-12}$ cm^2/s.

On the basis of these results Buttry and Anson [119] provided more refined conclusions than in their previous paper [110]. According to the new interpretation, $D_{2/3}$ for $Ru(bpy)_3^{2+}$ in Nafion is dominated by the physical motion of the ion through the hydrophilic phase within the Nafion coatings. It is larger than that for $Co(bpy)_3^{2+}$, because the cross-phase electron exchange between the concentrated but slower-moving ions in the hydrophobic phase and the ions in the hydrophilic phase is much faster in the case of $Ru(bpy)_3^{3+/2+}$ ions, thus it allows the latter to carry most of the diffusional motion. The value of k_e obtained for electron hopping between the redox sites of $Co(bpy)_3^{2+/+}$ was 2×10^3 dm^3/mol/s, which is much smaller than that reported in a homogeneous solution, $k_e > 10^8$ dm^3/mol s. This could be explained in terms of diffusion control of the hopping process, i.e., the diffusion rate constant observed is not a true self-exchange rate constant. The diffusion-controlled rate constant k_d, calculated by using the Smoluchowski equation, is 4×10^3 dm^3/mol s, which is very close to the observed k_e value and much smaller than k_e obtained in solution.

The results obtained by Rubinstein et al. [12], with the help of impedance spectroscopy for $Ru(bpy)_3^{3+/2+}$/Nafion system, were also consistent with the model that assumes two parallel diffusion paths in the film. However, the values obtained for D were different—$D_1 = 1.5 \times 10^{-10}$ and $D_2 = 2.3 \times 10^{-9}$ cm^2/s—while Buttry and Anson calculated $D_1 = 2 \times 10^{-11}$ and $D_2 = 9 \times 10^{-7}$ cm^2/s. This discrepancy was interpreted in terms of the differences in the types of experiments. In addition, Rubinstein et al. applied a model that assumes no coupling between the diffusional paths, while in the interpretation of Buttry and Anson a coupling exists between the two ways of diffusion in the case of $Ru(bpy)_3^{2+}$, and noncoupled diffusion was assumed only in the case of $Co(bpy)_3^{2+}$. (It should be mentioned, concerning the results of Rubinstein et al. [12], that equally good fitting to the experimental points can be obtained by assuming nonuniform film thickness [30], but the authors ruled out this possibility.)

It is important to state that the equations for D are different for coupled and noncoupled cases [12,139]: For coupled cases,

$$D = D_1 f_1 + D_2 f_2 \qquad (34)$$

whereas for noncoupled cases,

$$D^{1/2} = D_1^{1/2} f_1 + D_2^{1/2} f_2 \qquad (35)$$

where f_1 and f_2 are the fractions of the total incorporated reactants present in each phase.

On the basis of the results on the cooperative adsorption of ferricenium (Fc^+) and $Ru(bpy)_3^{2+}$, Rubinstein [136] suggested an alternative explanation for the concentration independence of D in the case of $Ru(bpy)_3^{2+}$. According to Rubinstein, the model of Buttry and Anson is erroneous, because $Ru(bpy)_3^{2+}$ ions are incorporated primarily by ion-exchange association with polymer sulfonic groups, the majority of which is found in the ionic clusters. In his opinion, the main factor is the structural change caused by the incorporation of $Ru(bpy)_3^{2+}$, which is reflected as changes in the value of k_e. In his model, k_e decreases with c for two reasons. First, the hydrodynamic mobility of the complex ions decreases with c due to partial dehydration, shrinking of the clusters, and increased cross-linking. Second, k_e depends on the dielectric permittivity inside the film, which decreases with c. Thus, the compensating effect of the opposite trends in c and k_e provides the explanation for the lack of concentration dependence.

The existence of two diffusion paths was also considered by Sharp et al. [139] for the interpretation of the results obtained for $Os(bpy)_3^{3+/2+}$/Nafion systems (see earlier). In their models, however, as the concentration is increased, the hydrophobic phase will be saturated by the redox species, and with the further increase of the concentration, $Os(bpy)_3^{2+}$ ions will incorporate into the hydrophilic region. In this way, an alternative diffusion (conduction) path will be accessible, resulting in a rapid increase of D. The $Ru(bpy)_3^{2+}$/Nafion [186], as well as Nafion containing [(trimethylammonio)methyl]-ferrocene (Cp_2FeTMA^+) and $Os(bpy)_3^{2+}$, were also investigated thoroughly by Bard et al. [187]. From potential-step chronoamperometric measurements $D = (5 \pm 2) \times 10^{-10}$ cm^2/s was determined for $Ru(bpy)_3^{2+}$/Nafion film. No dependence of D on the nature of counterions was found, which indicated that ionic migration was inactive. It was found that D is independent of the concentration in the loading range 0.7–70% (this denotes the percentage of the maximum stoichiometric amount). However, not all of the material in the film was found to be electroactive, while the fraction of the material that was electroactive increased with loading. It was emphasized that the thickness measurement is a major problem in obtaining quantitative data and the uncertainties of thickness and thickness distribution generate even greater uncertainties in D measured. The existence of different zones in Nafion was considered, and the dominating role of hydrophobic interactions as well as of the electrostatic forces was suggested. It was recognized that the results can be described by a model involving interactions between the electroactive groups as well as interconversions of different forms of $Ru(bpy)_3^{3+/2+}$ species. It was assumed that the electron hopping may contribute to the charge transport, and this contribution decreases in order of $Ru(bpy)_3^{2+} >> Os(bpy)_3^{2+} > Cp_2FeTMA^+$ [187], as indicated by the comparison of D with D_M

Charge Transport in Polymer-Modified Electrodes 129

obtained from permeation measurements. A concentration-independent D value (= 1.7×10^{-10} cm^2/s) was found in the case of Cp$_2$FeTMA$^+$, too. The lack of concentration dependences was attributed to a nonuniform distribution of Ru(bpy)$_3^{2+}$ within the film or, alternatively, to the cross-linking formation induced by the introduction of these ions. A smaller contribution of electron transfer for Cp$_2$FeTMA$^+$, as compared to Ru(bpy)$_3^{2+}$, was assigned to a smaller k_e value of the electron exchange reaction of the former system. The heterogeneous charge transfer rate constant for Ru(bpy)$_3^{2+}$/Nafion was also determined by Leddy and Bard [188]. They concluded that the k_s values for species in solution are typically 10^2–10^3 times higher than for virtually the same species in a polymer matrix, e.g., in H$_2$O solution k_s = 0.07 cm/s for Ru(bpy)$_3^{2+}$, while in Nafion membrane k_s = 10^{-4} cm/s. It was also pointed out that $k_s/D^{1/2}$ values for the polymer and solution species are, however, comparable, i.e., within a factor of 5. This kind of relationship between k_s and D has not been predicted by most of the usual models for electron transfer, either under homogeneous or heterogeneous conditions, when the reaction rate is not near the diffusion control. However, we will return to this problem later.

In some contrast to the results surveyed above concerning the dependence of D on the concentration of Ru(bpy)$_3^{3+/2+}$ and Os(bpy)$_3^{3+/2+}$ species in Nafion film, it was observed by He and Chen [137] that D_e increases with c, and this increase was found to be much faster than the increase of the concentration of the redox center, especially at high concentrations. The model elaborated for the explanation of this observation was discussed in Sec. III.C.1.

Along with the systems discussed above, the charge transport of other neutral and cationic species incorporated in Nafion film has also been investigated [136]. Rubinstein [136] studied the behavior of ferrocene/ferricenium (Fc^0/Fc$^+$) methyl viologen (MV$^{2+/+/0}$) and N,N,N',N'-tetramethyl-p-phenylenediamine systems. It was found that when neutral and charged forms are involved in the redox reaction, the electrochemical response is unique, in that two oxidation peaks can be observed. This can be explained in terms of the incorporated species in the different domains of the polymer. Viologen/Nafion systems have also been investigated by Gaudiello et al. [189] and Hodges et al. [190]. On the basis of the results obtained by simultaneous electrochemical and electron spin resonance (seesr) techniques, Gaudiello et al. [189] concluded that, in this film, the charge is carried mainly via the physical motion of ions. The high local mobility of the ions in Nafion is suggested by the presence of hyperfine structure in the esr spectra, while for redox species, covalently anchored to the polymer films, broad and featureless spectra were obtained, indicating that the redox centers are immobile on the time scale of the esr experiments [189, 191]. Gaudiello et al. [189] calculated the upper limit of k_e, and presented that $k_e = 8 \times 10^5$ dm^3/mol s at $c = 0.74$ and mol/dm^3 $\delta = 8 \times 10^{-8}$ cm. Because it gives a value for $D_e < 3 \times 10^{-9}$ cm^2/s and $D = 3 \times 10^{-9}$ cm^2/s was measured, the electron hopping may contribute to D, but the major contribution comes from D_0, taking into account that the value of k_e given above is an upper limit. Hodges et al. [190] used a combined electrochemical-spectrophotometric method for the study of three viologens, which differed in their overall charges. They indicated that the differing overall charges of these viologens have little effect on their diffusion rates. The diffusion coefficients measured decreased with increasing loadings in all cases. In agreement with the interpretation given by Gaudiello et al., it was concluded that electron exchange is a minor mechanism in these systems.

Tsou and Anson [192] investigated the diffusion coefficient of a heterobinuclear complex $(NH_3)_5RuC_5H_4NCH_2NHC(=O)$–CpFeCp (RuLFe, Cp = unsubstituted or substituted cyclopentadienide) within Nafion coatings. The three oxidation states exhibited large differences in their diffusion rates, as shown in Fig. 12. The possible origins of the differences and their concentration dependences were analyzed in terms of electrostatic cross-linking, electron transfer cross-reactions, the decrease of k_e as a consequence of the decrease in D_0 with c, and the different numbers of electrostatic binding sites required by the species having different charges. The possible role of the migrational motion was also considered. For the purpose of comparison, D values were determined at various concentrations for mononuclear complexes that were structurally analogous to either end of the binuclear RuLFe complex. The results are shown in Fig. 13. The concentration dependence of D observed was attributed to the effect of electrostatic cross-linking and hydrophobic interactions.

Pourcelly et al. [29] measured the influence of water content on the kinetics of counterion transport in perfluorosulfonic membranes. The variation of the membrane conductivity with the water content has two different origins. The first is a change in the rate constant of the elementary ion transfer reaction, due to the interactions between the mobile ions and the fixed sites, which depends on their hydration states. The second is connected with the change of the microstructure of the membrane material.

It might be expected that for ion-exchange films, where no such internal structure as in Nafion films is expected, the situation is much simpler. However, the results on poly(styrenesulfonate)-containing redox cations [82,138,193–197] or poly(4-vinylpyridine)-containing complex redox anions [106,198–206] and other systems [207,208] have not supported this idea.

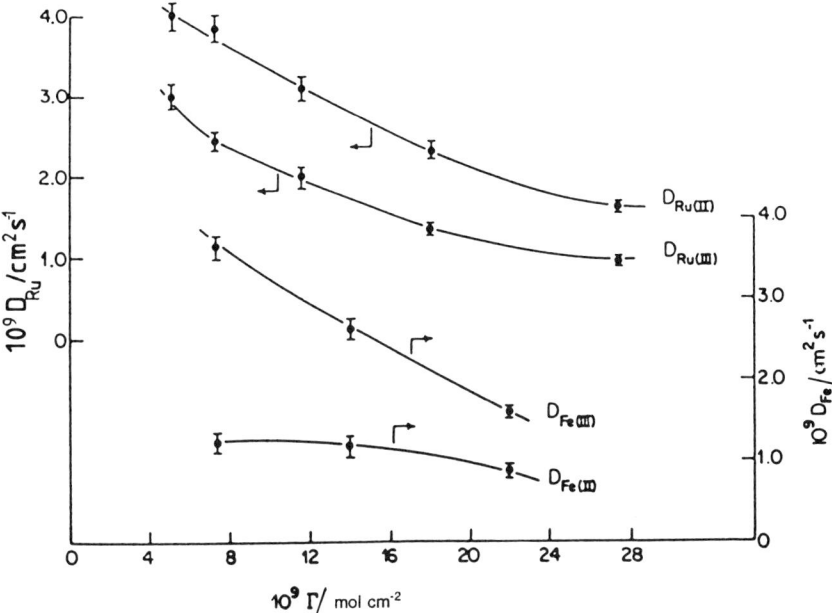

FIG. 12. Concentration dependences of the four diffusion coefficients evaluated for the binuclear RuLFe complex (see text). Supporting electrolyte = 0.1 mol/dm^3 CF$_3$COONa + 0.1 mol/dm^3 CH$_3$COOH at pH 4.5. (From Ref. 192, with permission of the American Chemical Society.)

Poly(styrene sulfonate) (PSS) with electrostatically bound Ru(bpy)$_3^{3+/2+}$ and Os(bpy)$_3^{3+/2+}$ was investigated by Faulkner et al. [86,138,193–195]. These studies were carried out in acetonitrile, because the film is water-soluble. Unfortunately, this prevents a comparison with the Nafion films, especially because in this case the film is weakly solvated and thus it is compact. Nevertheless, many interesting observations were made concerning the dependence of the film thickness on c, the electrochemically induced variation of polymer morphology, and the formation of domainlike substructure, in which Ru(bpy)$_3^{3+}$ centers cluster together to form tight, almost electroinactive zones. No concentration dependence of D was found. The use of luminescence as an indicator of electron exchange dynamics especially helped reveal some interesting features of the system. The quenching process was considered to be a bimolecular contact interaction, and on the basis of the rate of quenching, a diffusion coefficient for the motion of segments binding the Ru-centers and/or D_e was calculated. The D value determined from this

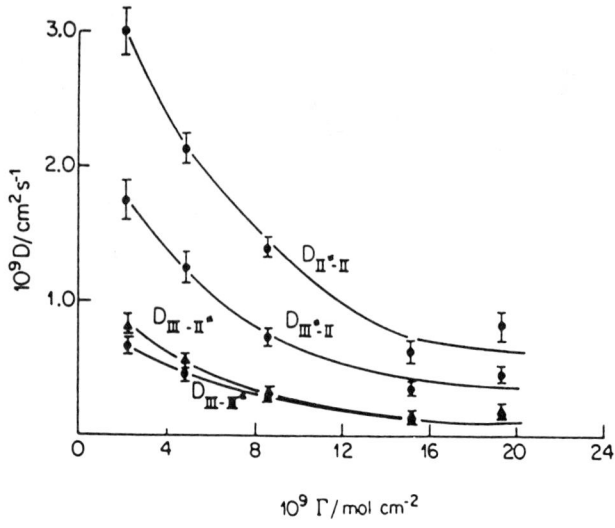

FIG. 13. Concentration dependence of diffusion coefficients for mononuclear (ferrocenylmethyl) trimethylammonium and ruthenium pentaamine pyridine complexes. (From Ref. 192, with permission of the American Chemical Society.)

experiment was found to be 40–80 times larger than the D values measured by chronocoulometry. The difference was discussed in terms of counterion dynamics in the two experiments. In order to overcome the solubility problem of PSS films in water, copolymerized [138], PSS codeposited with cationic polymers [196] or partially sulfonated films [197] were used.

Polycationic films, such as protonated poly(4-vinylpyridine) (PVP) [121], quaternized PVP (QPVP) [16,17,35,201–205] and poly(L-lysine) [207,208], are also widely studied matrices. In this case, complex anions capable of redox transformation ($Fe(CN)_6^{3-/4-}$, $IrCl_6^{2-/3-}$, $Mo(CN)_8^{3-/4-}$, $W(CN)_8^{3-/4-}$, $Ru(CN)_6^{3-/4-}$, $Co(CN)_6^{3-/4-}$, and $Fe(edta)^{1-/2-}$) were incorporated.

Oyama and Anson [198] investigated the behavior of the Ru^{3+}(edta)/PVP system as functions of film thickness, temperature, supporting electrolyte composition and solvent. The activation energy evaluated ($E_a = 19$ kJ/mol) was regarded as a measure of the barrier, faced by small segments of the PVP chains. It was also concluded that the swelling of the film affecting both the segmental and ionic motions is a key factor in the behavior of the film.

Shigehara et al. [199] evaluated k_e values for $IrCl_6^{2-/3-}$ and $Fe(CN)_6^{3-/4-}$ redox couples, and it was found that k_e decreases with the film thickness for both redox pairs. Oyama et al. [121] measured D values for PVP films containing $Mo(CN)_8^{4-}$, $W(CN)_8^{4-}$, $Fe(CN)_6^{3-/4-}$ species by the help of chronoamperometry, chronocoulometry, and chronopotentiometry. As Fig. 14 shows, D decreased with c for all the redox anions studied. This effect was attributed to the structural changes caused by cross-linking, which reaches a limit at c ~ 1 mol/dm³.

In order to achieve large diffusional rates, Inoue and Anson [200] used mixtures of PVP and several copolymers. The diffusion coefficients measured

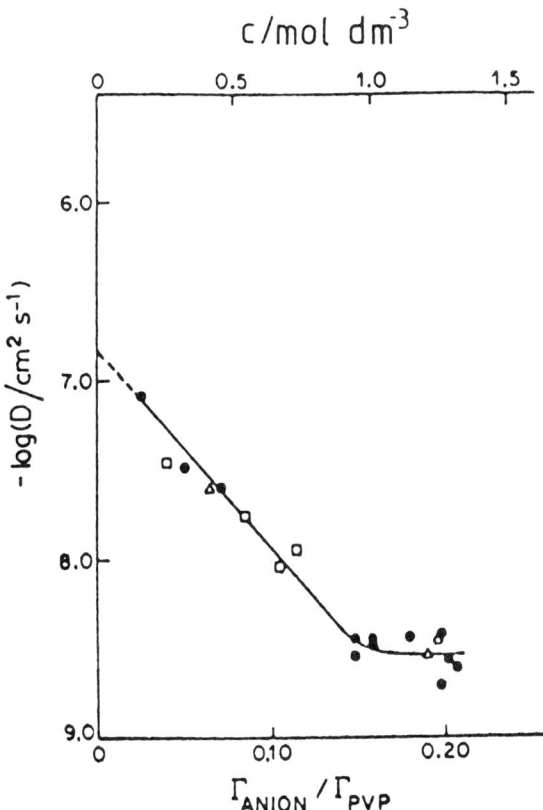

FIG. 14. Semi-logarithmic plot of D vs. the concentration of incorporated anion and of the molar ratio of incorporated anion, Γ_{anion}, to protonated PVP groups, Γ_{PVP}. Anion incorporated: (●) $Mo(CN)_8^{4-}$; (□) $W(CN)_8^{4-}$; (Δ) $Fe(CN)_6^{4-}$. Supporting electrolyte: 0.2 mol/dm³ CF_3COONa + 0.01 mol/dm³ CF_3COOH. (From Ref. 121 with permission of Elsevier Scientific Publishing Company.)

for $Fe(CN)_6^{3-/4-}$ decreased with c. This effect was explained in terms of the existence of different domains in the film and by the diffusion of the species through bottlenecks between adjacent hydrophilic clusters.

Doblhofer and Lange [201–204] concluded that in quaternized and cross-linked PVP, $IrCl_6^{2-}$ and $Fe(CN)_6^{3-}$ are bound by the organic matrix more strongly than their reduced counterparts. This leads to electrochemical rectification, i.e., the reduction of the oxidized species proceeds as fast as at uncoated electrodes, while stationary oxidation is inhibited. They have found that an extensive cross-linking occurs in the presence of $IrCl_6^{2-}$, which results in a low permeability. It was established that the charge transport proceeds predominantly by ion motion, and electron hopping plays no role. It should be mentioned, however, that the Ruff's analysis cannot be applied in this case, because the charge transport takes place by migration (no suppression of migration) and the concentration of neither redox species is constant.

Oh and Faulkner [205] investigated the electron transport dynamics in $Fe(CN)_6^{3-/4-}$/QPVP–modified electrodes applying transient and steady-state methods. The measurements were taken at different temperatures and concentrations of the redox centers. As the temperature decreased, an interplay between the surface and diffusional behavior was observed in the cyclic voltammetric studies. The chronocoulometric Q vs. $t^{1/2}$ plot was linear in the temperature range 0–25°C, suggesting a diffusion-controlled charge transport. The activation energy for this process depends on the concentration of the redox centers (E_a = 26.5 and 80 kJ/mol at c = 0.1 and 0.6 mol/dm^3, respectively) and on the identity of the electroinactive anions (E_a values are in the range of 26.5 and 54.4 kJ/mol at c = 0.1 mol/dm^3 for different electrolytes).

The diffusion coefficients slightly decrease with c, and there is no appreciable difference between the transient and steady-state values, as shown in Fig. 15. The permeation rate ($D_s\chi$) decreases more strongly with c, but except when c > 0.6 mol/dm^3, it is much higher than D. These observations led to a series of very important conclusions. First, the mobility of ionic species is larger than the electron diffusion rate, i.e., neither the diffusion of $Fe(CN)_6^{3-/4-}$ species nor the motion of counterions is a rate-controlling factor. Second, the electrostatic cross-linking between the highly charged $Fe(CN)_6^{3-/4-}$ ions and the positively charged polymer lattice greatly affects the charge transport rate in this films, because it diminishes the rate of all of the physical diffusion processes. Third, the effect of anionic species can be explained in terms of ion pairing and ionic hydration. In the presence of ClO_4^- ions, the film is compact and rigid, because ClO_4^- engages in a rather strong ionic interaction with the QPVP network, resulting in a small ionic activity

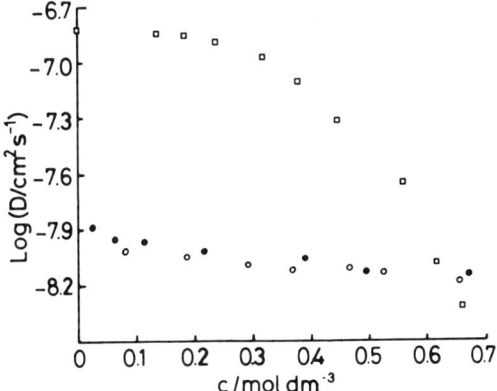

FIG. 15. Electron-diffusion coefficients and permeation rates vs. the concentration of $Fe(CN)_6^{3-/4-}$ loaded in QPVP. Electron diffusion coefficients measured by chronocoulometry (○) and steady-state voltammetry (●). Permeation rates ($D_s \cdot \kappa$) of $Fe(dmbpy)_3^{2+}$ are the squares. All data for films immersed in 0.1 mol/dm^3 KNO_3. (From Ref. 205, with permission of Elsevier Scientific Publishing Company.)

and a low water content in the film. The strong interaction between ClO_4^- ions and the polymer was also observed by Niwa and Doblhofer [203], because even $Fe(CN)_6^{3-/4-}$ ions were replaced by perchlorate ions by ion exchange. In nitrate solutions, the film is much more swollen (compare the behavior of the poly(vinylferrocene) electrode in Sec. IV.B.2).

The view of Oh and Faulkner on the charge transport contrasts with that set forth by Doblhofer et al. [201–204], who suggested a model based on the expulsion and reuptake of $Fe(CN)_6^{4-}$ ions. This difference may be explained by the fact that, in this case, partially loaded films were studied, while Doblhofer et al. investigated mostly fully loaded films, where the electroinactive counterions did not take part in the charge neutralization. Nevertheless, the views taken by the two groups still remain different, as Faulkner et al. state that besides the long-range physical diffusion, the short-range segmental motion coupled with fast electron exchange of colliding species is also operative.

Cross-linked PVP films containing either hexacyanoferrate (3−/4−) or hexachloroiridate (2−/3−) anions were investigated in CF_3COOH/CF_3COONa (pH2) buffer by Armstrong, Lindholm, and Sharp [16,17]. They applied the impedance technique and chronocoulometry. It was assumed that there is always an excess of supporting electrolyte present in these highly swollen films, even at high loading levels, and the mobility of the counterions does not affect the charge propagation. Different behavior was detected for the two systems. In the case of $Fe(CN)_6^{3-/4-}$, the rate of the heterogeneous charge transfer is much smaller than that for PVP films loaded with $IrCl_6^{2-/3-}$. The k_s value for the $Fe(CN)_6^{3-/4-}$ system was found to be $k_s = (2.7 \pm 0.7) 10^{-4}$ cm/s, which is an order of magnitude lower than at a bare electrode, $k_s = (4.6 \pm 3.3)10^{-3}$ cm/s. When Ir-complex was also present in the film, k_s for the Fe-complex decreased further by about an order of magnitude, depending on the loading and the relative concentrations of the different species. It should be mentioned that Oyama et al. [206] obtained k_s values higher by an order of magnitude for the $Fe(CN)_6$/PVP system, $k_s = 3.2 \times 10^{-3}$ to 1.6×10^{-4} cm/s, with the help of normal pulse voltammetry. In addition, Lindholm et al. [17,35] found no concentration dependence, while Oyama et al. [206] observed the decrease of k_s with increasing loading. The dependence of D, as a function of potential was analyzed for the oxidation and reduction processes on the basis of the Chidsey and Murray model [154]. In the case of $Fe(CN)_6^{3-/4-}$ system, an S-shaped D-E curve was found with $D_{ox} = 1.6 \times 10^{-8}$ cm^2/s and $D_{red} = 3.5 \times 10^{-8}$ cm^2/s. The behavior of $IrCl_6^{2-/3-}$/PVP electrode was more complicated. While no significant differences between D_{ox} and D_{red} were observed at medium and high concentrations, at low concentrations $D_{ox} < D_{red}$, and thus an S-shaped D vs. E curve was obtained at low c, but a curve with a maximum was found at medium and high c. It is of interest to note that D decreases, while interaction energy, i.e., the g value of Eq. (32) increases with increasing c. $D_{ox} = 2 \times 10^{-7}$ cm^2/s and $D_{red} = 4 \times 10^{-7}$ cm^2/s at $\Gamma_{Ir}/\Gamma_{PVP} = 1.4\%$, while $D_{ox} = D_{red} = 3.1 \times 10^{-9}$ cm^2/s at $\Gamma_{Ir}/\Gamma_{PVP} = 7.4\%$. It was concluded that the physical diffusion of the complexes plays a dominant role in the $Fe(CN)_6^{3-/4-}$/PVP system over the whole concentration range and in $IrCl_6^{2-/3-}$/PVP electrodes at low concentrations. For the latter system, at high loading levels, the contribution of electron hopping becomes substantial, because at high concentrations, owing to the increasing electrostatic cross-linking and the increasing repulsive interaction forces, which may cause compound formation, the redox couple will be immobilized on the binding sites. The decrease of D with c, however, refers to a diffusion-controlled electron hopping. At high concentrations, only a fraction of the total number of redox centers was electroactive, which causes uncertainties in the data evaluation.

Protonated poly(L-lysine) films containing anionic redox couples were studied by Anson et al. [207,208]. This material exhibits exceptionally high swelling in aqueous media, and thus allows very rapid physical diffusion. The diffusion coefficients obtained for the permeation, $\chi D_s = 6.6 \times 10^{-6}$ cm^2/s

Charge Transport in Polymer-Modified Electrodes

and for the diffusion $D = 2.8 \times 10^{-7}$ to 1.1×10^{-7} cm^2/s (decreasing with c) of Co(C$_2$O$_4$)$_3^{3-}$, respectively, are much larger than most values that have been reported for electrostatically bound ions in polyelectrolyte coatings. Similarly to the Nafion-based electrodes, two pathways were assumed. One of them, lying close to the fixed cationic sites, is characterized by $D = 1 \times 10^{-7}$ cm^2/s; the second pathway is in the solutionlike internal region of the film with $D = 2 \times 10^{-6}$ cm^2/s.

For the sake of showing the diversity of results and ideas, we discussed the different systems separately. It seems to be useful to give a short summary on the most often studied systems as well. Charge transport in Nafion films containing Co(bpy)$_3^{3+/2+}$ [110,119], Ru(bpy)$_3^{2+}$ [12,110,119,137], and Os(bpy)$_3^{3+/2+}$ [117,137,139] were investigated by potential step techniques [110,117,119,137,139,172] and electrochemical impedance spectroscopy (EIS) [12,137,139]. Beside the concentration dependence [12,110,117,119,137,172], the effect of film thickness [139], potential [12], and electrolyte [172] were studied. For the explanation of the experimental results, several models were applied, mostly successfully: Dahms-Ruff theory [110,119,139], He-Chen model [137,139], and Andrieux-Savéant theory for coupled electron hopping–ionic displacement [117,139]. However, because none of them gave a perfect description, other effects were also considered, such as partition of ions between two phases in the Nafion coatings and the existence of two parallel diffusion paths [12,119,139], ion association [117], structural changes, and the formation of cross-links with increasing loading [12,172]. The effect of the water content [29] and the problem of incomplete electroactivity [187] were also emphasized. Similarly, the results obtained on the charge transport in PVP or QPVP films could not be explained within the framework of a single theory without further assumptions. It also became evident that the experimental conditions substantially influence the rate and mechanism of charge transport. For the study of these systems, along with the mostly used potential step methods [17,35,121,199,201–206], EIS technique [16,17,35], steady-state measurements [205], and normal pulse voltammetry [206] also provided valuable data. It was found that the rate of charge transport strongly depends on the concentration of the redox couple incorporated in the film [16,17,35,121,199,201–206], on the temperature [198,205], on the potential applied [16,17,35,206], on the nature of supporting electrolytes [198,203,205], and even on the film thickness [121,199]. Two redox couples, Fe(CN)$_6^{3-/4-}$ [17,35,121,199,201–206] and IrCl$_6^{2-/3-}$ [16,121, 199,200–204], have been studied extensively. It has been established that strong interactions, not only between the positively charged polymer lattice and the electrochemically active counterions [16,17,201–206], but also between the QPVP network and the electrochemically inactive anions [198,203,205], have a determining influence on the behavior of these systems. They affect the rates of both the ionic charge transport and the heterogeneous charge transfer step [16,17,35,206]. At high loading the structural changes caused by cross-linking [121,198,201–206] come into prominence, but the ion pairing [205] and ionic hydration [205] also affect the charge transport dynamics in these films.

B. Fixed-Site Redox Polymers

In respect to the charge transport, there is a major difference between these systems, where the redox groups are covalently attached to the polymer chain, and the ion-exchange polymers in the mechanism of the physical diffusion of electroactive species. Namely, in this case the mobility of the redox ions is ensured by the physical motion of the segments and chains only. As we have mentioned earlier, it is most likely that the dominant charge transport process is the electron exchange reaction governed by the polymeric motions and coupled to the diffusion of electroactive counterions.

First, we will survey the properties of the systems where the films contain the same kind of redox groups—mainly inorganic complex ions—but are covalently and not electrostatically bound. Thereafter, we will deal with the organometallic and organic redox polymer electrodes.

1. Fixed-Site Redox Polymers Containing Inorganic Complex Ions

The early reports have been discussed in detail in Murray's review [5], thus we will focus our attention on the results achieved after 1984, so only two papers from the period 1981–82 discussing the D(c) function will be mentioned below.

As already discussed in Sec. III, a rather interesting concentration dependence of D was found for a copolymer containing $Os^{3+/2+}$ and $Ru^{3+/2+}$ complex redox sites [146]. This remains one of the most carefully designed experiments, because the dilution of Os(II) sites with structurally and electrostatically identical Ru(II) sites prevents major structural changes in the film upon varying Os(II)/Ru(II) ratio. Shigehara et al. [122] studied PVP-Fe(CN)$_5$ and QPVP-Fe(CN)$_5$ systems. An increase of D was observed with the quantity of redox sites, but D became independent of c at high concentrations, as shown in Fig. 16. The initial increase of D was attributed to the electron-hopping mechanism, but because the electroactivity of the film substantially decreased with the concentration, no reliable conclusion can be drawn.

Oh and Faulkner [143] investigated quaternized poly(4-vinylpyridine) films with coordinatively attached luminescent Re(CO)$_3$ (phen) (phen =

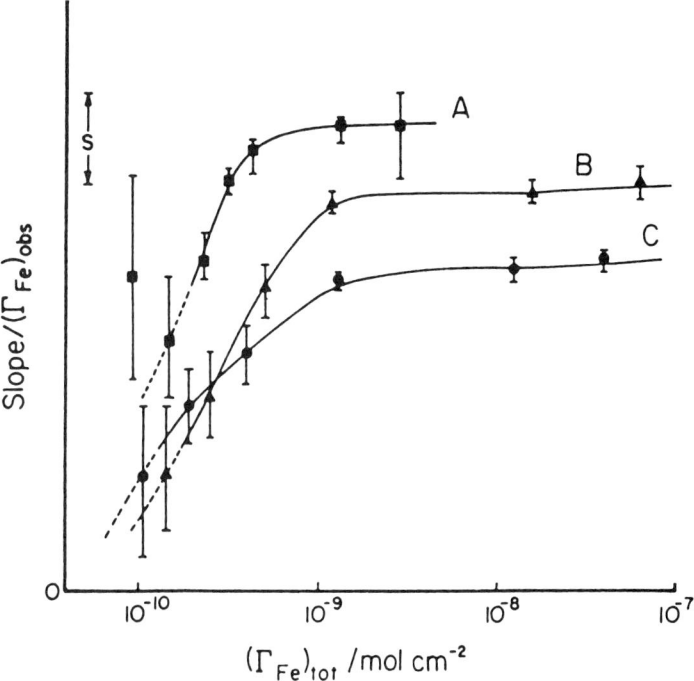

FIG. 16. Normalized slopes of chronoamperometric plots for films of PVP-Fe(CN)$_5$ as a function of the quantity of electroactive Fe(CN)$_5^{3-}$ groups in the films. The pyridine units contained in the films were (A) 5×10^{-8}, (B) 10^{-7}, (C) 5×10^{-7} mol/cm^2. The scale factor, s: (A) 0.5×10^{-5}, (B) 0.25×10^{-6}, (C) 0.25×10^{-7} A/mol s$^{1/2}$. (From Ref. 122, with permission of the American Chemical Society.)

1,10-phenantroline) and redox (Ru(bpy)$_3$Cl) probes. This study made it possible to draw a comparison between the ion-exchange and fixed-site films.

In accordance with the results obtained for QPVP films containing Ru(bpy)$_3^{3+/2+}$, a strong influence of anions was observed, as illustrated in Fig. 17. The rate of diffusion and permeation decreases simultaneously, while the activation energy increases in order of NO$_3^-$ > Cl$^-$ > p-toluenesulfonate > ClO$_4^-$, which was explained in terms of ionic hydration and ion-pair formation ability as previously. The luminescence results supported this idea. (It should be mentioned that Lyons et al. [209] found a similar, but weaker dependence of E$_a$ for PVP-Ru(bpy)$_2$Cl$^{2+/1+}$ system. By using the probe-beam deflection technique, Haas et al. [210] proved that the anions are responsible for the charge-compensation process in [Ru(bpy)$_2$ClPVP]Cl-coated electrode

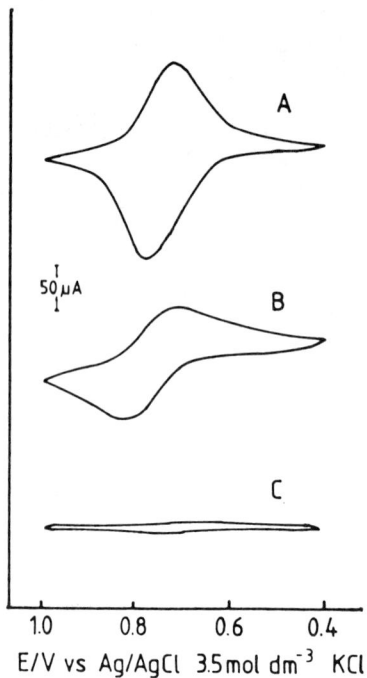

FIG. 17. Cyclic voltammograms of QPVP-Ru(bpy)$_2$Cl$^{2+/1+}$ films in different background electrolytes: (A) KNO$_3$; (B) K-p-toluenesulfonate; (C) NaClO$_4$. Electrolyte concentration: 0.1 mol/dm^3; scan rate = 50 mV/s; T = 23°C; c ≅ 0.5 mol/dm^3 calculated from the dry film thickness; d = 650 nm. (From Ref. 143, with permission of the American Chemical Society.)

immersed in 1 mol/dm^3 HCl.) The energy of activation does not depend on the site concentration, which contrasts with the behavior of QPVP/Fe(CN)$_6^{3-/4-}$ system, but increases with the degree of cross-linking (e.g., in 0.1 mol/dm^3 KNO$_3$ E_a = 25.41 and 59 kJ/mol at 4.8 and 12% cross-linked films, respectively). The difference between the systems can be readily understood, as in this case no electrostatic cross-linking is induced by the incorporation of redox ions.

Interestingly enough, D depends on c, as shown in Fig. 5. The explanation for this finding was provided by a model that emphasizes the dominance of local segmental motion in determining the electron diffusion rate. At low loading level, where D sharply rises with c, the rate is determined by the degree of occupation of the cages defined by the cross-links, because the charge transfer is possible only when a neighboring cage is filled in. At high loading, the rate is limited by the transfer of electrons from cage to cage, which is governed by the polymeric motion. This interpretation is supported by the fact tht the transition between the regimes occurs roughly at the point where the concentration of redox centers equals the cross-link concentration.

The [(PVP)$_5$-Ru(bpy)$_2$Cl]Cl system, in contact with 1 mol/dm^3 HCl, was studied by Gabrielli et al. using the impedance technique [10,32]. At the formal potential, $k_s = 1.6 \times 10^{-5}$ cm/s was obtained. It was found that $D_{ox} > D_{red}$, and the ratio increases as Γ decreases. The difference between D_{ox} and D_{red} was interpreted in terms of the Chidsey-Murray model [154], assuming attractive interactions between the redox sites. The analysis of two kinetic models has been offered. The first one is based on a diffusionlike charge transport process due to the diffusion of the electroactive species and electron exchange reaction. The second model assumes fixed redox sites and electronic conductivity.

Forster and Vos [211,212] have reported the effects of redox site concentration, electrolyte concentration, and temperature on charge transport and the electrode kinetics of electrodes modified with osmium-containing PVP films in sulfuric acid [211], as well as in sodium chloride and lithium perchlorate solutions [212]. The variation of D and the ratio of Os(bpy)$_2$/PVP units at different electrolyte concentrations is shown in Fig. 18. (In the calculation of the concentration, the authors also considered the change in the thickness with the loading level.) The activation energy in 0.1 mol/dm^3 H$_2$SO$_4$ remains approximately constant with c (18 + 6 kJ/mol), while in 1 mol/dm^3 H$_2$SO$_4$ E_a decreases with loading, E_a = 23 to 36 kJ/mol. The activation entropies are negative for all electrolyte/redox site concentration combinations, which was

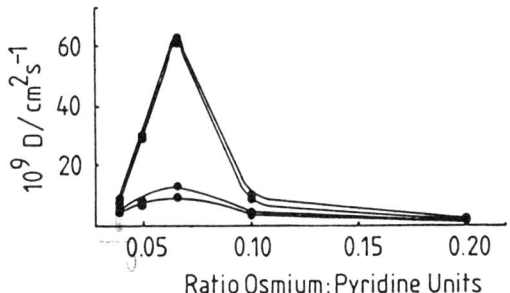

FIG. 18. The effect of H$_2$SO$_4$ concentration and redox site loading on D for [Os(bpy)$_2$PVPCl]Cl films. The electrolyte concentrations are, from top to bottom, 1.0, 0.8, 0.2, and 0.1 mol/dm^3. (From Ref. 211, with permission of Elsevier Scientific Publishing Company.)

attributed, following Murray's idea [213], to the rate-determining role of electron hopping or ion movement. The Gibbs energy term is, however, independent of c (ΔG_a = 46 kJ/mol). The analysis of the data obtained for the heterogeneous rate constant by using sampled current voltammetry was also carried out. The effects of the redox site and H_2SO_4 concentrations were very similar to those observed for D, as shown in Fig. 19. Negative activation and positive reaction entropies were perceived. In 0.1 mol/dm^3 H_2SO_4 the enthalpy and entropy of activation for heterogeneous electron transfer become more positive and less negative, respectively, with increasing osmium loading. In 1 mol/dm^3 H_2SO_4, this behavior is reversed.

Relying on the results described above, Forster and Vos stated that the electron exchange reaction is not involved or at least is not the rate-determining step of the charge transport in this polymer, and the polymer chain motion or diffusion of counterions are the most hindered processes. It was concluded tht the requirements of high electron transport rate and rapid ion diffusion, to a certain extent, are mutually exclusive. At high loading it is the ion diffusion, while at low loadings it is the polymer chain motion that limits D. This explains why the highest rates of charge transport were observed for intermediate loadings, where the film swelling allows a higher ion diffusion rate. It was also emphasized tht the electrode kinetics and the rate of homogeneous charge transport are strongly coupled, suggesting that processes such as ion diffusion influence both processes in similar ways.

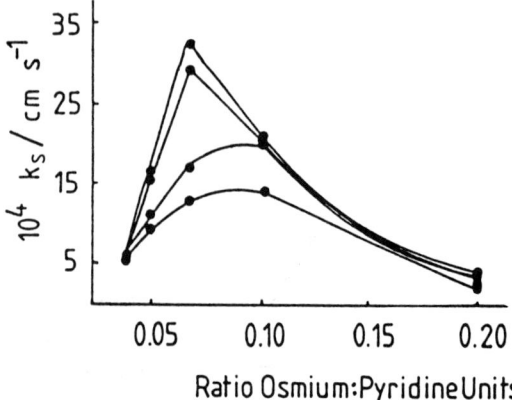

FIG. 19. The effect of H_2SO_4 and redox site loading on the heterogeneous electron transfer rate constant. The film and the conditions are as described in Fig. 18. (From Ref. 211, with permission of Elsevier Scientific Publishing Company.)

Charge Transport in Polymer-Modified Electrodes 143

Recent studies by the same authors [214] on poly(N-vinylimidazole) films containing pendant $[Os(bpy)_2Cl]^+$ moieties also supported these ideas.

The solvent transport during the redox reaction of the $[Os(bpy)_2PVP_nCl]^+$ film was also studied by the EQCM technique in aqueous and nonaqueous media [56]. In acetonitrile media, there is only a small amount of solvent ingress during oxidation. In an aqueous solution, the situation is different, because no solvent incorporation was observed in 0.1 mol/dm^3 NaClO$_4$, but in 0.1 mol/dm^3 sodium p-toluenesulfonate, 24 molecules of water accompany the anion movement into or out of the film during oxidation or reduction. This corresponds to a 13% swelling of the film in the oxidized state as compared to its reduced state.

In order to separate the electron and physical diffusion rates for redox polymers, Murray et al. [76–78] have carried out measurements making use of the sandwich-type electrode arrangements, where poly $[Os(bpy)_2(vpy)_2]^{2+}$, poly$[Ru(bpy)_2(vpy)_2]^{2+}$, and poly(viologen)$^{2+}$ films are contacted by electrodes on both sides. Two experimental modes were applied. In the first case, one of the electrodes was a porous gold, which allowed the entry of solvent and supporting electrolyte ions into the polymer film. The other mode, called solid-state voltammetry, in fact, hardly resembles any electrochemical measurement, because in this case there is no contact with electrolyte solution at all: the film is exposed to either an inert gas or a pure organic solvent atmosphere. Nevertheless, the current response on a voltage bias, applied to the two contacting electrodes, carries information on the effective electron diffusion coefficient. In the first mode, the potential of one electrode is usually kept constant, while a potential scan was applied on the other electrode.

A small sweep rate (v) and/or thin films (small d) give steady-state current (i) − potential (E) curves in which the steady-state concentration gradients of the redox sites are generated within the film. It is stated that, in this case, no net counterion transport occurs concurrently with electron transport, but the ionic atmosphere, which is microscopically local to the redox sites, may influence the kinetics of the electron hopping. When the potential of one electrode is sufficiently positive, the redox sites contacting it are always in an oxidized state, and the potential of the other electrode is swept into the direction where the contacting sites are in their reduced state. At equilibrium the current-potential curve can be described as follows:

$$i = \frac{nFADc^2}{\Gamma \{1 + \exp[(gnF/RT)(E^{0'} - E)]\}} \quad (36)$$

where g is a parameter ideally equal to 1 and $E^{0'}$ is the formal potential of the redox couple in the polymer.

The limiting current (i_L) occurs when the ratio of [ox] to [red] is 1, i.e., when all redox sites adjacent to the two electrodes are in their oxidized and reduced states, respectively, and a linear concentration gradient exists across the polymer. Then

$$i_L = nFADc^2/\Gamma = nFADc/d \tag{37}$$

In an ideal case, neglecting the migration effects, $D = D_e$.

A relatively good agreement regarding the $D^{1/2}c$ product was found between the data obtained by the sandwich electrodes and chronoamperometry for different Ru and Os polymers. (The $D^{1/2}c$ product was calculated only because the extent of swelling was not known.) The $D^{1/2}c$ products show some scatter as a function of Γ, but no significant trend can be observed.

For poly[Os(bpy)$_2$(vpy)$_2$]$^{n+}$ films, it was recognized that the effective electron mobilities are strongly dependent on the particular mixed-valent state generated. The Arrhenius plots for the different redox couples are shown in Fig. 20.

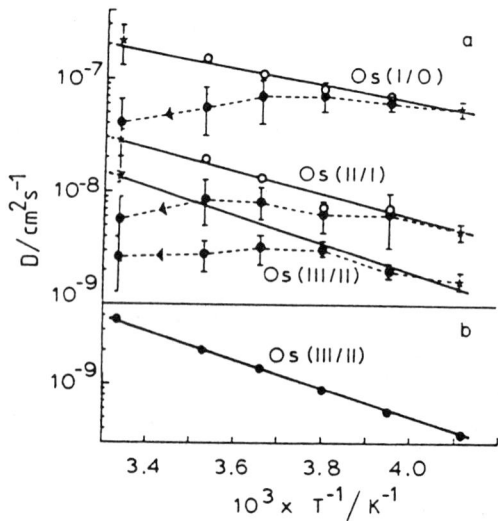

FIG. 20. Arrhenius plots for temperature dependence of D for [Os(bpy)$_2$(vpy)$_2$]$^{n+}$ sandwich electrodes with Γ ranging from 5.6 × 10^{-8} to 14.5 × 10^{-8} mol/cm^2. (●), Measurements made proceeding from low to high temperature (showing effect of partial polymer decay); (*) measurements on virgin electrodes; (○), data points corrected for polymer decay. Electrolyte: 0.1 mol/dm^3 TEAP/acetonitrile. (From Ref. 76, with permission of the American Chemical Society.)

Charge Transport in Polymer-Modified Electrodes 145

For Os(III/II) couple, $D^0 = 2.7 \times 10^{-4}$ cm^2/s and $E_a = 24.2$ kJ/mol were calculated. Because of the instability of the other species for the values of D^0 and E_a, a rough estimation was suggested, thus $E_a = 18$ and 13.8 kJ/mol, and $D^0 = 5.4 \times 10^{-5}$ and 4.5×10^{-5} cm^2/s for Os (II/I) and Os(I/0), respectively.

Using $D = 5 \times 10^{-9}$ cm^2/s, Murray et al. [76, 77] calculated $k_e = 1 \times 10^7$ dm^3/mol s for Os^{3+}/Os^{2+}. This value is close to the homogeneous rate (2.2×10^7 dm^3/mol s). They concluded that the diffusion process is controlled by the intrinsic barrier to the electron self-exchange between the redox sites. Following a similar line of argumentation, the same conclusion was drawn from the data obtained for poly[Ru(vpy)$_3$]$^{2+}$ where $D = 1.9 \times 10^{-9}$ to 6.2×10^{-10} cm^2/s were determined and $k_e = 4 \times 10^6$ dm^3/mol s in solution.

From the solid-state voltammetry, when the polymer films were previously converted to a mixed-valent form (see Ref. 77) for poly[Os(bpy)$_2$(vpy)$_2$](ClO$_4$)$_{2.5}$ polymer D_e(for Os(III)) = 2.1×10^{-8} cm^2/s and $D_i = 7 \times 10^{-13}$ cm^2/s, while for poly[Os(vbpy)$_3$(BF$_4$)$_{2.5}$] $D_e = 0.3-1.2 \times 10^{-8}$ cm^2/s and D_i spanning values from 3×10^{-11} to 3×10^{-15} cm^2/s were calculated.

In D_i values a dispersion was observed. In the author's view, in the first approximation the faster $D_i \cong 10^{-11}$ cm^2/s represents D_i for the bulk of BF$_4^-$ counterions diffusing during the electrolytic formation of Os$^{2+/3+}$ concentration gradients, while the smaller D_i values ($D_i = 10^{-15}$ cm^2/s) may reflect BF$_4^-$ counterions in those parts of the polymer film that are much less mobile, or alternatively may actually reflect a slow relaxation value of the electron diffusion constant, which is caused by the formation of the concentration gradients. Small values of D_i could also reflect a structural dispersion in the polymer or may result from slow changes in the polymer structure.

It is interesting that Murray et al. [77] detected an exponentially rising current-voltage curve when potential steps were applied to dry, mixed-valent osmium polymer films. This indicates that, in this case, the intersite potential reaches a high value ($\Phi \sim 10-50$ mV), which leads to the appearance of the exponential dependence of the rate of the electron exchange reaction on the free-energy difference between sites (i.e., $-nF\Phi$). Usually, it is assumed that the potential difference driving each hop is small, i.e., the equations can be linearized. It is generally so, but it serves as a good reminder that one should not forget that the equations we use to describe the electron-hopping process are linearized, and thus based on the assumption mentioned above. Recently Dalton and Murray [78] studied the electron self-exchange reactions for viologen (2+/1+) and viologen (1+/0) redox couples in a redox polymer formed by the oxidative electropolymerization of viologen monomer N,N'-

bis(3-pyrrol-1-yl-propyl)-4,4' bipyridinium tetrafluoroborate by sandwich electrode experiments. The results were analyzed in a manner applied in the case of osmium and ruthenium polymers [76]. They concluded that the concentration-gradient-driven electron self-exchange in the liquid acetonitrile-bathed viologen $(2+/1+)$ mixed-valent state of this polymer, $k_e = 8 \times 10^3$ dm^3/mol s, is much slower than that for the viologen $(1+/o)$ mixed-valent state, $k_e = 1.6 \times 10^5$ dm^3/mol s, which has a smaller activation barrier. The respective diffusion coefficients are $D_e(2/1) = 2.1$-3.4×10^{-11} cm^2/s and $D_e(1/0) = 6.1$–15×10^{-10} cm^2/s, and show no strong dependence on the nature of counterions (ClO_4^-, BF_4^-, PF_6^-), although in the case of tosylate anions a somewhat lower value was obtained. This is consistent with the absence of macroscopic counterion transport requirement in the steady-state experiments. The effect of tosylate may reflect either a forced increase in the viologen intersite distance (due to the size of the tosylate anion) or a specific ion-site (inner-sphere) interaction that contributes to the electron transfer barrier. Dalton and Murray also analyzed the counterion diffusivity in the case of this polymer. A comparison of the data revealed that the D values determined by chronoamperometry are almost equal to those obtained from the steady-state experiment for $D_{2/1}$, but they are twofold higher in the case of $D_{1/0}$.

The explanation for this was provided by the migration effect in terms of the theory of Andrieux and Savéant [141,145]. The analysis indicated that $D_i(2/1) = 1 \times 10^{-11}$ cm^2/s and $D_i(1/0) \cong 1.5 \times 10^{-10}$ cm^2/s, which means that $D_i(2/1) < D_e(2/1)$, while $D_i(1/0) < D_e(1/0)$. Therefore, the theory of Andrieux and Savéant holds in this case, and this result points to the fact that the D_e values determined by chronoamperometry may be inflated by electroneutrality coupling with a counterion that diffuses more slowly. The difference in the D_i values may refer to stronger ion pairing in the case of viologen $(2+/1+)$, which in turn could supply an inner-sphere barrier term, resulting in a smaller k_e value.

2. Organometallic Redox Polymers

The most important representatives of this group of polymers are the ferrocene-containing ones, notably the poly(vinylferrocene) (PVF) electrodes. They were chosen very early as a convenient model system of polymer-modified electrodes mainly because the behavior of the ferrocene/ferricenium couple was already well characterized, and also because the redox process is a simple, reversible one-electron transfer. It became clear, however, at an early stage of their investigation that the actual polymer structure, which strongly depends on the nature of the solvent and supporting electrolyte, as well as on temperature, affects its electrochemical behavior. For instance, their cyclic voltammetric and chronoamperometric responses, especially when the film is in contact with aqueous solution, are unusual, because the oxidative branch has a completely different shape from the reverse (reductive) branch with both techniques. Since ferrocene polymers have been discussed extensively by Murray [5], we will only refer to the most important papers published before 1984, and we will focus our attention on the new experimental findings and interpretations.

We have already mentioned the work of Peerce and Bard [104] in Sec. III.A. Leddy and Bard [188] examined the electrochemically deposited PVF polymer, as well as a copolymer of vinylferrocene and styrene in a ratio of 58:42 (PVF58) in acetonitrile solutions containing either $TBABF_4$ or $LiClO_4$, by convolution voltammetry. The values found for the diffusion coefficient of oxidized and reduced moieties, 3×10^{-9} and 1.5×10^{-9} cm^2/s, correlated well with the value determined by chronoamperometry, 2.8×10^{-9} cm^2/s. For the heterogeneous rate constant, the values $k_s = 3 \times 10^{-5}$ and 9.2×10^{-5} cm/s were obtained, which are much lower than those reported previously, and they are about two orders of magnitude smaller than that for the similar species in homogeneous solution. From chronocoulometric measurements, a somewhat higher $k_s = 10^{-4}$ cm/s was determined. For PVF58, $D = 7.8 \times 10^{-9}$ cm/s was obtained. The determination of the k_s value was not successful, because of the strong interactions in PVF58 films. The differences between the rates of oxidation and reduction—although they are not higher than the order of the error in the measurements—may be explained in terms of the square reaction scheme, introduced earlier for the interconversion of the sites [104].

Two important papers on the charge transport in PVF films have been published by Murray et al. [213,215]. For PVF films obtained by plasma polymerization, the diffusion coefficients and the activation energies were determined by chronoamperometry between -50 and $-84°C$ in 0.1 mol/dm^3 TBAP/butyronitrile electrolytes. Assuming that $c = 2.4$ mol/dm^3 $D^O = 10^{-8}$ to 2×10^{-9} cm^2/s and $E_a = 14$ to 19 kJ/mol were obtained, while the actual D values ranged from 10^{-13} to 10^{-12} cm^2/s [213]. The small D^O and D values were attributed to the very large and negative entropy of activation ($\Delta S_a = -142$ J/mol/K), which may reflect the reorganization of the polymer chains.

The diffusion-controlled electron exchange rate between the ferrocene sites was also calculated, $k_e = 102$ dm^3/mol s, which is much slower than the measured rate for ferrocene dissolved in methanol, $k_e = 8 \times 10^6$ dm^3/mol s.

This comparison also accentuates the importance of the polymer chain motion in controlling the electron hopping rate.

Chambers and Inzelt [216] also investigated the low-temperature behavior of PVF films. A decrease of the interaction parameter, r, with temperature was detected. It was attributed to the diminishing swelling at lower temperature, able, which leads to increased repulsive interactions between the ferrocene sites.

Daum and Murray [215] also realized that solvent swelling may play an important role in the electrochemistry of PVF films. They stated that in aqueous solutions the ferrocene sites exhibit a phaselike (constant) activity during oxidation, but in acetonitrile the ferrocene activity is proportional to the fractional film oxidation. They suggested a model of the film structure in aqueous solutions, in which the ferrocene and ferricenium sites are segregated into microscopic domains, differing both in solvent content and average state of oxidation, dynamically shrinking and growing, respectively, during film oxidation.

Since it was only later that it became clear that there is a strong counterion effect [54,217], it should be mentioned that Daum and Murray [215] used $LiClO_4$ electrolyte. Nevertheless, the conclusions drawn by Bard et al. [104,188] and by Murray et al. [123,213,215] share a common feature, i.e., potential-induced morphological changes occur during the electrochemical transformations of PVF films. Hillman et al. [46] confirmed this view on the basis of the data obtained by ellipsometric measurements. They indicated tht the reduced PVF film, which is homogeneous and compact, becomes more diffuse and inhomogeneous upon oxidation, and its polymer content decreases with the distance from the electrode. The swelling of the PVF films was measured by the electrochemical quartz crystal microbalance (EQCM) method [42,45,46,54,218,220]. Varineau and Buttry [42] detected practically no solvent sorption in the presence of PF_6^-, ClO_4^-, and BF_4^- anions. In contrast to this, Hillman et al. [45,46,218,219], as well as Inzelt and Bácskai [54], observed a substantial solvent incorporation into PVF films. According to Hillman et al. [45,46], the reduced film contains 6–9 H_2O molecules per ferrocene site, and the PVF oxidation is accompanied by the ingress of one anion and 4–5 H_2O molecules by redox sites at electrolyte concentrations, c_s < 1 mol/dm^3. Inzelt and Bácskai [54] discovered that the swelling of the PVF film strongly depends on the nature of anions and on the electrolyte concentrations. After the completion of the break-in process (see later), 2–3 and 10–12 moles of H_2O remain in the film in ClO_4^- and NO_3^--containing electrolyte, respectively. On the other hand, 4 and 6–8 moles of H_2O enter the film with 1 mole of ClO_4^- and NO_3^-, respectively, in the course of oxidation. The data given above refer to electrolyte concentrations of 0.1 mol/dm^3 NaClO$_4$ and 0.5–2.5 mol/dm^3 NaNO$_3$, where the maximum mass increase was observed. The anion dependence may be attributed to two effects. As one can see in Fig. 21, the shape and the position of the cyclic voltammograms, as well as the simultaneously obtained EQCM responses, depend on the nature of anions. The dependence of the peak potentials may refer to the different formation constants of the complexes or ion pairs [217] between the ferricenium sites and the anions. The differences in the frequency decrease (Δf) may be connected with the degree of hydration of the counterions studied [51,54]. It was noted that the hydration of ions and ion pairs formed during the electrochemical transformation, as well as the ion-site interaction, play a similar role in the electrochemistry of PVP polymers, as described previously [143,205]. The D values determined by chronoamperometry were D = 2.3×10^{-10}, 2.6×10^{-10}, and 3.1×10^{-10} cm^2/s in 0.5 mol/dm^3 K$_2$SO$_4$, LiClO$_4$, and NaNO$_3$ solutions, respectively, using c = 2.4 mol/dm^3 for the calculation [217]. Considering the uncertainty in the value of concentration, the differences between these D values are not significant.

Hillman et al. [218] made an important point concerning the transport rates of ionic and neutral species by analyzing the transient chronoamperometric/EQCM data. According to their observations, soon after a potential step, the counterions move faster than the water molecules, and the incorporation of solvent molecules exhibits a more substantial contribution at longer times. The long-time behavior is also associated by a slow polymer relaxation. The diffusion coefficient for the water in the film was ca. 10^{-10} cm^2/s.

Radiotracer studies accomplished by Inzelt et al. [63,65] have revealed that a significant amount of electrolyte is present even in the neutral film. These results indicate that the migration of co-ions should also be considered in the overall charge transport process. Figure 22 shows the change of the count rate originating from the radiation of labeled Ca^{2+} co-ions during a potential cycle. For the sake of comparison, the steady-state values are also illustrated, which demonstrate that during cycling there is no equilibrium with respect to the co-ion population. The chronoamperometric/radiotracer results, shown in Fig. 23, are even more demonstrative. As one can see, there is cation transport towards the electrode during reduction, and there exists a process in the opposite direction in the case of the anodic oxidation.

Inzelt and Szabó [217] investigated the effect of the electrolyte concentration. The variation of the peak potentials and peak currents of the cyclic voltammograms was explained in terms of a polyelectrolyte model. According to this model, with an increasing electrolyte concentration the oxidized film adopts a more compact structure, i.e., the film is less swollen due to the

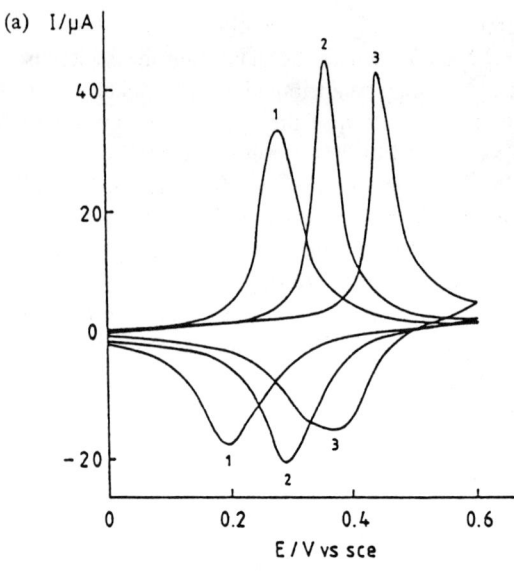

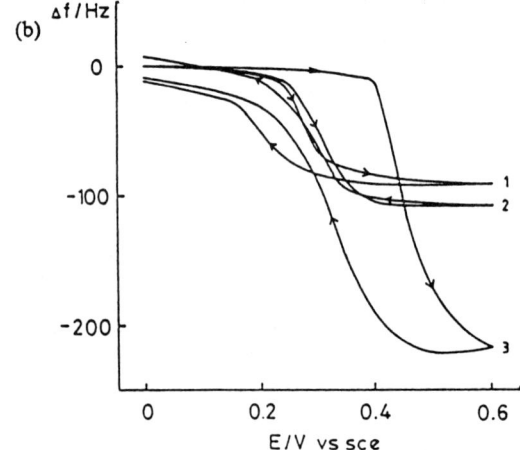

FIG. 21. (a) Cyclic voltammograms obtained for an electrochemically deposited PVF film ($\Gamma = 1.09 \times 10^{-8}$ mol/cm^2) in contact with NaClO$_4$ (1), NaNO$_3$ (2), and Na$_2$SO$_4$/H$_2$SO$_4$ pH 3.4 (3), respectively. Electrolyte concentration was 0.5 mol/dm^3. Sweep rate = 10 mV/s. (b) Frequency change-potential curves simultaneously obtained with the cyclic voltammograms shown in Fig. 21a. All curves are normalized to the initial frequency before each cycle to eliminate the effect of the change of the density and viscosity of the solutions. (From Ref. 54, with permission of Pergamon Press plc.)

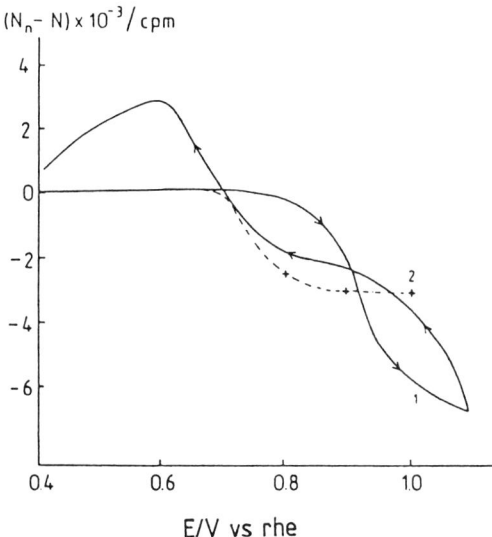

FIG. 22. The change of count rate (N) measured for a PVF film ($\Gamma = 1.1 \times 10^{-7}$ mol/cm^2) in the presence of labeled Ca^{2+} ions ($c_s = 2 \times 10^{-3}$ mol/dm^3 in 1×10^{-3} mol/dm^3 HClO$_4$) during a potential cycle with sweep rate 1.5 mV/s [full line (1)]. Dotted line (2) represents the results of a steady-state measurements. The change is referred to the count rate measured at a neutral film (N_n). (From Ref. 65, with permission of Pergamon Press plc.)

low water activity. The more compact structure, i.e., the increase in site concentration, enhances the electron-hopping reaction. However, at high concentrations, when the size of the solvent-filled cavities in the film becomes commensurate with the size of the moving counterions, the motion of ions becomes hindered. The two opposite effects result in a maximum charge transport rate at medium concentrations. We will discuss this effect in detail in connection with poly(tetracyanoquinodimethane) (PTCNQ) electrodes. This model was confirmed later by EQCM measurements [54]. The degree of swelling is higher and is more sensitive to the electrolyte concentration in NaNO$_3$ than in NaClO$_4$ solutions, which reflects a more extensive interaction between water and the charged ferricenium sites in the presence of NO$_3^-$-containing than in ClO$_4^-$-containing electrolytes. The electrolyte concentration effect was also investigated by Hillman et al. [45], and it was found that at $c_s > 1$ mol/dm^3 cations also enter the film, most likely in the form of ion pairs, due to the failure of permselectivity.

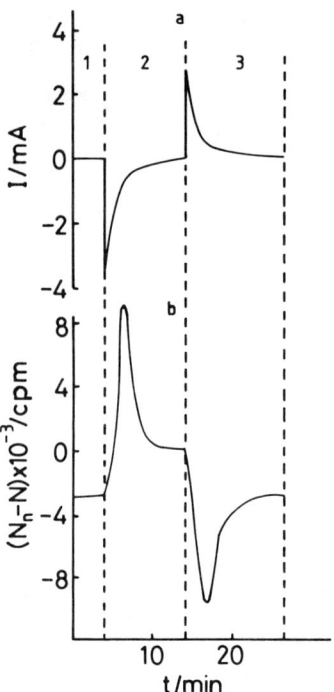

FIG. 23. Change of the current (a) and count rate (b) following potential switches in the case of a PVF film contacting with labeled Ca^{2+} solution (1) 1.0, (2) 0.3, and (3) 1.0 V. Other data as in Fig. 22. (From Ref. 65, with permission of Pergamon Press plc.)

It has been demonstrated [54,220] that the morphology of the polymer network strongly depends on the conditions of preparation. The electrochemically deposited thin films are more compact than thick ones. More stable and probably less diffuse films are formed if the films are deposited from more concentrated PVF/CH_2Cl_2 solutions and/or baked at elevated temperatures, while a solvent evaporation technique is used. These results indicate that the binding between the polymer chains and/or between the polymer and the metal substrate plays an important role, and the formation of these bonds is enhanced by the thermal treatment. It was also shown that reduced regions of PVF films in CH_2Cl_2 are less rigid than the oxidized ones due to the strong polymer-solvent interactions [220].

The effect of site concentration on the charge transport rate has been studied by Nakahama and Murray [123], as well as by Inzelt and Szabó [111].

Charge Transport in Polymer-Modified Electrodes 153

Nakahama and Murray used a copolymer of vinylferrocene and δ-(methacrylylpropyltrimethoxysilane) varying the ferrocene content of the copolymer from 38 to 88%. Unfortunately, the interpretation of the data was complicated by the incomplete electroactivity of the ferrocene sites and by a consequent ambiguity in the value of c. The D values determined by chronoamperometry increased with c, when the nominal ferrocene site concentrations (c_{NOM}) were used for calculation. When the concentration of the actually electroactive sites was substituted into the equation, no variation of D with c was observed.

The D values obtained for these copolymer films in contact with 0.1 mol/dm^3 LiClO$_4$/acetonitrile ranged between 1 and 30 × 10^{-11} cm^2/s and between 3.4 and 20 × 10^{-10} cm^2/s at ca. $-36°C$ when c_{NOM} and c detected were used, respectively. The energy of activation varied between 17 and 35 kJ/mol, decreasing with decreasing c, which indicates cross-linking at high concentrations.

Similar problems emerged when blends of PVF and PTCNQ polymers were investigated by Inzelt and Szabó [111]. In this case, the relative surface coverages of the polymers were varied, and in this way a mutual dilution of the TCNQ and the ferrocene sites can be achieved. Figure 24 shows that the redox transformation of TCNQ groups can be investigated separately from that of the ferrocene sites, because the former takes place between 0.2 and -0.2 V, while the latter occurs between 0.2 and 0.4 V (vs. sce). No segregation of polymers was observed. Some diffusional tailing was detected when the respective sites were diluted to a great extent. The dependence of the $D^{1/2}c$ products (derived from the Cottrell slope) on the concentration is presented in Fig. 25. When $D^{1/2}c$ values were plotted against the TCNQ molar fractions calculated from the amount of the polymer deposited (y_{TCNQ}), some deviation from linearity was observed, while when using the molar fractions determined by a slow-sweep rate cyclic voltammetry (x_{TCNQ}), practically linear plots were found for the reduction of PTCNQ and for the oxidation of PVF, respectively.

Assuming that the x value is proportional to c, no concentration dependence of D can be established. The temperature dependence of the chronoamperometric responses was also investigated. For the $D^{1/2}c$ vs. $1/T$ plots, straight lines were obtained over a temperature interval of 15–65°C. For PVF oxidation, $E_a = 16$ kJ/mol was obtained over the whole concentration range. For the TCNQ reduction, $E_a = 20$ kJ/mol was calculated, which was also constant between $x_{TCNQ} = 0.4$ and 1, showing that there is no interference between the redox processes of the different redox sites. Inzelt and Szabó [111] also examined the reasons for the incomplete electroactivity by using films of different thickness and replacing water by acetonitrile, which appears to be a better solvent for both PTCNQ and PVF. It was concluded that the

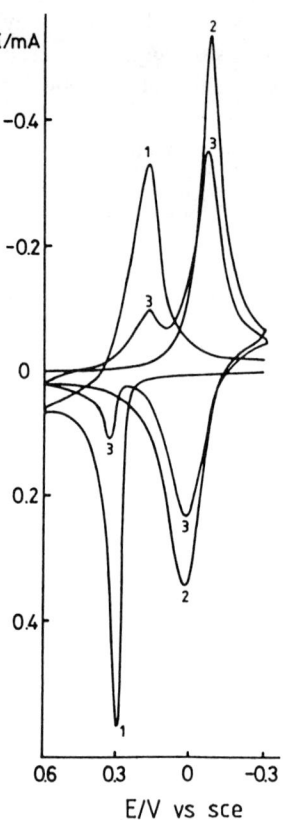

FIG. 24. Cyclic voltammograms for a PVF electrode ($\Gamma = 22$ nmol/cm^2) (1), for a PTCNQ film ($\Gamma = 28$ nmol/cm^2) (2), and for a PVF-PTCNQ blend electrode ($\Gamma_{total} = 55$ nmol/cm^2, $y_{TCNQ} = 0.56$, $x_{TCNQ} = 0.75$) (3) in contact with 2 mol/dm^3 LiClO$_4$ electrolyte. Scan rate = 40 mV/s.

most hindered step in the charge transport is the motion of the polymer chains. It was also assumed that the direct uptake of electrons by the redox groups from the metal substrate is also possible, because a swollen polymer network with rather weak and few connections between the polymeric chains renders a relatively fast segmental and chain diffusion possible, i.e., within the time scale of the experiment a substantial fraction of the redox sites reaches the metal surface by the physical diffusion of the polymer.

The rate of the charge transfer and charge transport processes occurring in PVF film in 0.1 mol/dm^3 TBAP/acetonitrile solutions were studied by Hunter

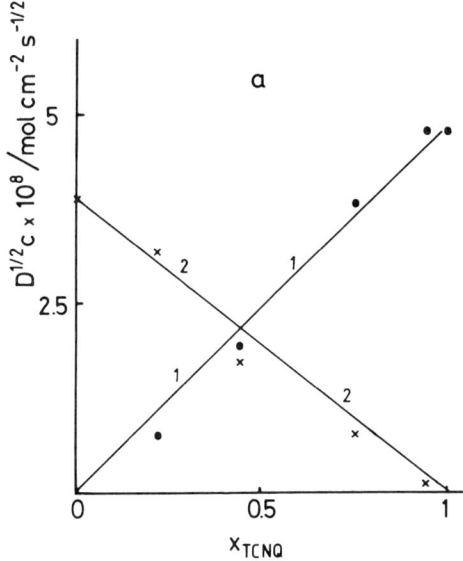

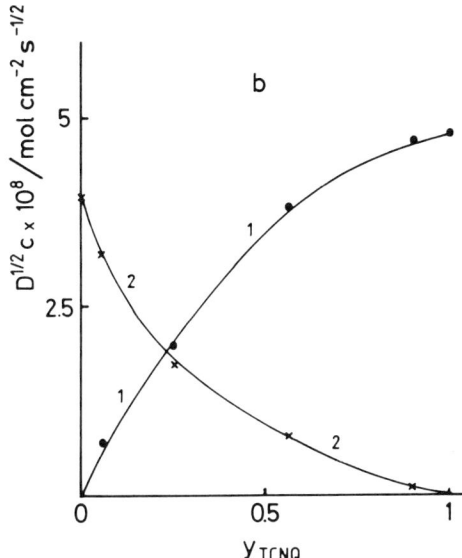

FIG. 25. Plots of $D^{1/2}c$ vs. molar fraction of TCNQ: x is for the observed, y for the applied mole fractions, curves (1) and (2) are for the $D^{1/2}c$ of the reduction of PTCNQ and oxidation of PVF, respectively.

et al. [18] using the impedance technique. The potential dependence of D shown in Fig. 26 was interpreted in terms of the Chidsey-Murray model [154], assuming attractive interactions between the sites. However, in the determination of D values, the theoretical variation of D exceeds the experimental error only slightly, and because of the solvent swelling during the oxidation of the film, one has to be very cautious when drawing a conclusion about the type and magnitude of the forces of interaction. The heterogeneous rate constant was found to be $k_s = (3.7 \pm 1.5) \times 10^{-5}$ cm/s, which is similar to that reported by other authors [188].

Kawai et al. [221] investigated the redox properties of poly(vinylferrocene sulfonic acid) (PVFS) films. This seems to be an interesting system, because it is not permselective, and thus both cations and anions may act as charge-compensating ions. The type of the charge compensating ion is variable, depending on the kind and concentration of the electrolytes used. The $D^{1/2}c$ values ranging between 0.9×10^{-8} and 3.6×10^{-8} mol/cm^2 s$^{1/2}$ were determined. The slowest rate was observed in the presence of $Mg(ClO_4)_2$ and

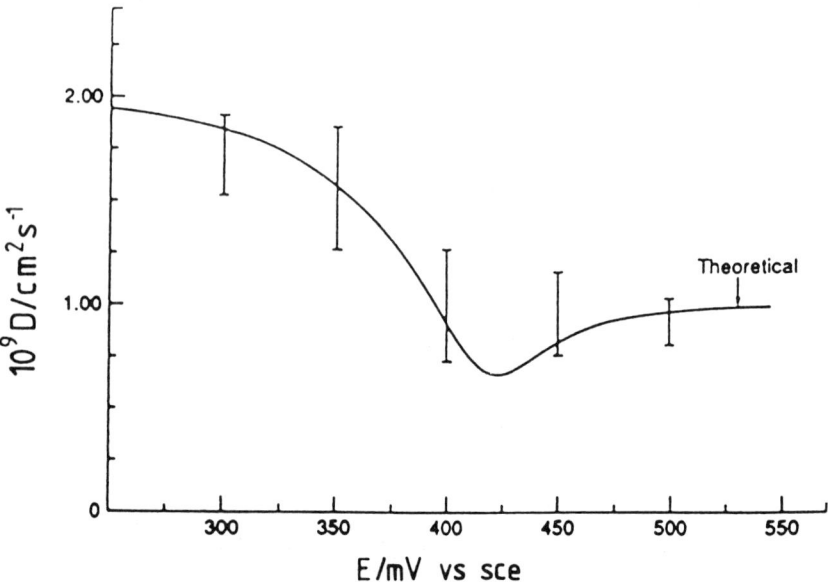

FIG. 26. Experimental (points) and theoretical (lines) values of D as a function of applied potential for a PVF film. Electrolyte: 0.1 mol/dm^3 TBAP/acetonitrile. Theoretical curve was calculated by the help of Eq. (28) assuming $k_e\delta^2 = 10^{-9}$ cm^2/s and $g/kT = -3$. (From Ref. 18, with permission of The Electrochemical Society, Inc.)

explained by the strong $Mg^{2+}-SO_3^-$ interaction, which retards the diffusion of ClO_4^- ions.

3. Organic Redox Polymers

Polymeric materials containing built-in organic redox couples have also been used to prepare polymer-modified electrodes. These systems are expected to show even more complicated electrochemical behavior, because in organic electrochemistry, the redox processes are usually coupled with protonation or other follow-up reactions, e.g., dimerizations and disproportionation.

Because of the polymer environment and the high concentration of redox sites, the effect of the chemical reactions may be stronger or weaker than in a solution. In this selection, we will survey the results that have been obtained on the selected representatives of this important class of polymer film electrodes.

a. Poly(tetrathiafulvalene)

Kaufman et al. [222–225] prepared polymers from a series of tetrathiafulvalene (TTF)-substituted polystyrenes and characterized their behavior as a function of the nature of counterions and polymer structure, using a combination of transient electrochemical techniques and uv-vis-near ir, as well as esr measurements. Although Murray [5] has already dealt with the fundamental properties of these PTTF polymers, we summarize some of the basic experimental findings and ideas of Kaufman and co-workers concerning the factors influencing the rate of charge transport, because they described several significant phenomena, such as break-in, dimerization, voltage-induced morphological changes, etc., for the first time. The results obtained by combining electrochemical and spectroscopic methods proved that beside the formation of TTF^+ and TTF^{2+}, dimeric species TTF_2^+ and TTF_2^{2+} are also formed in the course of the oxidation. However, the data indicated that mixed-valence conduction is less likely, because no formation of stable mixed-valence clusters takes place, and the electron transport occurs via hopping (or tunneling) by means of neighboring molecular group collisions. Solvent swelling plays a dominant role in determining the rate of the segmental motions and the diffusion of counterions, and thus in determining the overall D value. The degree of swelling depends on the potential, because the oxidation of the originally neutral polymer creates a polyelectrolyte, which can swell to a far greater extent than the neutral film. The variation of the value of the diffusion coefficient was interpreted in terms of a free-volume (density of packing of the macromolecule) theory.

The morphology (and the free volume) of the polymer is determined by the chemical structure of the polymer and by the experimental conditions (solvent, ions, temperature, potential etc.). The requirements for rapid electron transport (high density) and for a fast counterion transport (high free volume) are, to some extent, mutually exclusive.

D depends on the nature of the counterions: $D = 2.3 \times 10^{-10}, 6 \times 10^{-10}$, and 1.3×10^{-10} cm^2/s in 0.1 mol/dm^3 TEAP, TEAPF$_6$, and TEA-tosylate/acetonitrile solutions, respectively. This order does not correspond with a simple size relationship, therefore other effects (e.g., ion-pairing) should also be assumed. Thus, this also refers to limitations of the free-volume model.

b. Quinone Polymers

Degrand and Miller [226] investigated anthraquinone polymers adsorbed on a hanging mercury drop. The anthraquinone content of the polymer ranged from 20 to 75%. They reported that both k_s and D strongly depend on the pH. In addition, in acid media the charging of the coating is limited to the layer adjacent to the electrode, while in neutral and alkaline solutions the charging of a part of the bulk of the coating takes place. The rate of charge transport decreased with c. The electrochemical transformation of the sites was faster in alcoholic solution than in aqueous media. The chain motion was concluded to be crucial to charging the film, thus a good solvent, which swells the film and in this way plasticizes it, makes the chain motion less hindered. Similar observations concerning the solvent effect have been made for films of other quinoid polymers prepared by reacting acryloyl chloride with dopamine [227], as well as for poly(vinyl-p-benzoquinone) films [228,229]. Hoang et al. [229] determined the D values for cross-linked poly[p-(9,10-anthraquinone-2-carbonyl)styrene]-co-styrene in TEAP/dimethylsulfoxide solution. A value of $D = 5.1 \times 10^{-11}$ cm^2/s was determined by chronoamperometry, and no concentration dependence was found, but at high loading the incomplete electroactivity made the data evaluation problematical, especially in the case of thick films.

c. Poly(tetracyanoquinodimethane)

The polymers containing tetracyanoquinodimethane (TCNQ) units in the polymer chain have been thoroughly investigated in the laboratories of Chambers and Inzelt [30,31,51–53,64,65,111,191,216,230–239]. PTCNQ polymers are also a good model system to demonstrate the effects of protonation and film swelling on the kinetics and the equilibrium of charge transport and charge transfer processes.

The majority of the studies were carried out on the polymer synthesized by treatment of 2,5-bis(2-hydroxy-ethoxy)-7,7′,8,8′-tetracyanoquinodimethane monomer with stoichiometric amounts of adipyl chloride in N,N′-dimethylacetamide [230,240].

Charge Transport in Polymer-Modified Electrodes

On the basis of the results of combined electrochemical and esr [191,230], as well as uv-vis-near ir [231,232] techniques, it was established that the following reactions should be taken into consideration when the electron acceptor sites are reduced in neutral unbuffered or in alkaline media (pH > 9)

$$TCNQ + e^- + M^+ = TCNQ^- \cdot M^+ \quad E_1^o \quad (38)$$
$$TCNQ + TCNQ^- \cdot M^+ = TCNQ_2^- \cdot M^+ \quad K_1 \quad (39)$$
$$2TCNQ^- \cdot M^+ = TCNQ_2^{2-} M_2^+ \quad K_2 \quad (40)$$
$$TCNQ_2^- \cdot M^+ + e^- + M^+ = TCNQ_2^{2-} M_2^+ \quad E_2^o \quad (41)$$
$$TCNQ^- \cdot M^+ + e^- + M^+ = TCNQ^{2-} M_2^+ \quad E_3^o \quad (42)$$
$$TCNQ_2^{2-} M_2^+ + 2e^- + 2M^+ = 2TCNQ^{2-} M_2^+ \quad (43)$$

The existence of the dimerization equilibria was also confirmed by equilibrium electrochemical radiotracer [64,65] and QCM [52,53] measurements. The formal potentials of the redox couples and the dimerization constants were calculated by the treatment of Andrieux et al. [241].

In buffer solutions in the pH range ca. 4–9, the following protonation equilibria should be taken into account:

$$TCNQ^{2-} M_2^+ + H^+ = TCNQH^- M^+ + M^+ \quad (44)$$
$$TCNQH^- M^+ + H^+ = TCNQH_2 + M^+ \quad (45)$$

The dissociation constants were evaluated from the cyclic voltammetric data using Laviron's treatment [242], which was elaborated for systems, where the electron and proton transfer steps are coupled and the protonations can be considered to be at equilibrium. Since Laviron has predicted the decrease of the effective rate constants for electron transfer reactions in surface electrochemical processes with equilibrium protonations, it is of interest to investigate this question first.

A typical series of voltammograms obtained for PTCNQ films in the regions of pH 5.8–9.5 is shown in Fig. 27. It should also be mentioned that in a more acidic solution, pH < ca. 5, a more dramatic change takes place in the PTCNQ film voltammetry, because the film passivates in the 0.3 to -0.3 V potential region [232]. Although the film becomes electroinactive, it does not undergo an irreversible chemical transformation or dissolution process, because at higher pH its electroactivity can be restored. This behavior is reminiscent of the hydroquinone polymer films of Miller et al. [226]. Assuming $c = 3.6$ mol/dm^3, D values of 5×10^{-11} and 6×10^{-13} cm^2/s were derived by chronocoulometry for the first and second reduction processes, respectively, at pH 7 and 0.5 mol/dm^3 phosphate buffer. While the change of the peak currents and D values with the pH of the bulk solution is in accord with the theory of Laviron, when the buffer strength, the temperature, and the

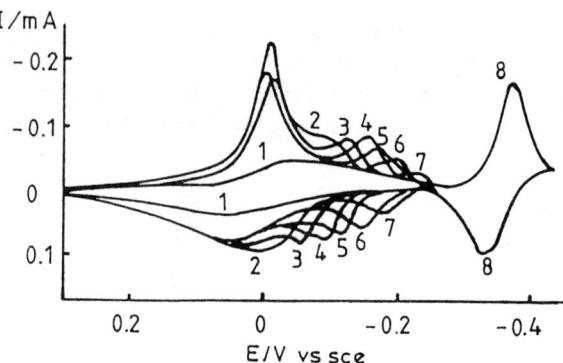

FIG. 27. Cyclic voltammograms of a PTCNQ film electrode in contact with 0.5 mol/dm^3 phosphate buffer solutions: pH (1) 5.8, (2) 6.2, (3) 6.4, (4) 6.6, (5) 6.8, (6) 7.0, (7) 7.2, and (8) 9.5, sweep rate = 5 mV/s.

extent of electrolysis were varied, further investigations revealed that the slow mass transport–controlled protonation of the reduced TCNQ sites by the buffer components is responsible for the effect observed. Although steady-state conditions may exist during the potential cycling, with regard to the extent of protonation, the protonation equilibrium may not be established inside the redox polymer films for two reasons. First, concentration-dependent pH gradients between the polymer film and solution exist as a result of the extensive consumption of protons by the high concentration of the redox sites during reduction or the release of protons during oxidation. In unbuffered solutions, the slower reduction of the film during the second electron transfer step is attributed to the mass-transport exclusion of hydroxide co-ions from the film. This statement is supported by the observation that the extent of neutralization depends on the sweep rate, temperature, and electrolyte concentration [236]. Second, the film pH and the buffer capacity are also affected by Donnan-exclusion of buffer co-ions. A simple membrane equilibrium calculation revealed that when the nominal pH of the solution is equal to 5.81, the film pH is 5.66 and 5.14 for 4.8 and 0.8 mol/dm^3 acetate buffers, respectively, or even less if the activity coefficient of the acetate ion in the film and the partition coefficient for the partitioning of acetic acid into the film is not unity. The shift in the peak potential of 37 mV was in agreement with the calculation. Similar behavior was observed in phosphate buffer, where both $H_2PO_4^-$ and HPO_4^{2-} species are co-ions, and thus the buffer capacity in the film is even more diminished due to the Donnan-exclusion. These effects seem to be general in polymer film electrochemistry.

Charge Transport in Polymer-Modified Electrodes 161

The problem that in polymer films the protonation equilibrium may not be established was also realized by Laviron [243,244]. O'Connell et al. [243] failed to find the predicted pH dependence of D for redox polymers containing monoquaternized bipyridine, N,N-dimethylthionine, thionine, and safranine, but in this case this was attributed to the slow counterion transport and/or to the slow movements of the polymeric chains.

No substantial differences in the electrochemistry of PTCNQ were observed as a function of the nature of counterions (Li^+, Na^+, K^+, Rb^+) or co-ions (Cl^-, ClO_4^-, SO_4^{2-}, CH_3COO^-, $H_2PO_4^-$, HPO_4^{2-}). On the other hand, the EQCM measurements attested that the film swelling depends on the nature of counterions in order $Li^+ > Na^+ > K^+$ [51]. It was detected that in not too concentrated solutions 12, 7, and 4 H_2O molecules are transferred by one Li^+, Na^+, and K^+ ion, respectively, which is in good accordance with the estimated hydration number of these ions. In the presence of bivalent cations (Mg^{2+}, Ca^{2+}, Ba^{2+}), a positive shift in the peak potentials was found, which reflects a stronger interaction between the bivalent cations and the reduced TCNQ sites, and most likely interchain cross-linking [235]. The radiotracer studies revealed that the concentration ratio of Ca^{2+}/Na^+ in a reduced film is significantly higher than that in the solution phase, which means that the reduced PTCNQ film behaves like a selective ion-exchange membrane [64,65].

Perhaps the most interesting observations were made when the voltammetric behavior of the PTCNQ films was studied at different electrolyte concentrations [30,31,51–53,112,234,235]. The characteristic cyclic voltammograms are presented in Figs. 28 to 30.

Independent of the composition of the supporting electrolytes, there are two characteristic features of the cyclic voltammograms. The first is the shift in the peak potentials (E_p) with the electrolyte concentration (c_s). A nernstian dependence of E_p was obtained in dilute solutions, i.e., the slopes of the E_p vs. log c_s plot were 0.06 and 0.03 V/dec for univalent and bivalent ions, respectively. However, for higher c_s a deviation from linearity was observed, and the E_p vs. c_s curve even exhibited a maximum. The incomplete Donnan exclusion of co-ions at high concentrations may have an influence, however, it cannot be responsible for this maximum dependence, as only a leveling of the curves is expected as a result of the imperfectness of the ion-exchange membrane. One should also rule out an activity effect. For the purpose of our subject, the variation of the peak currents is even more interesting. We have mentioned previously that a similar effect has been observed for PVF films, as well.

We attributed this effect to the swelling or shrinkage of the film referring to the behavior of polyelectrolyte gels. According to this model, as the

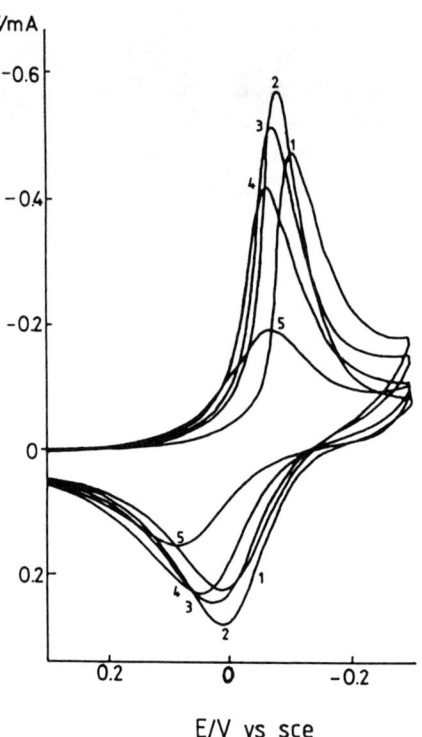

FIG. 28. Cyclic voltammograms of a PTCNQ electrode (Γ = 13 nmol/cm^2) in contact with lithium chloride solutions, concentration: (1) 0.625, (2) 1.25, (3) 2.5, (4) 5.0, (5) 10.0 mol/dm^3. Sweep rate: 60 mV/s. (From Ref. 112, with permission of Pergamon Press plc.)

concentration of electrolyte is increased, the ionic shielding of the polyelectrolyte increases, and the charged polymer film will adopt a more compact structure. In addition, at a high electrolyte concentration, the activity of the solvent is low, consequently the swelling of the film is less than that encountered in dilute electrolyte solutions. In a more compact structure, the concentration of redox sites are high, the free volume is low, i.e., an increase in the rate of electron hopping and simultaneously a deterioration of the film permeability concerning counterions are expected. Therefore, the maximum in the i_p vs. c_s curves is the result of the balanced effects of the enhanced electron-exchange process and the hindered counterion motion. Inzelt et al. [234] modeled these effects in an empirical fashion by scaling the concentra-

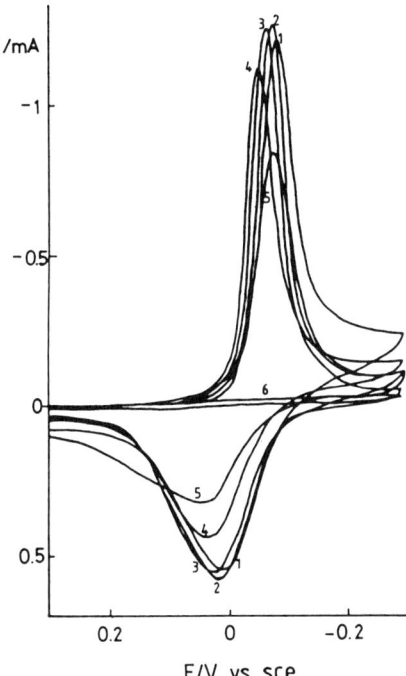

FIG. 29. Cyclic voltammograms of a PTCNQ electrode ($\Gamma = 21$ nmol/cm^{-2}) in contact with aqueous CaCl$_2$. Concentration: (1) 0.15, (2) 0.31, (3) 0.625, (4) 1.25, (5) 2.5, (6) 5.0 mol/dm^3. Sweep rate: 25 mV/s. (From Ref. 112, with permission of Pergamon Press plc.)

tion of electroactive sites in the polymer film and the effective charge transport diffusion coefficient with $c_s^{1/2}$.

By employing the empirical equations,

$$c = Z(1 + Bc_s^{1/2}) \qquad (46)$$

and

$$D = D^\circ(1 - H'c) \qquad (47)$$

a semi-quantitative description of the concentration effect on the peak currents and peak potentials has been obtained. Z, B, and H' are empirical parameters characteristic of the system under study. The values of these parameters depend on the nature of the solvent, of the counterions (their size and charge),

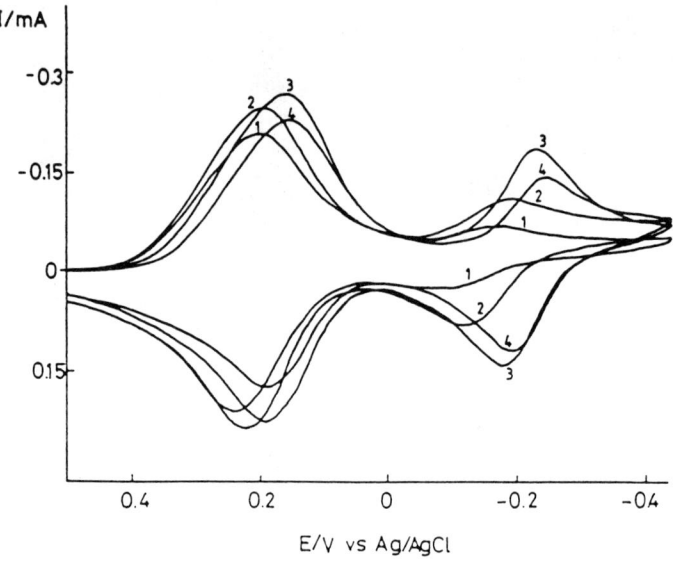

FIG. 30. Cyclic voltammograms of a PTCNQ electrode ($\Gamma = 12$ nmol/cm^2) in contact with an acetonitrile solution of TBAP, concentration: (1) 0.776, (2) 0.49, (3) 0.10, (4) 0.072, mol/dm^3. Sweep rate: 25 mV/s. (From Ref. 112, with permission of Pergamon Press plc.)

and of the polymer forming the film. The combination of Eqs. (46) and (47) with the Randles-Sevcik equation, also with the appropriate Nernst equation, gave the relationship

$$i_p = K_i[1 - H(1 + Bc_s^{1/2})]^{1/2}(1 + Bc_s^{1/2}) \tag{48}$$

and

$$E_p = K_E + \frac{RT}{zF} \ln \left\{ c_s \left[1 - H\left(1 + Bc_s^{1/2}\right) \right]^{1/2} \right\} \tag{49}$$

where

$$K_i = 2.69 \times 10^5 \, D_{ct}^{o1/2} A v^{1/2} Z, \text{ and } H = ZH'$$
$$K_E = E^o - \frac{RT}{zF} \ln K \pm 0.0285$$

and K is the formation constant of the salt, ion pair, or complex. The + and − signs can be applied to indicate oxidation and reduction, respectively, when the counterions have opposite charges. The most remarkable conclusion of these calculations is the fact that the variation of the i_p and E_p values with c_s can be described with the same set of parameters for a given system. In addition, the variation of Z, which is characteristic of the chemical structure of the film, B, which is in connection with the swelling (solvent-polymer and ion-polymer interactions) and H, which expresses the permeability of the film depending on the sizes of the penetrating ions and the solvent-filled cavities (free volume in the film), showed rather reasonable, systematic changes as the solvent was replaced with a better one or univalent ions were substituted for bivalent ones. In accordance with our earlier considerations, the concentration effect is even more remarkable in the case of bivalent ions [235]. These phenomena have been observed for several systems (see our examples, PTCNQ and PVF), and they have been found to be independent of the nature of the solvents (water and acetonitrile). However, comparing the cyclic voltammograms in Figs. 28 to 30, the effects, which may be produced by the nature of the solvent medium, can be observed. As acetonitrile is a better solvent for PTCNQ than water, the swelling of the film will be more extensive. By analyzing the wave shapes obtained in aqueous and acetonitrile media, we may conclude that in the better solvent the interactions between the polymer segments will be less strong, and there will be no difference in the swelling and deswelling rates during oxidation and reduction.

Joo and Chambers [237] demonstrated that in simple alcoholic solvents D decreases as the solvent size increases, but D × η remains constant. It suggests that in dilute solutions, where the free volume in the polymer is high, the counterion migration and the segmental motions are influenced by the viscosity (η) inside the film. Another feature worth mentioning is that the electrolyte concentration affects the second waves somewhat more strongly where dianion sites are formed, which are expected to be more sensitive to the shielding of coulombic forces. The electrolyte concentration effect for the modified Ni electrode differs from that of the polymer film electrodes [112,245]. The shift in the peak potentials as a function of c_s corresponds to a regular nernstian response, and the peak currents do not depend on c_s. As the nickel hexacyanoferrate layer is rigid by nature, solvent swelling plays no role. This may be considered an indirect proof of the assumption in regard to the polyelectrolyte swelling and its effects on the charge transport kinetics, as well as on the interactions inside the film.

The EQCM measurements provided direct experimental proof for a relation between the swelling and the electrolyte concentration [51–53]. We will attempt to discuss these results together with the temperature dependence and the break-in process.

Figures 31 and 32 show the temperature dependence of the cyclic voltammograms of a PTCNQ electrode in contact with LiCl solutions of different concentrations. The frequency curves obtained simultaneously with the cyclic voltammograms are shown in Figs. 33 and 34. (The curves in Figs. 33 and 34 are normalized to the initial frequency value obtained at the lowest temperature.) As we mentioned earlier, the frequency change (mass increase) is mainly due to the solvent sorption in the case of PTCNQ electrodes, especially when Li^+ ions are used as counterions [51].

One can see that the increase of electroactivity and solvent swelling with temperature is substantial, and the swelling is much less in more concentrated

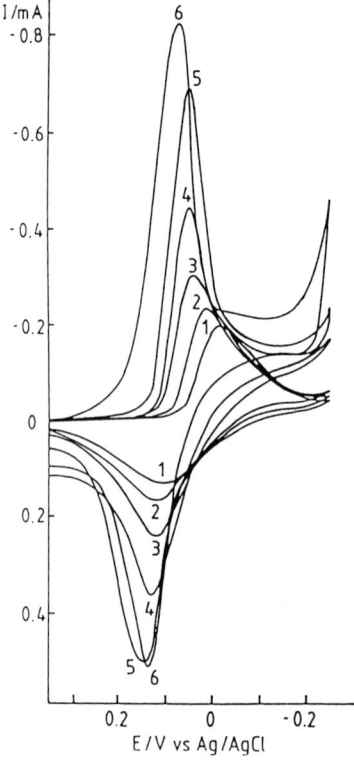

FIG. 31. Cyclic voltammograms of a PTCNQ electrode ($\Gamma = 6.34 \times 10^{-7}$ mol/cm^2) in contact with 2.5 mol/dm^3 LiCl solution at different temperatures: (1) 14.5, (2) 22.5, (3) 32, (4) 37, (5) 43.5, and (6) 55.5°C. Sweep rate = 10 mV/s. (From Ref. 52, with permission of Elsevier Scientific Publishing Company.)

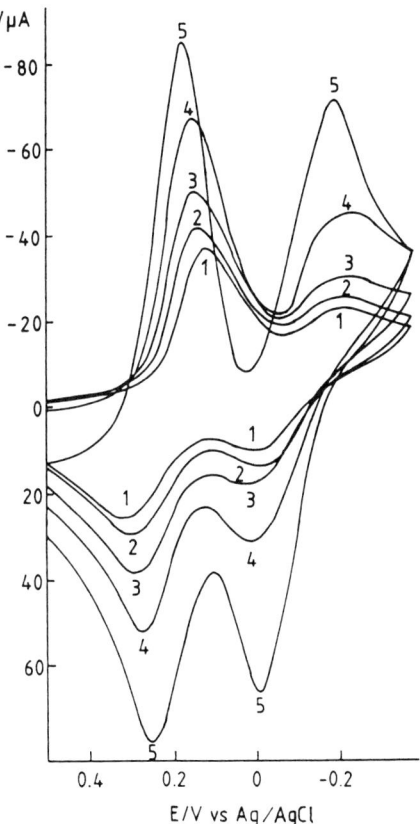

FIG. 32. Cyclic voltammograms of a PTCNQ electrode ($\Delta f = -6525$ Hz, $\Gamma = 5.6 \times 10^{-7}$ mol/cm^2) in contact with 10 mol/dm^3 LiCl solution at different temperatures: (1) 27, (2) 36, (3) 46, (4) 56, and (5) 70°C. Sweep rate = 2.5 mV/s. (From Ref. 52, with permission of Elsevier Scientific Publishing Company.)

electrolyte solution. The latter phenomenon is expected on the basis of the model described previously. (It should be mentioned that the possible maximum hydration number of Li$^+$ ions in 10 mol/dm^3 LiCl solution is 4.3 mol H$_2$O/1 mol Li$^+$, thus it is not surprising that in this solution only two water molecules are transferred by one Li$^+$ ion according to the EQCM data.) The increase of the electroactivity (total charge consumed during cycling) is most likely related to the enhancement of the chain and segmental motions, which are partially frozen-in at low temperatures, due to the higher stability of the

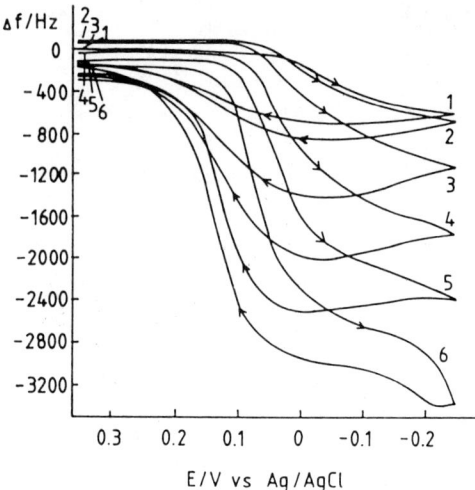

FIG. 33. Frequency curves obtained simultaneously with the cyclic voltammograms of Fig. 31 at temperatures (1) 14.5, (2) 22.5, (3) 32, (4) 37, (5) 43.5, and (6) 55.5°C. (From Ref. 52, with permission of Elsevier Scientific Publishing Company.)

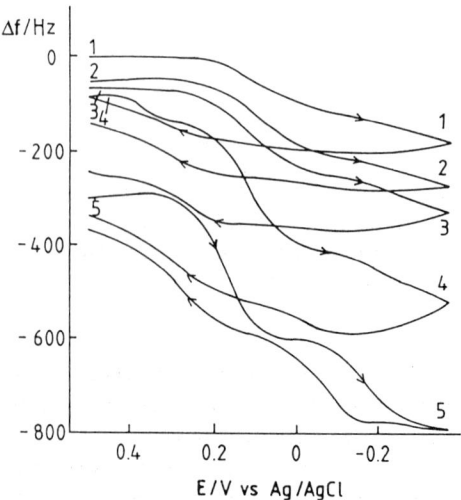

FIG. 34. Frequency curves obtained simultaneously with the cyclic voltammograms shown in Fig. 32 at temperatures (1) 27, (2) 36, (3) 46, (4) 56, and (5) 70°C. (From Ref. 52, with permission of Elsevier Scientific Publishing Company.)

Charge Transport in Polymer-Modified Electrodes

cross-linkages formed between the polymer chains by van der Waals forces. The behavior of the polymer films on electrodes shows a striking similarity to that of inhomogeneous gels [246,247]. In a simple model of these systems the network-forming process and the number of elastically active elements are assumed to be strongly affected by the swelling agent and by temperature.

The changes observed for PTCNQ films at different temperatures and electrolyte concentrations are reversible. However, when electrodes returned from elevated temperature to room temperature, a relatively long time (>10–20 min) was needed to restore the original voltammetric response, as illustrated in Fig. 35. Apparently, the polymer film adopts an extended, perhaps solvent swollen conformation at the elevated temperatures that requires a long

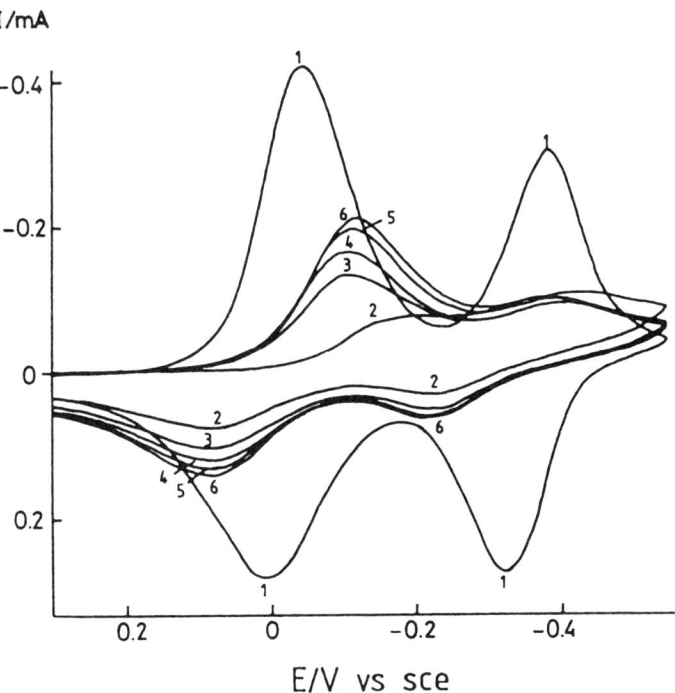

FIG. 35. Cyclic voltammograms obtained for a PTCNQ electrode ($\Gamma = 16$ nmol/cm^2) in contact with 10 mol/dm^3 LiCl at (1) 69°C and (2–6) after rapid cooling at 22°C recorded after delays of (2) 4, (3) 9, (4) 13.5, (5) 22.5, (6) 38.5 min. (From Ref. 112, with permission of Pergamon Press plc.)

relaxation time for the restoration of the room temperature structure. Such behavior is observed in the studies of polymer gels, when the variation of temperature results in the hysteresis of macroscopic polymer properties such as swelling, elasticity, turbidity etc. [112,246,247].

PTCNQ films in contact with 10 mol/dm^3 LiCl or CaCl$_2$ (c_s = 1 mol/dm^3) solutions are remarkably stable in the potential region of the second waves, even at a temperature as high as 70°C. This stability is a consequence of the decreased solvent swelling, if we consider that the dissolution is an infinite swelling (e.g., in 2.5 mol/dm^3 LiCl at elevated temperatures, a dissolution occurs in the potential range where the second electron transfer takes places). The consideration of these facts is equally essential in the design of electrodes for practical purposes. For application, we would need electrodes, which are stable, highly permeable for ions, but at the same time fulfill the conditions needed for a fast electron transfer. Unfortunately, as we have seen earlier, these requirements are, to some extent, mutually exclusive. However, optimum conditions can be obtained by varying the experimental conditions.

The effect of temperature and electrolyte concentration on the electrochemistry of PTCNQ electrodes was also investigated by impedance spectroscopy [30,31]. Complex impedance displays observed at different temperatures for a PTCNQ electrode in contact with 2.5 and 10 mol/dm^3 LiCl, respectively, are shown in Figs. 36 and 37. The variation of the ohmic resistance, with temperature and electrolyte concentration, follows the variation of the solution resistance, which indicates that the film resistance is negligible at the formal potentials of the redox couples of the polymer.

The temperature dependence of the double-layer capacitance was interpreted in terms of the classical double-layer models, but the variation of the effective charged area was also considered. The increase of the low frequency capacitance with decreasing electrolyte concentration and increasing temperature can be attributed to the enhancement of polymer chain motion, as well as to that of the motion of counterions bound to the charged sites of the polymer network, similar to the explanation given for the variation of the total charge consumed during the potential cycling. The charge transport rates were found to be $(1.7 \pm 1.2) \times 10^{-8}$ and $(1.3 \pm 1.1) \times 10^{-10}$ cm^2/s in 2.5 and 10 mol/dm^3 LiCl solutions, respectively. The uncertainty is due to the estimated variation of film thickness. Some variation of the D values with potential was detected, which may refer to attractive interactions between the charged sites, but this variation does not exceed the experimental error of the measurements, as we have seen in the case of PVF films.

The values of D obtained by the impedance analysis are two orders of

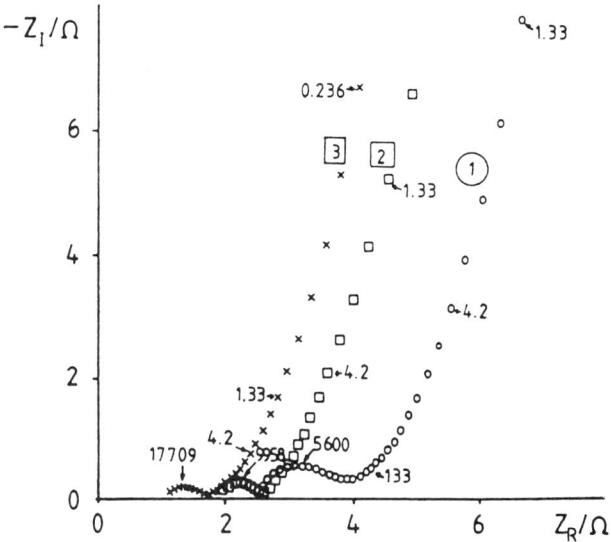

FIG. 36. Complex impedance display of a PTCNQ electrode ($\Gamma = 2 \times 10^{-7}$ mol/cm^2) in contact with an aqueous solution of 2.5 mol/dm^3 LiCl at a potential of 0 V vs. Sce at temperatures (1) 0, (2) 23, and (3) 41°C. Numbers refer to values of frequency in Hertz. (From Ref. 31, with permission of Pergamon Press plc.)

magnitude higher than those determined by potential step methods under similar conditions. This does not seem to be a unique problem, as it has been also observed for other systems [16,18,19,21,22]. It may be attributed to the differences in the experimental conditions. In the case of the potential step methods, when a neutral film is converted into a charged one, the incorporation of a large number of counterions is needed to maintain electroneutrality, while in the case of the impedance measurements, this problem does not arise, because the necessary ions are already in the film.

The following values for the activation energies were derived for the first reduction process $E_a = 12.9$ and 45.7 kJ/mol in 2.5 and 10 mol/dm^3 LiCl, respectively, and $E_a = 12.5$ kJ/mol for the second reduction step in 10 mol/dm^3 LiCl.

For the heterogeneous rate constant, the following values were calculated: $k_s = 1.13 \times 10^{-4}$ and 5.9×10^{-7} cm/s for the first electron transfer in 2.5 and 10 mol/dm^3 LiCl solutions, respectively; for the second electron transfer, $k_s = 2.5 \times 10^{-8}$ cm/s was found in 10 mol/dm^3 LiCl electrolyte.

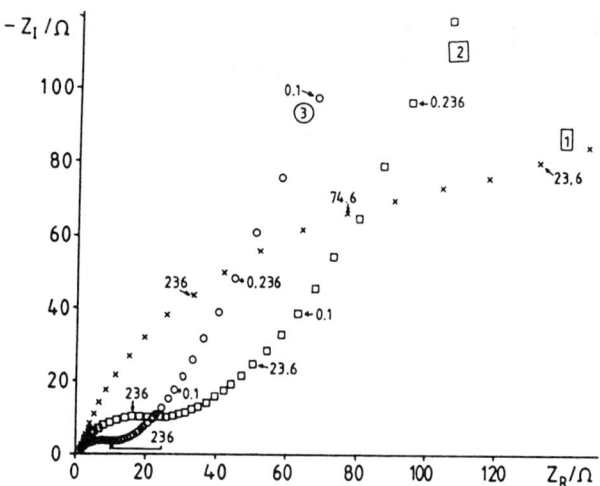

FIG. 37. Complex impedance display of a PTCNQ electrode ($\Gamma = 1.3 \times 10^{-7}$ mol/cm^2) in contact with an aqueous solution of 10 mol/dm^3 LiCl at a potential of -0.05 V vs. Sce at temperatures (1) 23, (2) 54, and (3) 67°C. (From Ref. 31, with permission of Pergamon Press plc.)

While the value of k_s obtained in 2.5 mol/dm^3 LiCl is similar to those obtained so far for several systems where the redox couples are in polymer-bound states [17,188,196,211,212], in very concentrated solutions the electron transfer is even more retarded. In all cases, the reactions proceeded more slowly in the polymeric environment than in solution. This has been attributed to several factors, such as the partially blocked nature of the electrode, the increased activation energy of the repolarization of the surrounding medium, or the slowness of the charge transport in the film [17]. From the temperature dependence of the exchange current density, the following energies of activation were derived: E* = 14.5 and 49.9 kJ/mol for the first electron transfer step in 2.5 and 10 mol/dm^3 LiCl, respectively; for the second electron transfer E* = 58.2 kJ/mol (10 mol/dm^3 LiCl) was obtained. The heterogeneous rate constant is much lower and the activation energy is much higher in concentrated solutions, which is in accord with the observations made in the case of cyclic voltammetric experiments. These data may also support the idea [17] that the counterionic sites and dipoles are attached to the macromolecular backbone and are thereby restricted in their motion. This hypothesis also involves the dependence of the reaction rate on the electrolyte concentration and on temperature. However, it may also be assumed that the dependence of the electron transfer rate on the electrolyte concentration and on temperature is due to the hindered chain and segmental motion of the polymer, because the frequency of "collision" between the metal substrate and the redox sites of the polymer is a function of the rate of these processes. In addition, the number of the active electron transfer sites accessible to the redox groups may be varied by temperature, because the fractional surface coverage may change with respect to the polymer segment. Similarly, the frequency of the site-site collisions within the film is a function of temperature and solvent content of the surface gel-like layer, and it manifests itself in the variation of D. Thus, it is not surprising that k_s and D vary with temperature and with electrolyte concentration in a very similar way, and also the energies of activation for the charge transfer and charge transport processes have almost the same values.

It has been observed for a range of neutral polymer films freshly deposited on metal substrates by the solvent evaporation technique that several potential sweeps are required for the films to become fully electroactive and reach a steady-state voltammogram [45,53,65,69,222,223,226,231,237]. This phenomenon has been referred to as the break-in effect [223]. The number of cycles required for this process depends on the film thickness, on the nature and concentration of the supporting electrolyte, on temperature and on the scan rate. A shorter break-in period can be observed for thin films at low scan rates and at elevated temperatures. More cycles are required when the concentration of the electrolyte solution is high and when the electrolyte consists of multivalent and/or large-sized counterions. Even if the film is broken in during a single sweep, the current-potential curve for the first half-cycle of a voltammetric experiment frequently differs from the curves for the subsequent cyclic scans. This phenomenon arises again when the film is in its neutral form [52]. Sometimes this latter phenomenon is also referred to as the break-in effect [45]. Both the latter phenomenon (also called a secondary break-in effect) and the break in of the virgin films are attributed to the incorporation of solvent molecules and ions into the film phase in the course of electrolysis, as well as to potential-dependent morphological changes [222]. (However, in some cases presumably the difference between the first and subsequent cycles is due to a chemical reaction, beside film swelling. For instance, in the course of the oxidation of virgin lutetium diphthalocyanine (Pc_2Lu) films in acidic aqueous electrolyte, Pc_2Lu^+ cations are formed in the film, a process accompanied by the incorporation of anions and water molecules into the film from the bulk electrolyte. On reverse and successive scans, a reversible two-electron peak develops, which reflects the formation of a protonated species, HPc_2Lu. In this process, the charge is compensated by movement of protons into/out of the film [248–251].)

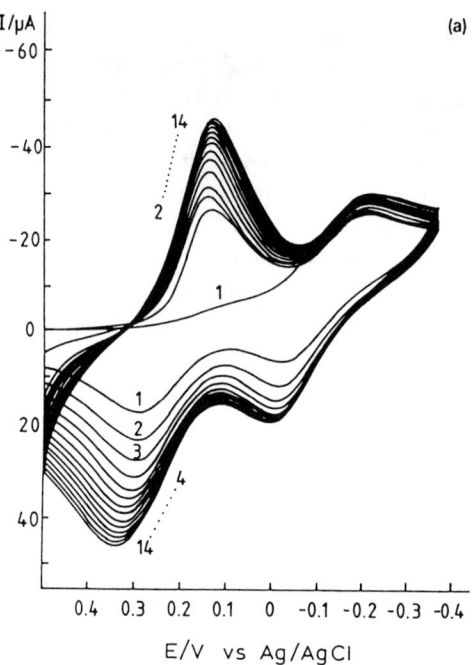

FIG. 38. (a) Cyclic voltammetric break-in of a freshly deposited PTCNQ electrode ($\Delta f_p = -6525$ Hz) in contact with 10 mol/dm^3 LiCl solution. Sweep rate = 6 mV/s. Delay time between the first and second, as well as between the second and third cycles is 1 min. (b) Frequency curves obtained simultaneously with the cyclic voltammograms shown in (a). Δf_p represents the frequency decrease caused by the deposition of the polymer layer, $\Delta f_p = 1000$ Hz corresponds to ca. 8.6×10^{-8} mol/cm^2. (From Ref. 53, with permission of Elsevier Scientific Publishing Company.)

Typical cyclic voltammetric and EQCM break-in patterns are shown in Figs. 38 to 40. Similar pictures were obtained for PTCNQ and PVF films during the break-in period, when the sorption of counterions was followed by in situ radiotracer method (Figs. 41 and 42) [65,69]. No penetration of species from the supporting electron into the virgin films prepared by solvent evaporation technique occurs, i.e., the incorporation of solvent molecules and ions takes place during the charging of the film only. However, broken-in films always contain the species of the supporting electrolyte even in their neutral state. These effects are nicely illustrated on Figs. 38 to 42. For films of similar thicknesses, both the extent of swelling and the duration of the break-in process strongly depend on the concentration of the supporting electrolyte.

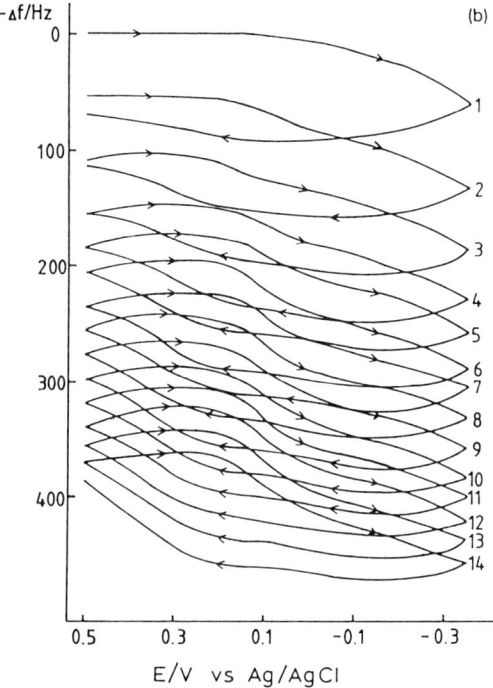

For thinner films the break in, i.e., the swelling and wetting of the polymer structure, is completed much faster, as shown in Fig. 39. In the case of very thick films and/or when concentrated electrolyte solutions are used, the state of full electroactivity is never reached during the cycling. A comparison of the cyclic voltammetric results with those of potentiostatic experiments in the radiotracer and the EQCM studies revealed that no equilibrium can be attained during the cyclic voltammetric experiments, even if the sweep rate is low. This phenomenon is shown in Figs. 43 to 46. In Figs. 43 and 44, the sorption of Ca^{2+} ions during cycling and under equilibrium conditions is presented. In Fig. 45, the cyclic voltammograms and the simultaneously obtained EQCM curves are shown for a thick PTCNQ film. For a very thick film, the mass change during one cycle is about one third of that obtained under potentiostatic conditions (Fig. 46). In addition, the full electroactivity and the maximum sorption were not reached during cycling. These results confirm the supposition concerning the swelling dependent polymer and ionic motions.

The effect of the polymer film composition on the TCNQ electrochemistry in acetonitrile media has also been studied by Chambers et al. [238,239].

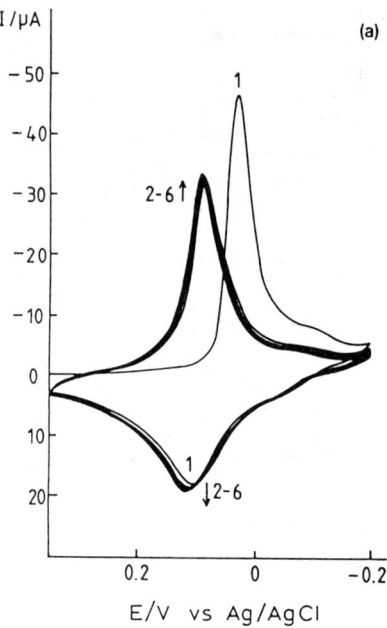

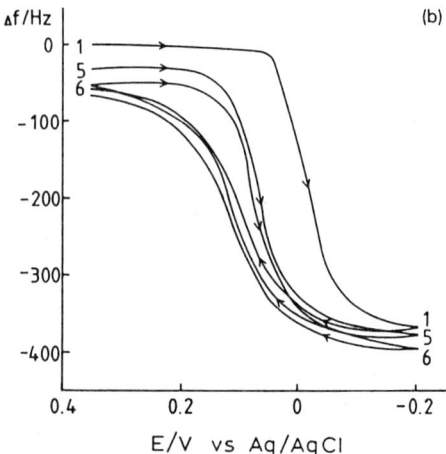

FIG. 39. (a) Cyclic voltammetric break-in of a PTCNQ electrode ($\Delta f_p = -743$ Hz) in contact with 2.5 mol/dm^3 LiCl solution. Sweep rate = 6 mV/s. (b) Frequency curves obtained simultaneously with the cyclic voltammograms shown in (a). (Because curves 2–6 are very similar, only curves 1.5 and 6 are shown.) (From Ref. 53, with permission of Elsevier Scientific Publishing Company.)

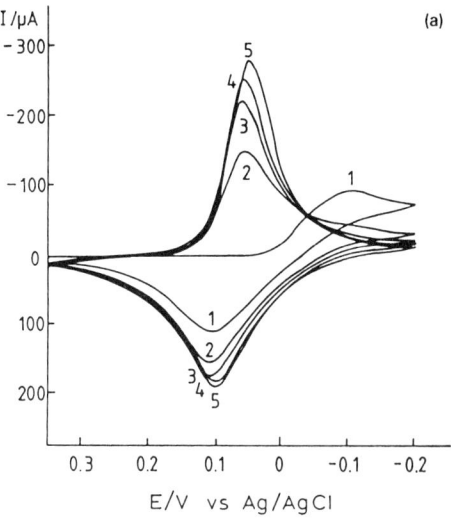

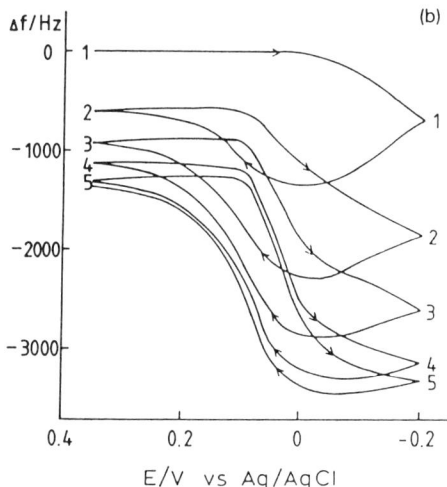

FIG. 40. (a) Cyclic voltammetric break-in of a PTCNQ electrode ($\Delta f_p = -4754$ Hz). Electrolyte: 2.5 mol/dm^3 LiCl. Sweep rate = 6 mV/s. (b) Frequency curves obtained simultaneously with the cyclic voltammograms shown in (a). (From Ref. 53, with permission of Elsevier Scientific Publishing Company.)

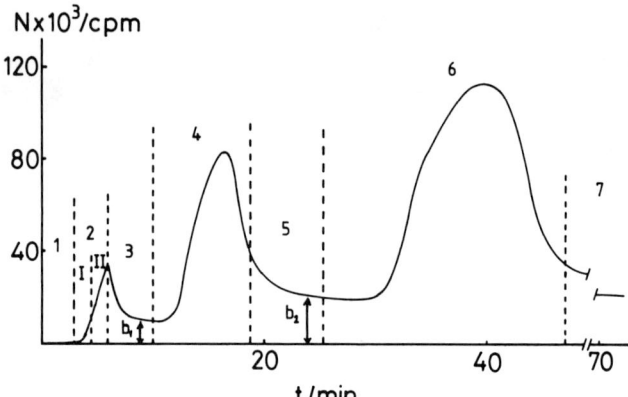

FIG. 41. Study of the behaviour of a virgin PTCNQ film in the presence of labeled Ca^{2+} ions (c_{Ca}^{2+} = 8 × 10^{-5} mol/dm^3 in 5 × 10^{-2} mol/dm^3 CH_3COONa + CH_3COOH; pH = 7). (1) Addition of labeled Ca^{2+} ions to the solution phase under open-circuit condition. (2) First two sweeps (I and II) between 0.9 and 0.25 V (v = 15 mV/s). (3) A pause at 0.9 V following the first two cycles. (4) Third cycle, v = 2.5 mV/s. (5) A pause at 0.9 V following the second cycle. (6) Fourth cycle, v = 1 mV/s. The change of the count rate at 1.0 V following the fourth cycle. (From Ref. 65, with permission of Pergamon Press plc.)

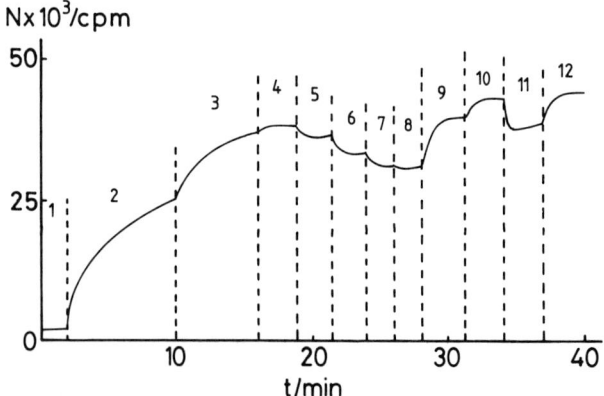

FIG. 42. The change of the count rate (proportional to Γ counterion) with time during subsequent potential switches in the case of a PVF electrode in contact with $Na_2^{35}SO_4$ (1 × 10^{-2} mol/dm^3). (1) 0.5, (2) 0.9, (3) 1.0, (4) 0.9, (5) 0.8, (6) 0.7, (7) 0.6, (8) 0.5, (9) 0.9, (10) 1.1, (11) 0.5, and (12) 1.0 V vs. rhe. (From Ref. 65, with permission of Pergamon Press plc.)

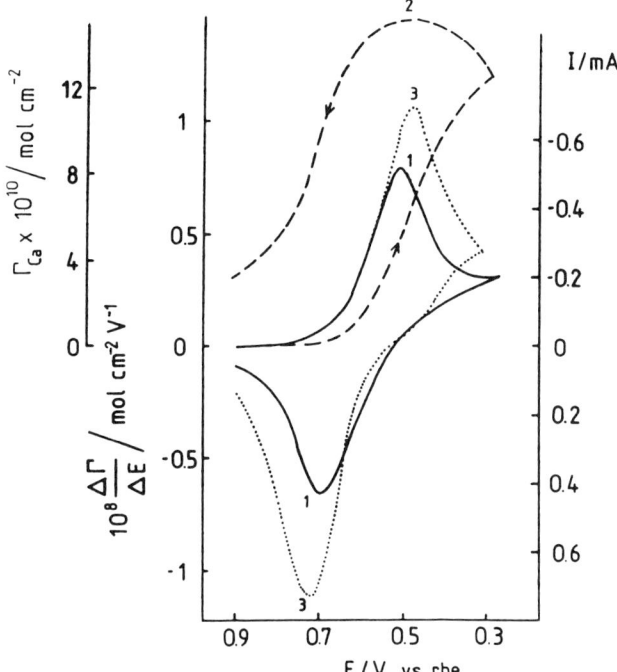

FIG. 43. Cyclic voltammetric (1), Γ vs. E (2), and $d\Gamma/dE$ vs. E (3) curves for a PTCNQ film ($\Gamma = 9 \times 10^{-8}$ mol/cm^2) obtained simultaneously by combined voltammetric and voltradiometric measurements. Sweep rate: 1 mV/s, $c_{Ca}{}^{2+} = 8 \times 10^{-5}$ mol/dm^3. Supporting electrolyte: 5×10^{-2} mol/dm^3 CH$_3$COONa + CH$_3$COOH (pH = 7). (From Ref. 65, with permission of Pergamon Press plc.)

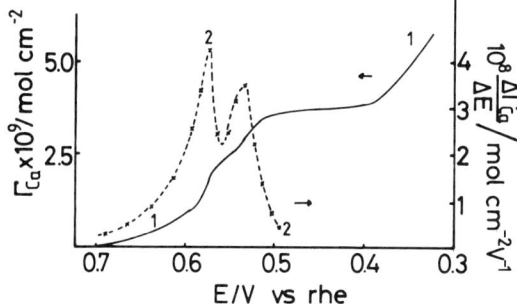

FIG. 44. Curve 1: potential dependence of the equilibrium Ca^{2+} sorption in a PTCNQ film ($\Gamma = 9 \times 10^{-8}$ mol/cm^2) at 3.2×10^{-4} mol/dm^3 Ca^{2+} concentration in 5×10^{-2} mol/dm^3 CH$_3$COONa + CH$_3$COOH (pH = 7). Curve 2: $\Delta\Gamma/\Delta E$ vs. E curve obtained from experimental data. (From Ref. 65, with permission of Pergamon Press plc.)

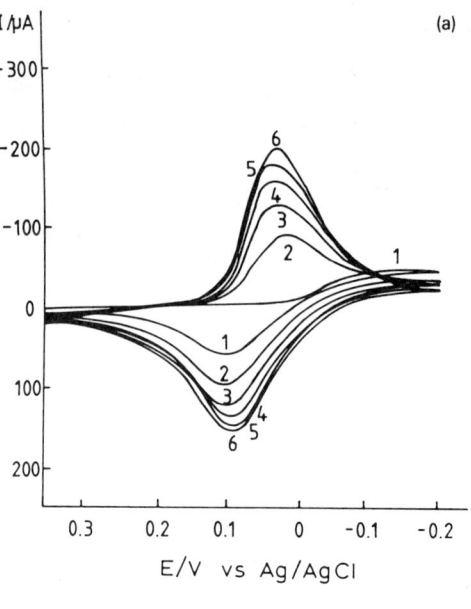
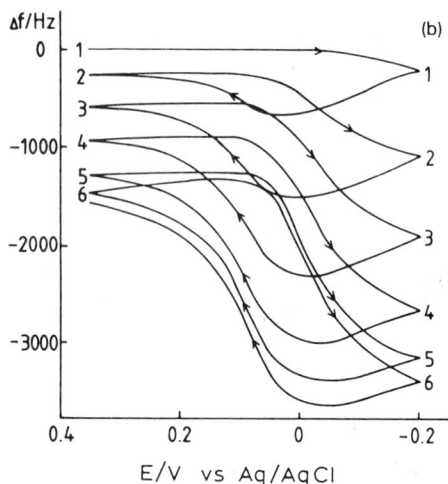

FIG. 45. (a) Cyclic voltammetric break-in of a PTCNQ electrode ($\Delta f = -13925$ Hz). Electrolyte: 2.5 mol/dm^3 LiCl. Sweep rate = 6 mV/s. (b) Frequency curves obtained simultaneously with the cyclic voltammograms shown in (a). (From Ref. 53, with permission of Elsevier Scientific Publishing Company.)

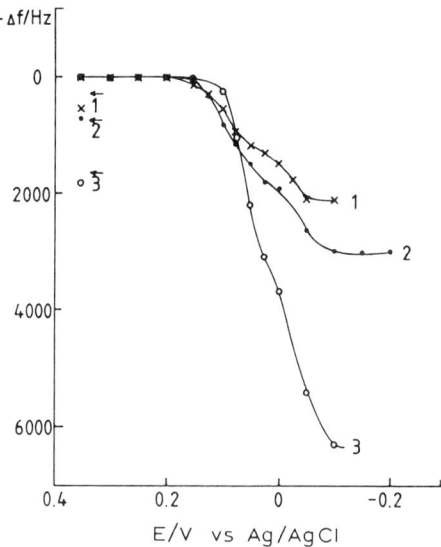

FIG. 46. Equilibrium frequency-potential curves obtained potentiostatically for freshly prepared PTCNQ films with a delay time of 15–21 min at each potential. Electrolyte: 2.5 mol/dm^3 LiCl. Δf_p values are -5809 (1), -7888 (2), and -13615 Hz (3). Points labeled with arrows (1, 2, and 3) were obtained for the reoxidized films after a 20-min delay time. (From Ref. 53, with permission of Elsevier Scientific Publishing Company.)

The variation of the film composition was achieved by the synthesis of a TCNQ terpolymer and by casting blended films of the TCNQ polyester and an isomorphic polyester containing a benzene ring in place of the TCNQ structure [238]. An increase of D with the site concentration was found when the nominal TCNQ concentrations were used, i.e., assuming similar solvent swelling for the different polymers. However, it was also observed that the charge transport rate constant increases with the film thickness (see Ref. 104 for PVF). This phenomenon, which contradicts the expected effect of uncompensated resistance, may be explained by a model that assumes a more ordered, less swollen film near the metal surface and less compact, highly swollen bulk polymer region. Activation energies of 30 and 39 kJ/mol were found for the TCNQ polyester and terpolymer, respectively. The difference in E_a values was interpreted in terms of a counterion-controlled charge transport rate, assuming that increased energy is required for the ion to move from site to site in the terpolymer, where the intersite distance is greater than in the TCNQ polyester.

The polymer composition also varied in different TCNQ polyurethanes [239]. The charge transport through the films was a sensitive function of the solvent and of the flexibility of the polymer chain backbone.

d. Viologens

Several polymers containing N,N'-alkylated bipyridines have also been studied. (We have already mentioned the sandwich experiments of Dalton and Murray [78], as well as the behavior of viologen ions in ion-exchange membranes [190,196].) These systems have been investigated since 1980 [251] and have continued to be the subject of charge transport studies. The viologen polymers are attractive materials because of their chemical stability, their relatively uncomplicated electrochemical behavior of dication/monocation redox couple, and their possible practical applications due to their electrochromic properties.

Bookbinder and Wrighton [251,252] reported D values in the range of $0.4–3 \times 10^{-10}$ cm^2/s, depending on the nature and the concentration of the electrolyte solutions. For instance $D = 0.37 \times 10^{-10}$, 1.4×10^{-10}, and 2.1×10^{-10} cm^2/s in 0.1, 1.0, and 4.0 mol/dm^3, respectively. The self-exchange rate constant of $k_e > 10^5$ dm^3/mol/s for dication/monocation centers was estimated.

The most detailed studies on poly(styrene-co-chloromethylstyrene) pendant viologens (PMV) have been carried out by Oyama et al. [253] using cyclic voltammetry and normal pulse voltammetry. Not all viologen sites in the PMV coating were electroactive. About two-thirds of the viologens sites were assumed to be located in the unswollen regions of polymers, which are inaccessible to the species of supporting electrolyte (0.2 mol/dm^3 KCl, pH 3). Both Γ (observed) and the ratio of Γ (observed)/Γ(total) decreased almost linearly with the loading ratio. Both the wet and dry thicknesses were measured, and the degree of swelling was found to depend on the viologen content of the polymer, which is understandable considering that the viologen sites are hydrophilic, in contrast to the styrene and chloromethylstyrene moieties. These experiments are worth mentioning because the wet thickness is seldom measured. The maximum increase observed in the thickness due to the swelling did not exceed a value of 30%. The actual concentration was very carefully calculated considering both the swelling and the incomplete electroactivity. The resulting D vs. c (observed) plot is shown in Fig. 47. Although D increases with c, the dependence is not completely linear, as predicted by the Dahms-Ruff theory. Because the swelling of the film depends on c, and consequently the motions of the polymer chains, solvent, and counterions

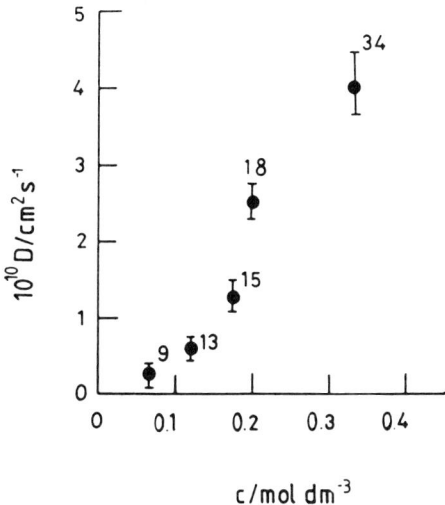

FIG. 47. Plot of D vs. c for viologen films (see text). The percent loadings of viologen sites in the polymer are indicated on each point. (From Ref. 253, with permission of the American Chemical Society.)

become easier when c increases, it was predicted that D_0 and k_e in Eq. (5) would also increase with c. Although this is a reasonable explanation for the observed D vs. c function, the theory of Buck [147–151] and Savéant [141,145] on the coupling of electron hopping and counterion displacement may be applied in this case, too. With the assumption of $\delta = 1$ nm, a $k_e = 0.4–2 \times 10^5$ dm^3/mol s was calculated; this value is much smaller than the electron exchange rate constant (8×10^6 dm^3/mol s) of the monomeric $MV^{2+/+}$ redox couple in solution.

The heterogeneous rate constant increased almost linearly with increasing c and ranged between 3×10^{-5} and 8.9×10^{-5} cm/s. D also varied with supporting electrolyte: for sodium salts in the order p-toluenesulfonate (PTS) > Cl$^-$ > ClO$_4^-$ > poly(styrenesulfonate) (PSS) or, when counterions of the chloride salts were changed, in the order Cs$^+$ > Na$^+$ > tetraphenylphosphonium$^+$ (TPP).

On the basis of these results, Ohsaka et al. [254] noted that the rate-determining step of the charge transport process within PMV coatings is the counterion motion, which is inconsistent with the results and explanations given later in Ref. 196. The values of k_s in the presence of NaCl or CsCl are 2 $\times 10^{-4}$ cm/s, and those in the solutions containing TPPCl, PTSNa, NaClO$_4$, or PSSNa are 1–9×10^{-5} cm/s. There was no explanation for this effect, but specific interactions between the sites and the ions or the variation of potential distribution in the double layer are considered.

Ohsaka et al. [196, 254] determined the dependences of the charge transfer and charge transport rates on the site concentration for such complex systems as PMV-PSS and PMV-Nafion. A parallelism was observed in the change of D and k_s, although the concrete variation of each parameter with the site concentration is very complicated and depends on the supporting electrolyte applied. On the basis of these results, it was suggested that the factors affecting polymer morphology contribute in a similar manner to both the rate-determining process of the heterogeneous electron transfer reaction at the electrode/film interface and the transport of charge through the polymer.

Tsou et al. [255] determined the diffusion coefficients for electroactive polymeric surfactants with pendent long-chain viologen redox groups coated on a glassy carbon surface. The difference between $D_{ox} = 1.4 \times 10^{-8}$ cm^2/s and $D_{red} = 1.0 \times 10^{-8}$ cm^2/s was attributed to the morphological changes accompanying the redox process, i.e., the reduction of the viologen groups leads to a more compact structure in the polymer.

C. Conducting Polymer Films

As we pointed out in Sec. III.E, despite many publications on conducting polymer films, a general model for the behavior of these systems is yet to be discovered. There are still not enough reliable data concerning the rate of charge transport in these films. This insufficiency of information is attributed to the peculiarities of this type of electrode, especially to the large change of their conductivity in the course of the electrochemical transformation and to the fast electronic charge transport, which is outside the range accessible by our usual techniques.

In this section, we discuss the results obtained on the charge transport for polymers containing conjugated aromatic rings, such as poly(pyrrole), and poly(thiophene), and for polymers containing aromatic rings connected by polarizable atoms, such as poly(aniline), poly(aniline) derivatives, and poly(o-phenylene diamine).

1. Poly(pyrrole)

Poly(pyrrole) films have been widely studied since their electrical properties were first reported by Diaz et al. [256]. In the early studies thin films of this polymer were shown to be capable of being driven repeatedly between the conducting and insulating states. This reaction was assumed to involve the oxidation of the extended Π system of the polymers, and it was thought that the partial positive charge of the pyrrole units is balanced by incorporated anions. The results are interpreted either in terms of the polaron-bipolaron model [23,165] or by a chemical model involving localized redox species with two possible conformations of the polymer [24].

The electrochemical behavior of the poly(pyrrole) films is influenced by many factors, such as the conditions of film preparation, the nature of the solvent and electrolyte salt, and the electrolyte concentration, which makes it difficult to compare the results [87,88]. We will survey the most important results on the electronic and ionic charge transport in poly(pyrrole) films, including the effect of the experimental conditions. We will also deal with the widely investigated problem of the distinction between faradaic charge transfer and nonfaradaic double-layer charging (capacitive current).

From the very beginning, one of the central problems concerned the mechanism of electron transport, including the question of the nature of charge carriers responsible for the high conductivity of the oxidized films. Obviously, this is not an electrochemical problem, but it is of the utmost importance to the mechanism of charge transport. One of the most reliable methods for obtaining information in this area is the combined electrochemical esr technique. Unfortunately, the results obtained so far are somewhat contradictory and allow different conclusions. In addition, no detailed studies on the temperature dependence of the esr signal have been carried out, which would have helped in the distinction between Curie spins (magnetic susceptibility due to individual charges) and Pauli susceptibility (assigned to metallic polaron bands). Street and co-workers [165] reported that esr absorption is due to such defects, which vanish during potential cycling, and the absence of paramagnetism in the metallic state is discussed within the framework of bipolaron formation. In a later paper, it was suggested that after the charge injection, polarons are formed that very rapidly recombine into bipolarons [257], and the bipolarons are responsible for the charge transport in oxidized form. The theoretical calculation of the energetics of polaron and bipolaron formation in PP also suggests that the formation of a bipolaron is more favorable than the formation of two polarons. However, all further studies [23,175,176] demonstrated that stable paramagnetic species are formed in the course of oxidations. A good one-to-one correlation between the number of spins and number of electrons (or charge consumed) is found at the early stage of oxidation. The film conductivity [23] also correlates with the esr signal intensity, which means that the polarons (or cation radicals) are the charge carriers. At higher potential or at a higher degree of oxidation, the spin concentration decreases, which can be interpreted in terms of either bipolaron [23,164] or localized dication formation [24]. The potential-dependent bipolaron-polaron equilibrium model was proposed by Genoud et al. [176].

Nowak et al. [258] pointed out that the equilibrium between polarons and bipolarons is determined by the oxidation level (concentration). Polarons are favored at low concentrations due to the increased entropy. The data obtained from simultaneous impedance/esr experiments also indicate that the conduction within PP films cannot be solely attributed to the bipolaron, and the polaron is the dominant species at a low doping level [23]. Albery et al. [24] are of the opinion that the delocalization involves three monomer units at most, because of an entropy decrease caused by removal of the free rotations between the monomer units. However, Hamnett [259] argues that this cannot hold as an argument against delocalization, because free rotation about the C-C bond joining neighboring pyrrole units is extremely improbable, even in the neutral form. Li and Albery [178] found evidence by in situ esr measurements that two radical cations disproportionate to form one neutral monomer and one dication. At low radical cation concentrations, the formation of dications via disproportionation or direct oxidation is not likely, but it is substantial at high oxidation levels.

Impedance spectroscopy has been widely used [11,15,21–24,28] to obtain data on the rate of charge transfer and charge transport processes, as well as to determine the potential dependence of the capacitance. Bull et al. [11] established that PP films are characterized by extremely large capacitances and conductances, which strongly depend on the supporting electrolytes; for example, both values are 5–10 times higher in 1 mol/dm^3 KCl than those in 0.5 mol/dm^3 Na$_2$SO$_4$. Tanguy et al. [15] analyzed the problem of the capacitive current. The impedance measurements in 1 mol/dm^3 LiClO$_4$/propylene carbonate indicate the existence of two types of ionic trapping sites in PP films. The amount of the deeply trapped ions, which cannot follow the ac signal, increases during the film charging, i.e., the few ions that have entered the film remain in a quasi-free state. Near the redox potential, the amount of quasi-free ions is maximal, giving a maximum film capacitance. In this interpretation, there are two types of current in the cyclic voltammetry: a capacitive current without hysteresis effect and a noncapacitive current arising from the deeply trapped ions, giving a large hysteresis responsible for the broadening of the reduction peak. This explanation contrasts with Feldberg's model [155], which assumes a potential-independent differential capacitance, and with the porous electrode model [22,129], where the capacitance is attributed to the double-layer formation at the surface developed in a porous electrode.

Waller and Compton [23] presented a complex plane impedance plot for a PP electrode in 0.1 mol/dm^3 TBAP/acetonitrile, which exhibits two forms, dependent upon the degree of doping (Figs. 48 and 49). The observed

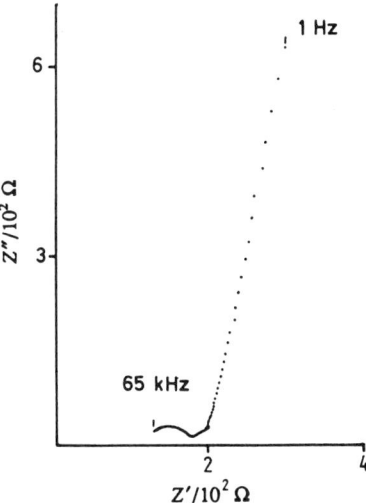

FIG. 48. The complex plane impedance plot for a PP-coated electrode, with a coat thickness 40 nm, measured in 0.1 mol/dm^3 TBAP/acetonitrile. The potential was -0.5 V vs. sce. (From Ref. 23, with permission of the Royal Society of Chemistry.)

impedance plot consisting of two semicircles proved to be consistent with a series of two parallel combinations of resistance and constant phase-angle admittance, which was interpreted in terms of the theory of rough, fractal electrodes. At higher doping levels, the complex impedance spectra obtained resemble those obtained for redox polymer films, which can be described by using a modified Randles equivalent circuit. Waller and Compton attribute the increase in the low frequency capacitance to a double-layer effect, and the Warburg section is interpreted in terms of a porous electrode model, instead of a plane electrode diffusion model. However, the existence of the Warburg section can be queried, and the straight line with an angle of 45° does not fit the accepted models elaborated for porous electrodes. Albery and co-workers [24] found no Warburg region on the impedance plot for poly(pyrrole) in contact with an unidentified aqueous electrolyte (Fig. 50). They used a transmission line equivalent circuit for the modeling of the polymer film. This type of equivalent circuit was introduced by Rubinstein et al. [13], but in their work, they allowed each rail to have its own resistance in order to model the different mobilities of the charge carriers in the polymer and of the counterions in aqueous pores. The distributed capacitance between the two lines describes the type of capacitance proposed by Feldberg [155]. The resistance

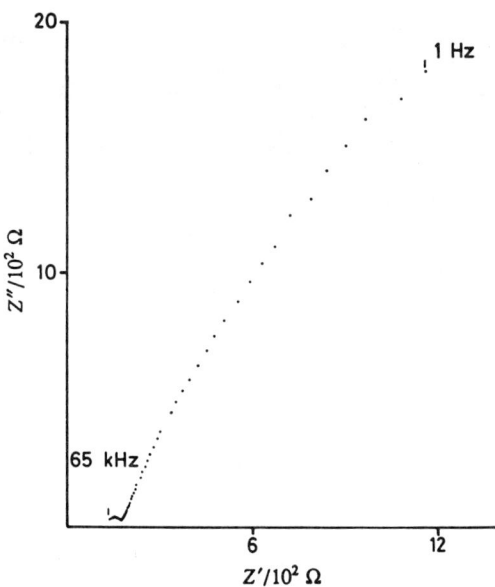

FIG. 49. The complex plane impedance plot for the PP-coated electrode described in Fig. 48. The potential was +0.5 V vs. sce. (From Ref. 23, with permission of the Royal Society of Chemistry.)

of the polymer line, R_p, was found to be equal to that of the aqueous line, R_s, in the transmission line model, which indicates that the motion of the polymer charge carriers and that of the counterions are coupled on the micro scale. This observation was interpreted in terms of the theory elaborated by Buck [147–151] and Savéant [141,145] (see Sec. III.C), but it was also assumed that the rate of the motions of electrons and ions might be limited by the polymer segmental motion.

This explanation fits the model proposed by Albery et al. [24,25], who on the basis of the esr results assume localized redox species, and thus an electron propagation that occurs via molecular exchange reactions. According to this model, which is believed to be valid not only for poly(pyrrole), but also for poly(aniline) and poly(thiophene), one form of the polymer is the stable conformation in a reduced state, which becomes metastable when the film is oxidized. The other conformation is stable when the film is conducting. The change between the two conformations follows first-order kinetics.

Albery and Mount [26] investigated another transmission line model in which the intrachain electron transport is assumed to be very fast and the interchain electron hopping is the rate-controlling process. Although several

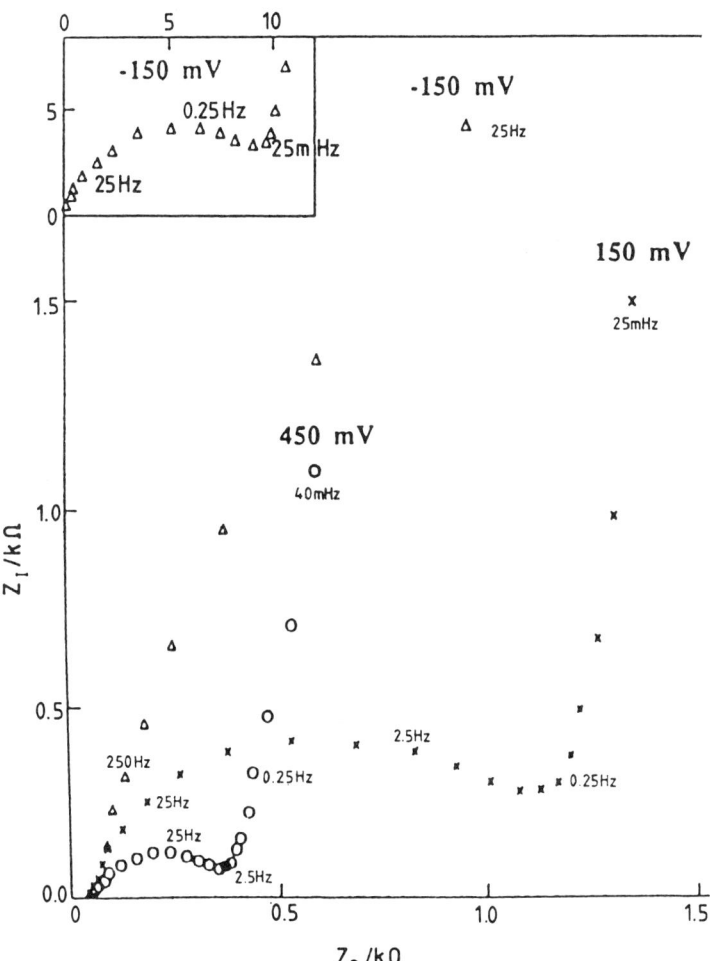

FIG. 50. Impedance plots for poly(pyrrole) at different potentials; the inset shows the high-frequency results at −0.15 V. (From Ref. 24, with permission of the Royal Society of Chemistry.)

cases with varying values of resistances and capacitances were studied, no satisfactory description of the experimentally observed impedance plot was obtained.

Two problems arise concerning the model of Albery et al. [24]. First, the localized redox site model does not give a reasonable explanation for the high conductivity of these systems. Second, the conclusions that were drawn on the basis of the equality $R_p = R_s$, which is obviously model-dependent, may not be valid. The appearance of the Warburg region may depend on many factors, especially on the film thickness and temperature, therefore, further experimental work should be devoted to this question, and further comparison of the experimental data and the results obtained by simulation based on different models should be made in order to gain a better insight into the nature of this complex system.

Penner and Martin [22] determined the D values for reduced poly(pyrrole) films in contact with 0.2 mol/dm^3 TBABF$_4$/acetonitrile by using impedance spectroscopy. The diffusion coefficients range from 1.4 to 3.4 × 10^{-9} cm^2/s and from 0.4 to 1 × 10^{-9} cm^2/s obtained from finite and semiinfinite diffusion regimes, respectively. However, as Fig. 51 shows, no well-defined Warburg region was obtained, and the uncertainty in the D values obtained from the low-frequency data is very high. A simple electron-hopping counterion diffusion mechanism with a rate-determining ionic diffusion is suggested. They found that the modified Randles-type equivalent circuit [6-10] gives a good description of the behavior of the reduced PP films, while for oxidized films a transmission line equivalent circuit is applicable to interpret the data.

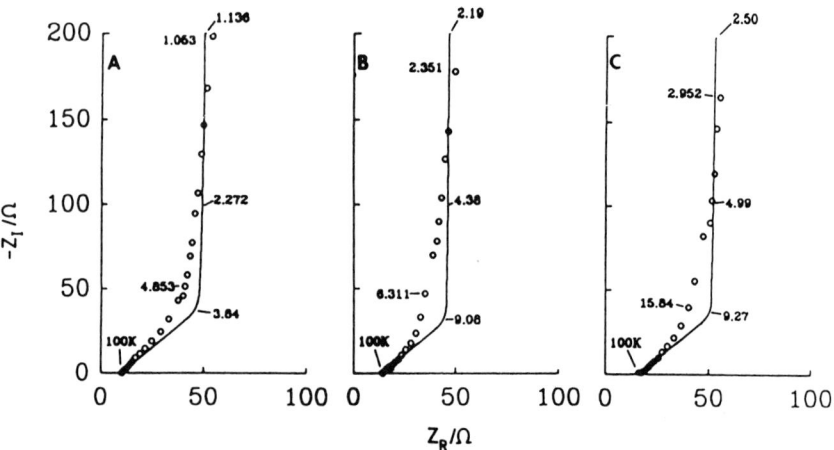

FIG. 51. Comparison of experimental (points) and simulated (solid curves) Nyquist plots for poly(pyrrole) films at three different potentials: (A) −0.33 V, (B) −0.35 V, and (C) −0.361 V vs. sce. Numbers to the right are simulation frequencies; numbers to the left are experimental frequencies. (From Ref. 22, with permission of the American Chemical Society.)

Charge Transport in Polymer-Modified Electrodes

From impedance data, with a modified Randles equivalent circuit, $D = 1.3 \times 10^{-9}$ cm^2/s was estimated for the anion diffusion coefficient in poly(pyrrole) film in LiClO$_4$/propylene carbonate solution by Panero et al. [21]. From galvanostatic pulse transients $D = 2 \times 10^{-10}$ cm^2/s and 7.5×10^{-0} cm^2/s for the oxidation and reduction, respectively, were given in the text by the same authors, while in a table they list $D_{red} = 1$–2.6×10^{-10} cm^2/s and $D_{ox} = 1$–2.6×10^{-11} cm^2/s without any explanation. The results obtained with impedance and pulse techniques differ substantially. This was attributed to the fact that the electrode area is involved in the calculation of D from pulse data, while in the impedance data evaluation from the limiting low-frequency capacitance and resistance, this area is not involved. However, this explanation does not seem to be entirely adequate. The thickness dependence was also considered for the apparent discrepancies of the data reported. The differences between D_{ox} and D_{red} were interpreted in terms of a diffusion model in which the mechanism of the oxidation is different from that of the reduction. During the oxidation, only anions participate as charge-compensating ions, while the reduction involves the incorporation of cations, as proposed originally by Kaufman et al. [40]. Here we should mention that Genies et al. [260] found $D_{ox} = 6.2 \times 10^{-10}$ cm^2/s and $D_{red} = 1.2 \times 10^{-9}$ cm^2/s in aqueous perchlorate solutions. In moist LiClO$_4$/acetonitrile, $D = 1.2 \times 10^{-10}$ cm^2/s was obtained, and a not too substantial variation of D was detected when ClO$_4^-$ was replaced by BF$_4^-$.

Electropolymerized poly(pyrrole)/poly(azulene) composite films in contact with LiClO$_4$/propylene carbonate show a somewhat unexpected behavior. The D value derived from the diffusion-controlled region of the impedance spectra, shown in Fig. 52, depends on the azulene content nonlinearly. A sharp decrease from $D = 1 \times 10^{-8}$ to 7.5×10^{-8} cm^2/s occurs between 25 and 50% azulene concentration. In this concentration range, a noticeable change of the low-frequency capacitance and the charge consumed during the cyclic voltammetric experiments can also be observed. This effect is attributed to the morphological change of the film caused by the incorporation of the more bulky azulene into the PP matrix. At low azulene content, the film is compact and the charge transport controlled by the diffusion of counterions is slow. At high azulene concentrations, the composite film has a more open structure and charge-transfer limiting is assumed. The D value for pure PP is much higher than that reported by Panero et al. [21]. Unfortunately, the comparison of D values is very difficult, because they depend on the equivalent circuit applied to model the electrochemical process. Even in this case, when identical equivalent circuit and data evaluation methods were used, the differences in D values may be related to the potential dependence of the

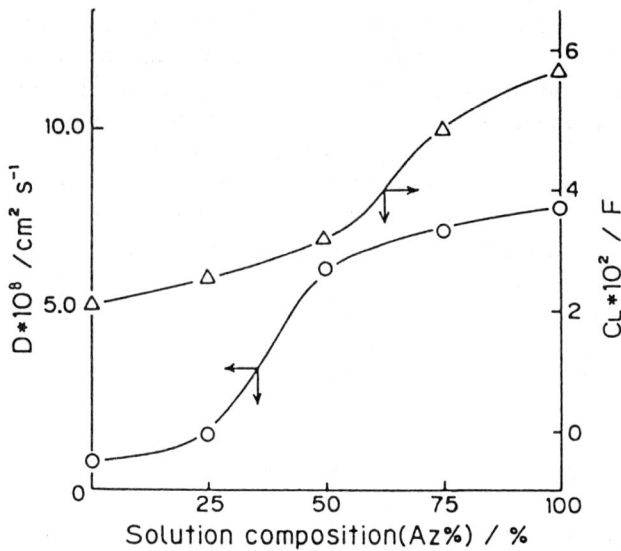

FIG. 52. Correlation of diffusion coefficient (D) and redox capacitance (C_L) for various poly(pyrrole)/poly(azulene) composite films against azulene concentration. (From Ref. 28, with permission of The Electrochemical Society, Inc.)

diffusion coefficient. Naoi et al. [28] calculated the D values at the respective anodic peak potentials, while Panero et al. [21] did not report the potential at which the measurements were taken; they indicated only that the experiment was carried out on PP electrodes doped to 5 mol% ClO_4^-.

Ion transport in pyrrole-based polymer films was investigated by Pickup and co-workers [129,181]. They emphasized that chronoamperometric data for PP should be analyzed with the help of a migration model (e.g., a single-pore model where a porous conducting electrode is treated as a finite transmission line in series with the uncompensated solution resistance), rather than by the commonly used diffusion model (Cottrell equation). Although the authors were aware of the complicated nature of the charging process, because several possible relations in respect to polymer and solution resistances as well as the possible role of the potential dependence of the capacitance were considered, eventually they decided to use a very simplified equation for the finite-difference simulation. Therefore, the data treatment method used seems invalid, because no simple double-layer charging occurs in the case of PP films, but a faradaic current also flows in the so-called double-layer region. These facts were neglected in the model. However, the authors argue that this model aids in the accumulation of more reasonable data, which correlate with those obtained by other methods. For instance, in a 1 mol/dm^3 KCl solution, the D value extracted from the migration model is 3.5×10^{-8} cm^2/s, while the D obtained by using a diffusion model is 1.1×10^{-6} cm^2/s. From ionic resistance measurements, D = 3.3×10^{-8} cm^2/s was derived, which is very close to the value (D = 1.2×10^{-8} cm^2/s) obtained by Burgmayer and Murray [261,262]. (It should be noted that the two resistance measurements differ from each other. In the experiment conducted by Pickup et al., two ssce electrodes were positioned on either side of the polymer film to measure the potential drop across the film caused by a constant current between the platinized Pt electrodes. From the resulting potential difference, the film resistance can be estimated if the solution resistance is known. Burgmayer and Murray used an arrangement where the PP films contained an embedded gold minigrid electrode.)

However, a comparison of data coming from these two sources cannot be considered a proof, as Genies et al. [260] obtained even smaller diffusion coefficients by using a diffusion model. Pickup et al. feel that the results reported by Genies et al. are erroneous, because of the uncompensated resistance. For perchlorate ions in acetonitrile, D = 2.3×10^{-9}, 1×10^{-9}, and 5×10^{-11} cm^2/s were found in the thickness ranges 2–12 μm, 1–2 μm, and 0–2 μm, indicating that the layer is more compact when it is close to the metal substrate. This is in agreement with the results of other studies [180]. For Cl$^-$ ions in water, D = 3.5×10^{-8} cm^2/s in PP and 1×10^{-7} cm^2/s in poly-[1-methyl-3-(pyrrol-1-ylmethyl)pyridinium] (poly-MPMP+) [181]. However, the incorporation of a high concentration (5.6 mol/dm^3) of permanent cationic sites has a negligible effect on the permeability in acetonitrile. This effect indicates that the solvation and the swelling of the polymer are much more significant in water than in acetonitrile. The chronocoulometric results suggest a strong anion dependence of the electrochemical transformation of PP films in acetonitrile. The charging/discharging process is relatively fast in the presence of Cl$^-$ and ClO$_4^-$, and much slower for benzenesulfonate- and toluenesulfonate-containing films. The D values obtained are 3.1×10^{-9} cm^2/s (Cl$^-$ and ClO$_4^-$) and 4.3–4.8×10^{-10} cm^2/s (toluenesulfonate and benzenesulfonate). Vork et al. [180] assumed that the diffusion of counterions is the rate-limiting step, but the discharging process in the presence of Cl$^-$ and ClO$_4^-$ is initially controlled by the transition of the polymer from the oxidized (conducting) state to the neutral (insulating) state. A substantial solvent effect was observed for ClO$_4^-$ diffusion in PP films—D = 4.2×10^{-8}, 3.6×10^{-10}, and 3×10^{-13} cm^2/s in aqueous, acetonitrile, and propylene carbonate solutions, respectively—which was attributed to the effect of swelling [263].

Inzelt and Horányi [66] investigated the sorption of SO_4^{2-} and Cl^- ions into poly(pyrrole) film by an in situ radiotracer method under potential cycling and steady-state conditions, using labeled H_2SO_4 and HCl. Similarly to the redox polymer films, the variation of the counterion content in the surface layer follows the potential changes as illustrated in Fig. 53. The mobility of sorbed sulfate ions was studied by the exchange of labeled sorbed species with nonlabeled species added to the solution phase in great excess. A relatively rapid exchange was observed at the beginning, as shown in Fig. 54. The results of this experiment attest that the sulfate ions embedded into the film during its preparation are mobile species, i.e., no irreversible uptake of SO_4^{2-} ion occurs. The radiotracer studies revealed that, in contrast to the observations for redox polymers (PVF, PTCNQ), no saturation with respect to the amount of the sorbed counterions can be observed in the case of polypyrrole films, as shown in Fig. 55. Below 200 mV, there is no change with the potential. Above this value, an almost linear relationship exists between the amount of sorbed SO_4^{2-} ions and the potential. This indicates that the penetration of anions into the film proceeds unhindered at potentials far beyond the peak potential of the voltammogram (E_{pa} = 420 mV [rhe]). This observation, however, is in accordance with the theoretical consideration outlined by Feldberg [155]. He suggested that the transformations during the charging and discharging processes involve both faradaic and capacitive electron transfer components. The situation is the same for Cl^- counterions,

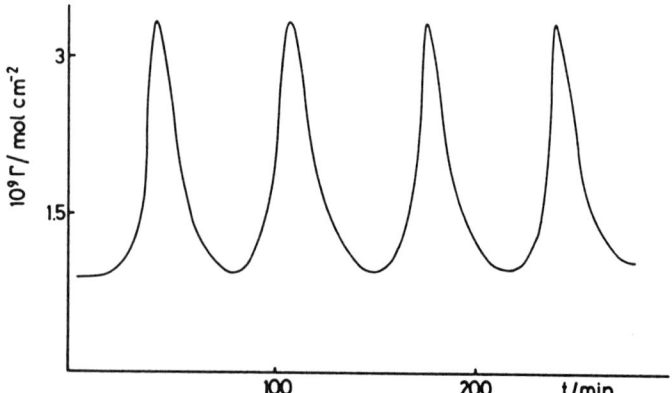

FIG. 53. Periodic change of the sorption of Cl^- ions during four consecutive potential cycles. Poly(pyrrole) electrode in contact with solution of 2×10^{-4} mol/dm³ Cl-36 labeled HCl. Scan rate: 0.4 mV/s.

Charge Transport in Polymer-Modified Electrodes 195

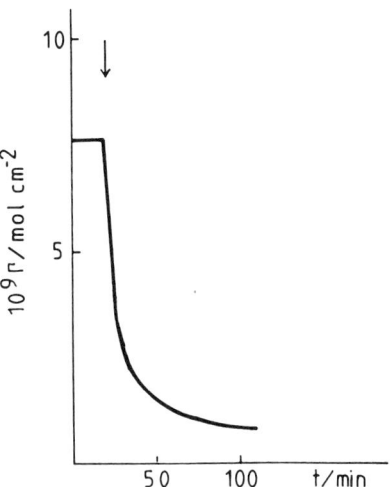

FIG. 54. Study on the mobility of sorbed counterions in PP films. The exchange of labeled embedded SO_4^{2-} ions with nonlabeled species added to the solution phase at the moment indicated by the arrow at E = 0 mV. Initial H_2SO_4 concentration: 10^{-5} mol/dm^3. Final concentration: 2×10^{-2} mol/dm^3. (From Ref. 66, with permission of Elsevier Scientific Publishing Company.)

as seen in Fig. 55. This figure also shows that the amount of the embedded SO_4^{2-} ions decreases significantly in the presence of ClO_4^- ions in great excess, yet only a partial displacement of SO_4^{2-} ions occurs. Thus, the interaction between the film and SO_4^{2-} ions is much stronger than that between the film and perchlorate ions. The potential step experiments revealed that the average number of chloride ions involved in the sorption process during the transfer of one electron is about 0.25. If we assume that each positive site formed by oxidation requires a compensating anion, we conclude that the oxidation process does not result in the formation of an equivalent amount of positive sites. This situation may occur when the oxidation involves simultaneous deprotonation, i.e., neutral radicals or molecules are formed and the protons leave the film.

When pyrrole is electropolymerized in the presence of poly(styrene sulfonate) the resulting poly(pyrrole) contains permanently trapped PPS [178]. In association with small cations (K^+) and favorable anions (NO_3^-, Cl^-, Br^-), the charge transport is reversible and fast, with $D = 6 \times 10^{-8}$ cm^2/s. In a sulfate solution, the kinetics of the charge transport is much slower due to the formation of cross-links, which leads to a more compact structure.

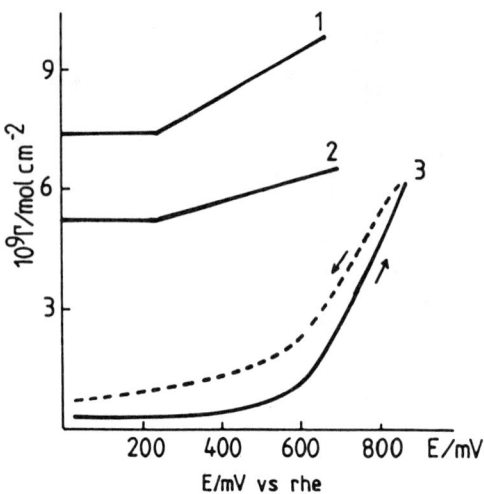

FIG. 55. Amount of sorbed ions as a function of potential in the case of PP films (steady-state measurements). (1) 1×10^{-4} mol/dm^3 H$_2$SO$_4$, (2) 1×10^{-4} mol/dm^3 H$_2$SO$_4$ + 1×10^{-2} mol/dm^3 HClO$_4$, (3) 2×10^{-4} mol/dm^3 HCl. (From Ref. 66, with permission of Elsevier Scientific Publishing Company.)

Naoi et al. [60] have demonstrated by the EQCM method that in PP films prepared in the presence of large polymeric anions (PPS and poly(vinylsulfonate)), mostly cations and solvent molecules move into and out of the film. Mostly anion motion occurs when small anions (ClO$_4^-$ or BF$_4^-$) are used, but cation sorption becomes significant for higher oxidation state.

The charge and mass transport in poly(pyrrole)/poly(styrene sulfonate) composites have also been studied by Baker et al. [61] using the EQCM method. The mass transport is cation specific, the frequency responses increase with the increasing cation mass at Li$^+$ = Na$^+$ < Cs$^+$ < TEA$^+$ < TBA$^+$. The reductive charge transport processes are faster than the oxidative transport process during switching. The origin of this effect may lie in the fact that, during the initial stages of the reduction process, the film is electrically conducting and thus can easily support the charge transport. The chronocoulometric D values range between 10^{-9} and 10^{-11} cm^2/s for mobile hydrated Na$^+$ ions. However, the concentration of accessible redox sites decreases with increasing film thickness, which makes the derivation of D from the $D^{1/2}c$ product uncertain.

2. Poly(thiophene)

The properties of poly(thiophene) are similar to those of poly(pyrrole) [89,264]. Because substituted thiophene and thiophene dimer (bithiophene) give higher quality films than the monomer and the substituted film is oxidized much more easily, these systems are more suitable for electrochemical studies than the films formed by the electropolymerization of the unsubstituted thiophene monomer. Below, we discuss some recent results concerning these systems.

Poly(3-methylthiophene) was synthesized and characterized in acetonitrile solutions of different anions and cations [265]. Marque et al. [265] found that the polymer morphology strongly depends on the nature of anions used in the electropolymerization process, causing a "memory effect" in the course of further electrochemical studies, when the original counterion is replaced by any other anion. The findings suggest that both cations and anions are present in the neutral polymer film; in the oxidized state, an excess of anions exist to ensure electroneutrality. The differences in the oxidation potentials between films cycled in lithium and tetrabutylammonium electrolytes show that cations also participate in the oxidation/reduction process. The polymer swelling depends on the degree of the solvation of cations, thus the incorporation of the largely solvated Li^+ ions also makes the anion motion easier. The anion effect seems to be influenced by the shape and size of the anions. The $D^{1/2}c$ values are approximately three times higher for the cathodic process than for the anodic process. (The results are presented in $D^{1/2}c$ form because of the difficulty in determining the accurate c values.) The highest values were obtained in the presence of PF_6^-, $D^{1/2}c = 8.4 \times 10^{-5}$ and 20.9×10^{-5} mol/cm^2 s$^{1/2}$ for the oxidation and reduction, respectively, when Li^+-salt was used. In TABPF$_6$ electrolyte, $D^{1/2}c = 7 \times 10^{-5}$ and 25×10^{-5} mol/cm^2 s$^{1/2}$ are derived.

These observations were confirmed by Servagent and Vieil [50], who established, on the basis of the EQCM results, that film oxidation takes place by the simultaneous insertion of ClO_4^- counterions, while during the cathodic process the incorporation of Li^+ ions is preferred to the release of perchlorate ions.

A similar conclusion was drawn by Hillman et al. [46] for poly(bithiophene) (PBT) films in 1 mol/dm^3 TEABF$_4$/acetonitrile solution. A mechanism involving both anion ingress and cation egress, together with some solvent transfer, upon the oxidation of poly(bithiophene) was suggested, making use of the results obtained on the mass changes during the redox transformation of film, as well as the nonnernstian dependence of the peak potentials on BF_4^- concentration. The statement that the counterions are not the solely transferred species was confirmed later [49], when PBT films in contact with acetonitrile solutions of $TEABF_4$, TBAP, TEAP, $TEAPF_6$, and NH_4ClO_4 were studied, and the electrolyte concentration was varied over the range of 0.01–1 mol/dm^3. The mass and charge fluxes correlated during oxidation, but not during reduction. The reduction of PBT takes place in two stages. It is assumed that beside the counterions, salt and solvent molecules are also incorporated into the film, and this provided the somewhat surprising observation that the salt and solvent are transferred in the opposite direction. Isotopic substitution of the solvent (using CD_3CN instead of CH_3CN) proved that one solvent molecule is bound to one redox site.

3. Poly(aniline)

The electrochemical behavior of poly(aniline) is more complex than that of other conducting polymer films. There is no doubt that at least two kinds of redox reactions are involved in the electrochromic reactions (yellow → green → blue (violet) color changes can be observed), and at least three forms of poly(aniline) should be taken into consideration:

1. Closed valence-shell reduced form (benzenoid structure)
2. Radical cation intermediate form (combination of quinoid and benzenoid structures)
3. A closed valence-shell oxidized form (quinoid structure)

By analogy to the aniline octamer, we may speak of leucoemeraldine, emeraldine, and pernigraniline forms, but we may assume that there is actually a continuum of oxidation states, ranging all the way from the completely reduced leucoemeraldine to the completely oxidized pernigraniline species. The idealized formulas of poly(aniline) are shown in Fig. 56. However, as we mentioned in Sec. III.E, the situation is more complicated, because all species of different oxidation states can be expected to be protonated to some extent. The degree of protonation depends on the protonation (dissociation) constant of the species (not known for all these species), consequently it varies with the pH and the potential.

Since both oxidation and proton addition may result in the partial depopulation of the Π system, unlike all the other conducting polymers, the conductivity of poly(aniline) depends on two variables instead of one, namely, the degree of oxidation and the degree of protonation of the poly(aniline) [89,168,266]. In addition, a hydrolysis reaction of the fully oxidized species in acid media resulting in soluble and insoluble (electroactive and non-electroactive) degradation products occurs [68,70,267]. Another important

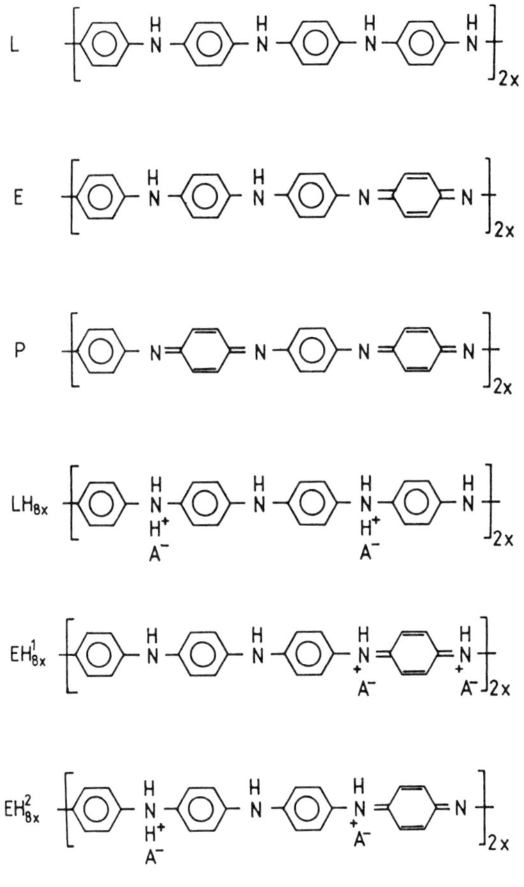

FIG. 56. The idealized formulas of poly(aniline) at different oxidation and protonation states. L, leucoemeraldine, E, emeraldine, P, pernigraniline forms; LH_{8x}, partially protonated leucoemeraldine; EH_{8x}^1 and EH_{8x}^2, protonated emeraldine forms.

feature of poly(aniline) electrochemistry seems to be that the films become passive above ca. pH 5. Thus, it is not surprising that many possible reaction schemes have been suggested [67,157,168,171,172,268].

As we have seen, the electrochemical esr measurements are very instructive in respect to the identification of the charge carriers. Kaya et al. [164] carried out in situ esr measurements on PANI films of different oxidation states in 0.1 mol/dm^3 H_2SO_4 and 1 mol/dm^3 $LiClO_4$/propylene carbonate solutions. (The esr signals were measured at open circuit conditions.) At the initial stage of oxidation, the spin concentration increases in a nernstian way, but at high positive potentials it tends to decrease gradually.

Genies and Lapkowski [171,172] used a combined cyclic voltammetric/esr method to characterize PANI films in $NH_4F/2.3HF$ medium. In both redox processes, as the potential is increased the spin concentration increases until a potential corresponding to three fourths of the voltammetric peak current is reached, and then it decreases to almost zero.

A redox mechanism has been proposed in which during the first electron transfer reaction, some of the nitrogen atoms are transformed from the amine to the imine form and, during the second reaction, the oxidation of the imine form to a new conjugated polymer takes place. It is distinct from the alternative mechanism, which assumes the formation of cation radicals at the potential region of the first wave and dication formation at the second wave as final and not as intermediate products. A model involving two polaron-bipolaron systems was used to interpret the results of the in situ esr measurements. According to this model, two types of polarons characterized by different g values are formed in the first and second oxidation steps, respectively. For the first redox system, a polaron is favored as opposed to a bipolaron, while for the second reaction, the situation is the contrary. The different g values indicate two different structures in the polymer involving conjugated aromatic rings, respectively. Also, on the basis of spectroelectrochemical measurements, Genies and Lapkowski [172] rendered the existence of two polaron and bipolaron states probable.

Glarum and Marshall [163] carried out the same experiment, but in 0.5 mol/dm^3 H_2SO_4. The increase of the spin concentration correlates with the injected charge. The number of spins increases simultaneously with the current in the cyclic voltammetric experiment, but irreversible changes takes place when the potential enters the second oxidation wave. On the basis of the esr and impedance measurements, they explained the electrochemical behavior of poly(aniline) films in terms of a one-dimensional electron band model. In the course of the oxidation, this band passes from a fully occupied condition to an empty state, with electronic conduction occurring only when the band is partially filled. The diffusion coefficient for the charge carriers in poly(aniline) estimated from the impedance measurement, $D = 10^{-6}$ cm^2/s, was almost the same as the spin diffusion coefficient, $D = 4 \times 10^{-6}$ cm^2/s, derived from the esr data. The thickness dependence was explained by the structural inhomogeneity of the film, i.e., at ca. 150 nm, a dense structure is replaced by a fibrous one. This change in the film morphology was established by ellipsometry and electron microscopy by Carlin et al. [158]. Because of the protonation dynamics, the poly(aniline) structure is not static, and this fact should be taken into account. As we have already discussed concerning the poly(pyrrole) electrodes, Albery et al. [24] critized the band model and interpreted the results in terms of a chemical model.

Rubinstein et al. [13] studied the impedance spectra of poly(aniline) films in contact with 2 mol/dm^3 HCl. An equivalent circuit consisting of a parallel combination of two discrete elements—a capacitor and a finite transmission line (repetitive combination of a capacitor and resistor) in series with a polymer and a solution resistance—was successfully applied to simulate the experimental data. Two models, a two-phase model and a double layer/faradaic model, were used to interpret the potential-dependent processes represented by the parallel capacitor and the transmission line. (These models are also discussed in Refs 24–26, 155, and 269.) According to the two-phase model, although there are conducting and insulating phases in the film, the exact nature of the processes occurring in the film is unclear. In the other model, a single homogeneous polymer phase is assumed, and the parallel capacitance and transmission line represent some combination of the double-layer capacitance and the faradaic reaction, both being potential dependent.

Within the framework of the latter model, two interpretations may be considered. The double-layer capacitance increased gradually upon switching of the polymer due to the increasing electrode area, and the oxidation/reduction of the film is a diffusion-controlled faradaic process. Alternatively, the capacitor is a pseudo-capacitance associated with a fast electron-transfer process confined to the film, while the transmission line represents the double-layer capacitance of a porous electrode. Both interpretations are consistent with the potential dependence of the data obtained. The diffusion coefficient increases during the oxidation of the film from 10^{-10} to 10^{-8} cm^2/s.

Poly(aniline) films in aqueous acidic media and in neutral organic solutions were also investigated by Fiordiponti and Pistoia [19]. Very high, thickness-dependent exchange current densities, i_0, were found, although it is not exactly clear how they determined the i_0 values, in the absence of well-defined semicircles in half-oxidized state. From the low-frequency capacitance and resistance, $D = 10^{-7}$ cm^2/s was calculated using the measured film thickness. It was, however, stated that D may be much lower ($D = 10^{-10}$ cm^2/s) because the ion diffusion requires a much shorter path than the geometrical thickness on the account of the high polymer porosity.

Osaka et al. [33] studied poly(aniline) films in propylene carbonate and in mixed organic solvents by impedance spectroscopy. The D values were estimated from the low-frequency resistance and capacitance on the basis of a modified Randles equivalent circuit (finite diffusion model). When the measured film thickness, d, was used to derive D from d^2/D, $D = 10^{-5}$ cm^2/s was obtained. They also used the average particle size (ca. 0.8 mm) instead of d to estimate D; then the D values range from 3×10^{-8} to 6×10^{-9} cm^2/s, depending on the nature of the solvent applied in the electrodeposition and/or in the study.

Deslouis et al. [210] dealt with the theory of electrohydrodynamic impedances for modified electrodes in contact with a solution containing a redox couple. They developed a model that led to impedance expressions in which the diffusion of electrons expressing charge transport through the film appears. Diffusion time constants, $d^2/D_e = 1-10^{-2}$ and 10^{-6} s for redox and conducting polymer, respectively, were considered.

La Croix and Diaz [182] showed that for poly(aniline) films in acid solutions, the rate of switching is strongly dependent on the anodic voltage step, and the response does not exhibit simple Fickian behavior. The response is initially linear with time and changes to a $t^{1/2}$ dependence at longer times. Thus, if the switching rate is controlled by mass transport, then it is limited by a process involving a moving front across the thickness of the film and not by the diffusion of ions. This picture is consistent with the linear kinetics observed. The constant velocity of an advancing boundary may be the result of a chemical reaction or a phase transition. La Croix and Diaz considered the effect of cell resistance as an alternative explanation. They proposed that the interpretations based on mass transport and the reported D values may be in error. The calculated D values are 1.4×10^{-10} cm^2/s (10^{-3} mol/dm^3 H$_2$SO$_4$ + 0.1 mol/dm^3 Na$_2$SO$_4$), 1.9×10^{-10} cm^2/s (0.1 mol/dm^3 H$_2$SO$_4$) and ca. 5.7×10^{-10} cm^2/s (1–2 mol/dm^3 H$_2$SO$_4$). However, in solutions more acidic than pH 1, the charge is linear with time in the initial period, while in 10^{-3} mol/dm^3 solutions, the $Q - t^{1/2}$ dependence is linear. Although the position and width of the cyclic voltammetric peaks are sensitive to the solvent/electrolyte combination of the solution, the rate depends on the acidity of the solution but is independent of the nature of the counterions.

Although it was realized at a very early stage of the research into conducting polymer film electrodes that the kinetics of charge transfer and charge transport are very fast, only recently did the improvement of the experimental techniques make it possible to demonstrate that the switching process of these polymers is indeed extraordinarily rapid.

Lacroix et al. [161] used a PANI electrode with an area of 0.15 mm^2 and applied partial ohmic drop compensation (only the solution resistance was compensated). They concluded that the switching time for the reduction is controlled only by the RC time constant of the cell, which is associated with the series of combination of solution resistance and the double-layer capacitance. Using the actual film thickness, d = 250 nm, a very high ionic diffusion constant was estimated, D = 2.5 × 10^{-6} cm^2/s, which is orders of magnitude higher than those reported previously.

Peter and co-workers [73,74] performed cyclic voltammetric and potential step experiments using a fast operational amplifier potentiostat equipped with positive feedback IR compensation to study the rate of switching in poly(aniline) films in 1 mol/dm^3 H$_2$SO$_4$. Excellent resolution was obtained by using a microelectrode with a diameter of 10 μm even at scan rate of 100 Vs^{-1}. The first voltammogram differs from the subsequent ones due to the imperfect rereduction. The potential step studies also attest that the oxidation rate is very sensitive to the presence of residual oxidized material, and the second oxidation transient is much faster than the first one. The change of the current transients, when the potential limit was varied, suggests a two-dimensional phase formation/transformation process; at least such a model shows the best fit. Neither a model of the linear propagation [182] nor a model based on the changes of film resistance during the electrochemical transformation [184] seems to be appropriate. The very fast switching indicates a rapid proton transfer through the film to be the charge-compensating process, rather than anion transport. Despite the use of a microelectrode, there is still a small ohmic drop distortion. The switching process is controlled by the RC time constant only in the reducing cycle, whereas the oxidizing cycle is slower and kinetically controlled.

Andrieux et al. [75] demonstrated that the use of ultramicroelectrodes of 5–20 μm diameter and a fast potentiostat with ohmic drop compensation makes the observation of the electrochemical response of conducting polymer films possible at scan rates as large as 364000 Vs^{-1}. Very thin (d = 5 nm) poly(aniline) films in contact with 1 mol/dm^3 H$_2$SO$_4$ exhibit the same behavior in fast scan rate experiments as that observed at low v values, i.e., two well-defined waves, and a secondary break-in effect, as shown in Fig. 57. The electron transfer kinetics is slower for the first cycle as compared with the subsequent cycles. A heterogeneous rate constant, k_s = 10 s^{-1}, was extracted. (The k_s is defined by an appropriate form of Butler–Erdey-Grúz–Volmer equation, where Γ was used instead of c, thus the dimension of the heterogeneous rate constant is s^{-1}, instead of cm/s.) Slower charge transfer kinetics was found for poly(3-methylthiophene) and poly(3,4-dimethylpyrrole) in TEAP/acetonitrile solutions.

Although the cyclic voltammograms obtained by Andrieux et al. [75] are very impressive and convincing, at the same time they are somewhat puzzling. The good correspondence between the low and very high sweep-rate voltammograms may indicate that it is solely the uncompensated resistance of the film and the solution that is the key factor determining the behavior of the

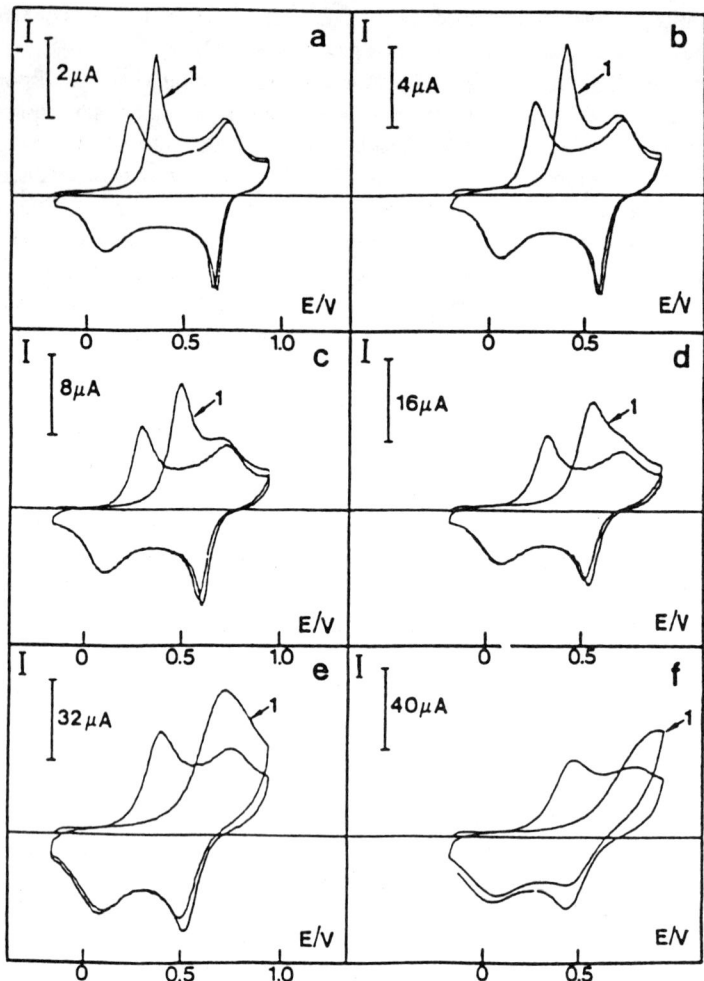

FIG. 57. Cyclic voltammograms of a poly(aniline) film in 1 mol/dm³ aqueous sulfuric acid solution on a 5-μm-diameter platinum disc working electrode. Delay time at the starting potential before the first scan: 200 ms. Scan rates: (a) 8,900, (b) 22,000, (c) 43,000, (d) 93,000, (e) 190,000, (f) 364,000 V/s. Potentials are versus sce. (From Ref. 75, with permission of Elsevier Scientific Publishing Company.)

conducting film. Hence, if the iR drop is properly compensated, we should obtain the same response. At the same time, it means that not only the charge transfer, but also the charge transport, as well as the phase transition are very rapid. Andrieux et al. [75] used a film with a thickness of 5 nm, which is not much higher than monolayer coverage. Thus, on the basis of these results, one should not draw any conclusions with respect to the charge transport process in thick films, especially because of the thickness-dependent morphology of PANI films. For somewhat thicker films (d = 36 nm), Peter and co-workers got similar results, but at a much smaller sweep rate (100 V s^{-1}). In this case, a rapid proton transfer through the film was assumed as a charge-compensating process. However, the main problem is that the salt/base equilibria depending on the pH involve the transfer of both protons and counterions. The proton motion, being the only ionic motion, may be considered only when all reduced and oxidized forms of the poly(aniline) are unprotonated, which is in disagreement with the results on the protonation equilibria in acid medium [171–173,268,270]. If there is a phase transition and/or a phase separation depending on the degree of oxidation and on the degree of protonation, it should be either very fast or very slow compared to the time scales of the fast and slow cyclic voltammetric experiments. A slow reorganization may be assumed due to the polymeric nature of the systems, but in this case, the two-phase model elaborated to explain the hysteretic redox behavior becomes invalid.

Since it is impossible to understand the charge transport in PANI films without having any idea about the sorption of protons and counterions, we will review the most important results accumulated in this field. For the sake of simplicity, we will consider only three forms of poly(aniline) mentioned earlier, which correspond to the respective participating initial forms and products formed during the first and second redox processes, i.e., we will neglect the intermediate radical species.

On the basis of the pH dependence of the peak potentials in cyclic voltammetric experiments (Figs. 58 and 59) [67,74,173,270], the redox processes can be described as illustrated in Fig. 60. This scheme was confirmed by the counterion concentration dependence of the peak potentials [270].

It is not a simple task to present precise pH and concentration dependence measurements. First, one should bear in mind that the changes in the pH and concentration of electrolyte refer to the solution phase, and thermodynamic relationships with respect to the film can only be obtained if an equilibrium exists between the film and the solution phase. Second, constant ionic

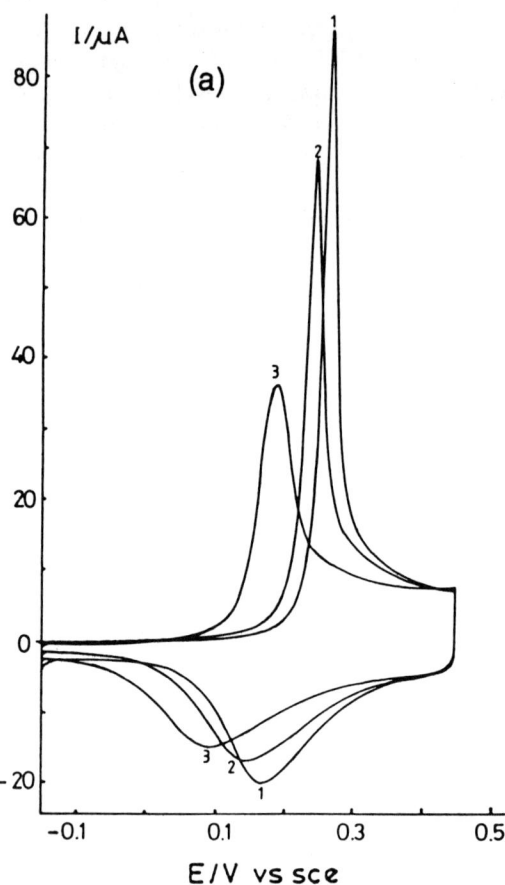

FIG. 58. Cyclic voltammograms of a PANI electrode in contact with aqueous solution of: (1) 5 mol/dm^3 HCl; (2) 0.5 mol/dm^3 HCl + 4.5 mol/dm^3 NaCl, (3) 0.05 mol/dm^3 HCl + 4.95 mol/dm^3 NaCl. Scan rate = 6 mV/s. Reference electrodes are: (a) sce, (b) rhe. (From Ref. 270, with permission of Elsevier Scientific Publishing Company.)

strength should be maintained during the determination of the pH effect, and constant pH is necessary when the influence of the counterion concentration is measured. It is of importance to eliminate the effect of junction potentials appearing in the measured potential values in the course of changing the pH and the electrolyte concentration. The role of the junction potential and the ionic activity in voltammetric response were analyzed in detail in Ref. 270.

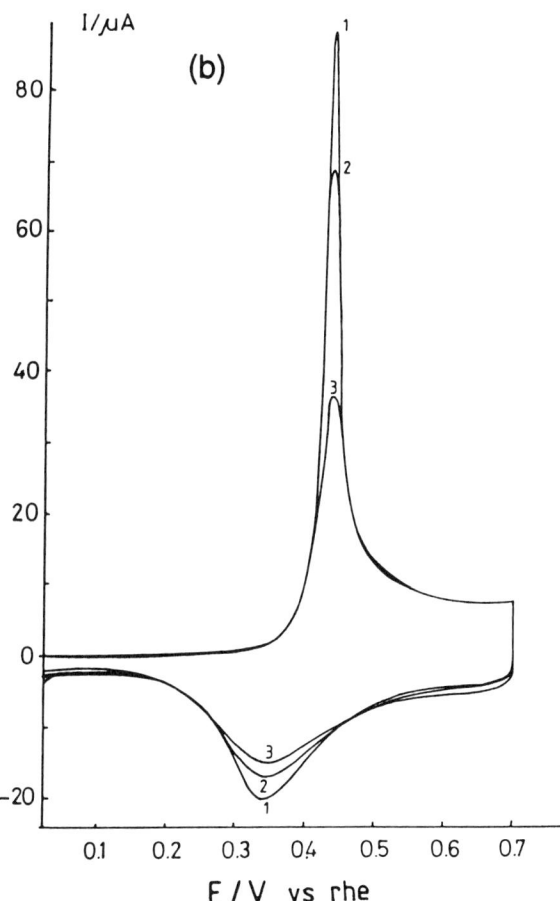

For instance, in Fig. 58a there is a deviation from the expected 60 mV shift of the peak potential as the HCl concentration is increased by an order of magnitude. This deviation can be ascribed to the simultaneous role of liquid junction potential and the effect of hydrogen ion activities, thus it can be eliminated by the use of hydrogen reference electrode immersed in the actual solutions, i.e., presenting the voltammetric response on rhe scale (Fig. 58b). In this presentation there is no shift in the peak potentials, which corresponds to one proton per electron process. The discussions of these problems have led to contradictory results and consequently to some confusion in the literature. For instance, the use of Hammett acidity functions, H_o [173], is unsuitable

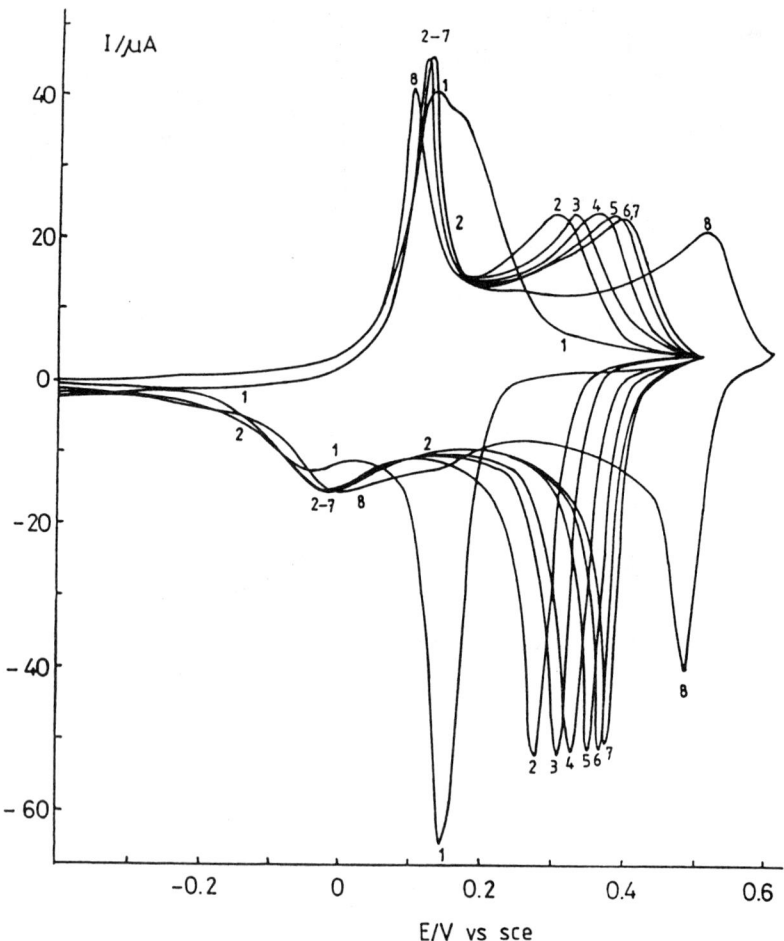

FIG. 59. Cyclic voltammograms of a PANI electrode in contact with acetic acid, sodium acetate + hydrochloric acid solutions of pH: (1) 4.65, (2) 3.4, (3) 3.1, (4) 2.95, (5) 2.75, (6) 2.58, (7) 2.48, (8) 1.6. Total concentration of CH_3COO^- + Cl^- anions is 0.6 mol/dm^3. Scan rate = 6 mV/s. (From Ref. 270, with permission of Elsevier Scientific Publishing Company.)

when one wants to correlate the peak potentials and the pH, due to the substantial junction potential effect. The effect of pH and electrolyte composition on the rate of the redox reaction has recently been highlighted by Peter and co-workers [74]. (Fast scan rate voltammetry at microelectrodes seems to be a useful method, because no degradation at the second wave occurs due to

Charge Transport in Polymer-Modified Electrodes

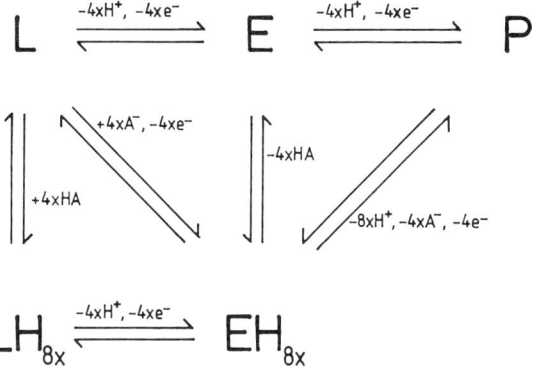

FIG. 60. Scheme of redox transformations and protonation equilibria of poly(aniline).

the relative slowness of the hydrolysis reaction [74,271].) The results suggest that acid/base equilibrium is reached in 1 mol/dm^3 H$_2$SO$_4$, although this may not be the case at higher pH values. The electroinactive fraction increases with increasing pH, which indicates that the emeraldine base is reduced more slowly than the emeraldine salt. The electroactivity of PANI is clearly influenced by the size of the anion. The size effect may be related to the smaller diffusion coefficient and mobility of the larger ions, but in the case of large anions, a pore size exclusion effect may also be relevant. Isotopic substitution (deuteration of sulfuric and hydrochloric acids) causes a shift in the peak potential of the first wave into the direction of more positive potentials, which may be explained by the higher mobility of H$^+$ compared D$^+$.

The larger overpotential observed on the first scan is probably due to the need for nucleating and growing emeraldine centers. Peter et al. [73–75] assumes a very rapid proton migration at short times, followed by equilibration of the oxidized phase by the diffusion-controlled movement of the acid. This would explain the time lag in anion movement observed in quartz microbalance [41] and radiotracer [67–72] studies of PANI, and in fact, it is in agreement with the models involving the formation of radical cation intermediate products. The fast proton migration may occur through different mechanisms such as transfer between nitrogen atoms along the polymer chain or prototropic motion through the structured layer of adsorbed water. There have been several attempts to measure the proton transport directly at PANI films [272,273].

Pfeiffer et al. [272] established that hydrogen and oxygen evolution reactions take place at the metal/polymer interface, therefore it allows a good estimation of the rate of proton migration and water diffusion in PANI films. A proton conductivity $X_{H+} > 1.6 \times 10^{-5}$ Ω^{-1}cm^{-1} was estimated. Although these measurements supply some information on the rate of these transport processes, their accuracy is insufficient. The formation of gas bubbles may cause an ohmic drop effect, and it may also destroy the film and the oxygen evolution at potentials where the film irreversibly decomposes.

Shimazu et al. [273] studied the proton transport accompanying the redox switching of PANI films using a long optical path length thin layer cell combined with an ion-selective FET sensor. Various acid-base indicators of low concentration were added to monitor the change in the pH of the solution spectroscopically. Absorbances at the absorption maxima of the acid and base forms and at the isobestic point of the indicator were recorded simultaneously during the potential cycling. The results provided evidence for the injection of protons into the poly(aniline) film upon reduction and the release of protons upon oxidation in $HClO_4$ + $LiClO_4$ solutions of pH -0.3 to 0.3. The proton concentration increased linearly with the charge which passed during the electrochemical transformation of the film, however, it remains constant at higher potentials where the capacitive current flows. By using a proton-selective FET sensor in the absence of an indicator in the solution, the potential and charge dependences of proton concentration obtained spectroscopically were reproduced. The proton per electron ratio linearly decreases with the solution pH value from 2 at pH 0.3 to ca. 0.2 at pH 3.

On the basis of EQCM measurements, Orata and Buttry [41] concluded that the initial oxidation process is accompanied by proton expulsion at low pH values, indicating the partial protonation of the PANI amine groups in the reduced form, with the loss of these protons on oxidation. This study also reveals the significance of anions. The incorporation of anions starts when a substantial portion of poly(aniline) film has already been oxidized, and this is the case even in very acidic solutions. The mass increase induced by the incorporation of anions and solvent molecules also continues in the double-layer charging region. The amount of solvent sorbed depends on the nature of anions and is larger for the strongly hydrated Cl^- ions than for sulfate ions. During the second oxidation process, a mass decrease was observed, indicating that anions together with protons decrease the film. These results were confirmed by Daifuku et al. [55], who measured the EQCM responses in aqueous $NaClO_4$ + $HClO_4$ solutions. However, in $LiClO_4$/acetonitrile solutions anion sorption was detected during oxidation at both redox peaks. The anion insertion takes place without solvent transfer.

Cordoba-Torresi et al. [59] studied the ion-exchange phenomenon in 1 mol/dm^3 HCl, HNO_3, and $HClO_4$ solutions by combined impedance/EQCM technique. They observed that the anion-exchange capacity is independent of the potential, but the cation-exchange capacity is larger at lower potentials, which is in accord with the results of other studies [273]. It was also concluded that the kinetics of proton expulsion and of anion incorporation processes may be different.

Bácskai et al. [274] also used EQCM technique to study the influence of pH and solution composition on the electrochemical behavior of PANI films. It was found that there is a significant accumulation of solution species in the film in all solution compositions and in the whole potential range studied. In 10 mol/dm^3 HClO$_4$ or H$_2$SO$_4$ solutions, the films become highly swollen, which eventually leads to the dissolution the films. However, a thin, compact electroactive layer remains on the platinum or gold substrate, which shows regular CV and EQCM behavior. Therefore, it was reasonably assumed that no chemical decomposition occurs, but physical dissolution of the less dense, outer part of the PANI film does. The thickness dependence of the swelling of the films was also explained by assuming the change of the film density (morphology) from a more compact structure near to the electrode surface to a less dense structure with increasing distance from the metal substrate [41,58]. (The film morphology strongly depends on the method of preparation [271,275], thus it is not surprising, that there are contradictions between the results reported for PANI films.) Significant mass changes during the oxidation of the leucoemeraldine to emeraldine form were observed only in the interval between pH 1 and 3, where the reduced poly(aniline) is practically unprotonated, while the half-oxidized material is protonated, which is in accordance with the scheme presented in Fig. 60.

Anion sorption accompanying the film formation and electrochemical transformations of PANI film has also been studied by an in situ radiotracer technique [67–72]. There is significant sorption of sulfate counterions even in the reduced film due to the partial protonation of the film in the pH range between 1 and 3. The sorbed labeled species can easily be exchanged for nonlabeled species. As one can see in Figs. 61 and 62, the anion sorption follows the film oxidation with some lag and continues to increase in the double-layer charging region.

As mentioned earlier, this type of behavior is characteristic of conducting polymer films. The results of the equilibrium sorption measurements attest that in contrast with redox polymer system, there is no saturation of the amount of sorbed counterions (Fig. 63). The reversible increase and decrease in the anion sorption at potentials where the second redox reaction takes place is connected with the deprotonation and dehydrogenation of the polymer film. A decrease in the amount of sorbed ions can also be observed in the case of

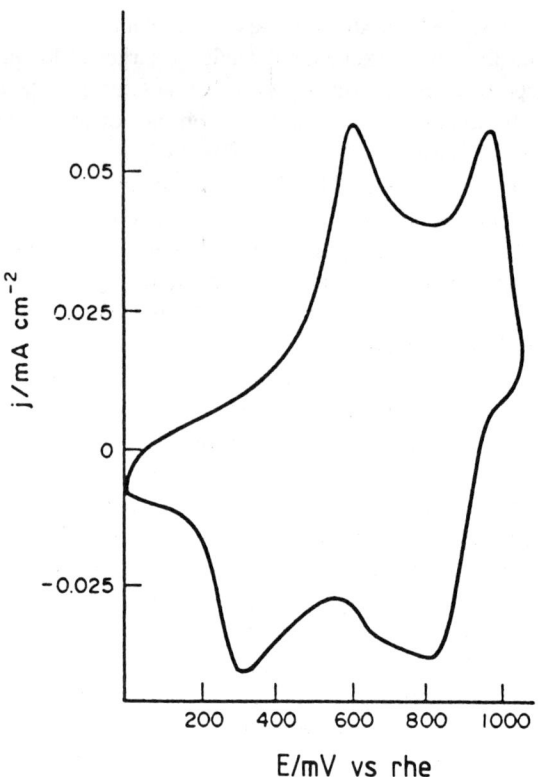

FIG. 61. Cyclic voltammogram of a PANI electrode in the presence of 10^{-2} mol/dm^3 labeled H_2SO_4. Scan rate: 0.4 mV/s. (From Ref. 67, with permission of Plenum Press plc.)

continuous potential cycling or potentiostatic experiments, when the upper potential limit is higher than the peak potential of the second wave, which is attributed to the overoxidation of the film (hydrolysis of quinoidal structure). The anion sorption increases with the concentration of the acid, indicating a higher degree of protonation of the film.

Barbero et al. [62,276] used the probe beam deflection (PBD) method to monitor the ion exchange between polymer film and bulk electrolyte during the electrochemical transformation of poly(aniline) and poly(N-methylaniline) (PNMAI) films. The cyclic voltammogram and the deflectogram obtained simultaneously for a PANI film in 1 mol/dm^3 HCl solution are shown in Fig. 64. The deflectogram can be explained by assuming a proton release preced-

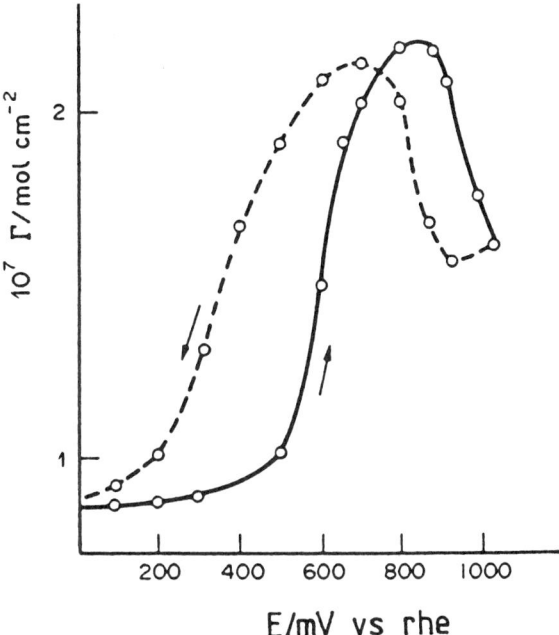

FIG. 62. Cyclic voltradiometric curve corresponding to the experimental conditions indicated in Fig. 61. (From Ref. 67, with permission of Plenum Press plc.)

ing the anion uptake during the oxidation scan and consecutive anion release, and proton insertion during the cathodic scan; this corresponds to the interpretation of the EQCM and radiotracer results. These investigations have revealed that the degree of protonation of PANI films depends on the nature of anions—at least this is the most reasonable explanation for the phenomena observed—when PANI films are oxidized in the presence of HCl, $HClO_4$ (1 mol/dm^3), and H_2SO_4 (0.5 mol/dm^3) solutions, respectively. In the case of Cl^- ion, the protons start to be involved at a pH of 0.5, while for HSO_4^- ion a pH of -0.5 is required. It was also demonstrated that the blocking of the imine nitrogen by methyl groups in PNMANI affects the second oxidation step, inasmuch as no deprotonation occurs, only anions are exchanged.

Genies et al. [156] have investigated the double-layer charging region of the cyclic voltammograms as a function of the anions and pH of the solution. The differential capacitance of PANI increases in order of $HClO_4$ < HCl < H_2SO_4 and is inversely proportional to the square root of proton concentration between pH = -1 and pH = 1. Different sites of PANI films are assumed to

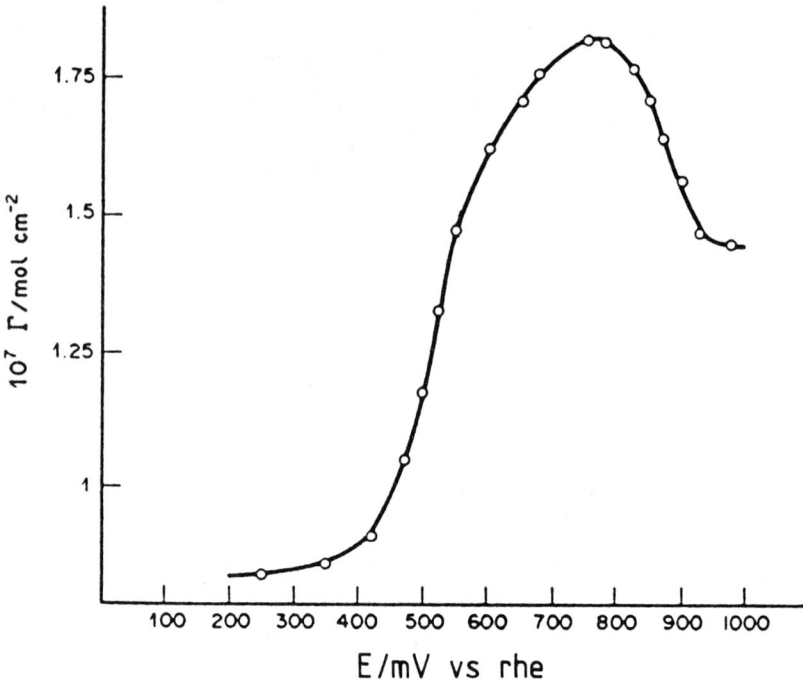

FIG. 63. Equilibrium (steady-state) sorption of counterions in a PANI film in contact with 10^{-2} mol/dm^3 labeled H$_2$SO$_4$. (From Ref. 67, with permission of Plenum Press plc.)

be responsible for the capacitive and faradaic processes. The capacitive process is reversible and independent of the pH, whereas the reversibility of the faradaic processes decreases as the pH of the solution increases.

Kalaji and Peter [34] have demonstrated that the simultaneous analysis of the periodic electrical and optical responses of poly(aniline) films offers a powerful approach to the characterization of the system and to establishing the distinction between the faradaic and nonfaradaic charging processes, inasmuch as the time-dependent absorbance changes can be related to the faradaic component of the total current. Apparently the ac response of PANI is dominated by faradaic processes, but the problem of the discrepancy between capacitance values derived from cyclic voltammograms and small-amplitude ac measurements remains unsolved.

Oyama and co-workers [183] have determined the charge transport diffusion coefficients and the heterogeneous rate constants for poly(*o*-phenyl-

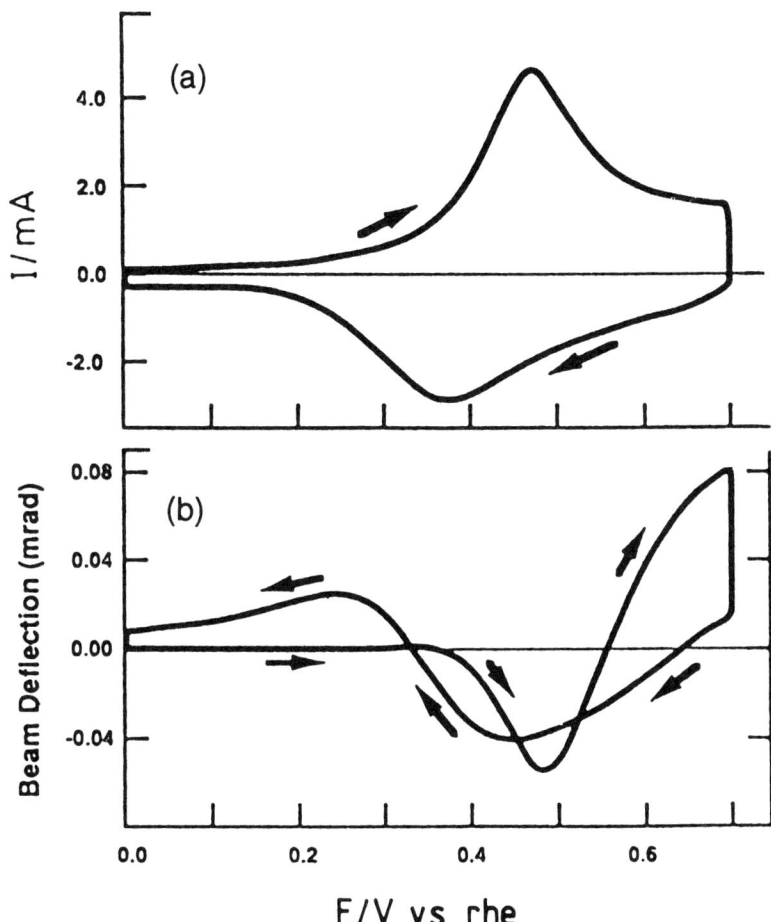

FIG. 64. Cyclic voltammogram (a) and cyclic deflectogram (b) for a poly(aniline) film ($\Gamma = 1.4 \times 10^{-8}$ mol/cm^2) in 1 mol/dm^3 HCl. Scan rate = 50 mV/s. (From Ref. 276, with permission of The Electrochemical Society, Inc.)

enediamine), poly(N-methylaniline), and poly(N-ethylaniline) films in pH 1 perchlorate solutions by normal pulse voltammetry. Somewhat surprisingly the normal pulse voltammograms are similar to those obtained for redox polymer films. The values for D_{ox} and D_{red} for each film are almost the same and range between 1×10^{-8} and 4×10^{-8} cm^2/s. The k_s values for these films are ca. 5×10^{-4} cm/s.

The temperature dependence of the poly(aniline) film voltammetric response in aqueous and nonaqueous media has been investigated by Inzelt [162]. Cyclic voltammograms obtained for PANI films in aqueous solution at different pH values as a function of temperature are shown in Figs. 65 and 66. Only a very slight increase of the peak currents with increasing temperature can be observed, i.e., the energy of activation, in this case, is much smaller than that observed for redox polymer films. Because a decrease in the rate of electronic conduction with temperature is expected, the results may be attributed to the increased rate of ionic charge transport, i.e., the motion of H_3O^+ and anionic counterions. It is also possible that two opposite effects—the diminishing electronic conduction and the enhancement of ionic charge transport—are balanced. The film stability increases with decreasing temperature, as one can see in Fig. 65; the film decomposition, due to the hydrolysis

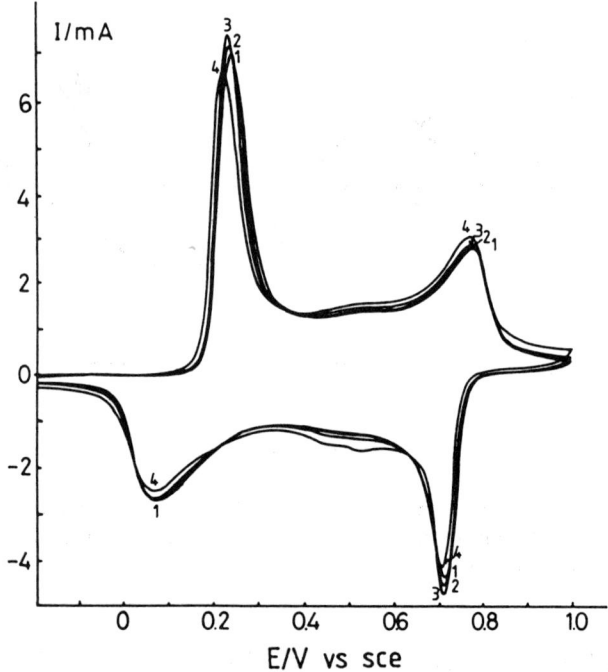

FIG. 65. Cyclic voltammograms of a poly(aniline) electrode in contact with 1 mol/dm^3 solution (pH 0) at different temperatures: (1) −1.5, (2) 4, (3) 13, and (4) 29°C. Sweep rate: 60 mV/s. (From Ref. 162, with permission of Elsevier Scientific Publishing Company.)

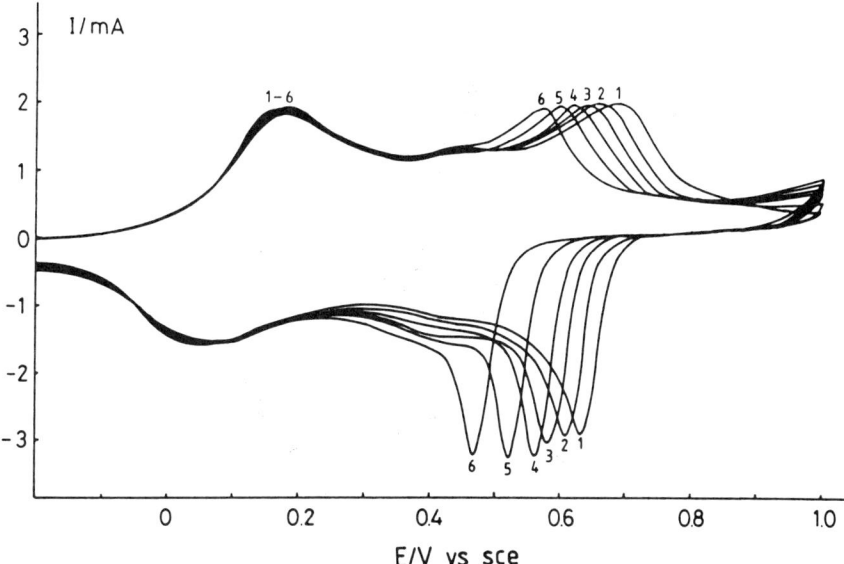

FIG. 66. Cyclic voltammograms of a poly(aniline) electrode in contact with 0.9 mol/dm^3 Na$_2$SO$_4$ + 0.2 mol/dm^3 NaHSO$_4$ solution (pH 2.65) at different temperatures: (1) 19, (2) 29, (3) 36, (4) 46, (5) 58, and (6) 66°C. Sweep rate: 60 mV/s. (From Ref. 162, with permission of Elsevier Scientific Publishing Company.)

of the reaction product of the second oxidation process, starts only at the highest temperature.

While there is practically no shift of the peak potentials with temperature at pH 0, the picture changes drastically at higher pH values, in that the peak potential of the second wave shifts to the direction of the more positive potentials as the temperature is increased (Fig. 66). This phenomenon is most probably due to the temperature dependence of protonation constants of the film functional groups. Because in the pH region 1–4 there is no participation of protons in the first oxidation step but the second oxidation step takes place with deprotonation, the variation of pK_a values will affect the position of the second wave only.

Cyclic voltammograms, obtained in moist acetonitrile media containing acid, attest that the electrochemical response is still fast at temperatures as low as −42°C (Fig. 67). This study has revealed that there is a change in sign of the reaction entropy at −18°C, which may indicate a phase transition or

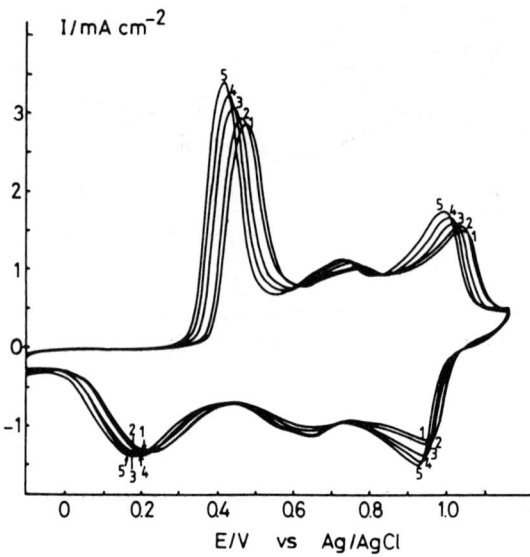

FIG. 67. Cyclic voltammograms of a poly(aniline) electrode in contact with moist acetonitrile containing 0.1 M H_2SO_4, 0.1 M TBAP, and 0.01 M HCl (water content 1.5%) at different temperatures: (1) -42, (2) -36.5, (3) -31, (4) -26.5, and (5) $-21°C$. Sweep rate: 25 mV/s. (From Ref. 162, with permission of Elsevier Scientific Publishing Company.)

morphological change of the film, but further studies are necessary to clarify the situation.

Depite the fact that by varying the temperature important activation and thermodynamic quantities can be determined, such studies are very scarce in the literature of polymer film electrodes [31,52,111,123,143,205,211–213, 216,236], and the temperature dependence study discussed above has been the only one carried out on conducting polymer film electrodes so far.

V. EFFECT OF FILM MORPHOLOGY ON CHARGE TRANSPORT IN POLYMERS

Film morphology, including changes with potential and experimental conditions, plays a major role in determining the rate of charge transport. We consider this question briefly from the aspect of theories elaborated for polymeric systems.

A. Theories of de Gennes on the Adsorption, Diffusion, and Charge Transport of Polymers

First we will deal with the works of de Gennes [95,277–281], which are very important in two respects: they provide a description of the structure of polymer layers adsorbed on solid surface [88,259–263] as well as a theory of charge transport in polymeric systems, including the transport through an adsorbed layer [279–281]. The adsorption model of de Gennes is based on the observation that in many cases the polymer sticks to the wall and cannot be desorbed by washing with the pure solvent. This is expected when the surface tension of the pure polymer melt is lower than the tension of the pure solvent.

Equilibrium concentration profiles calculated for a polymer solution near a solid surface consist of three distinct regions, as shown in Fig. 68. In the proximal region ($z \sim a < d$) the short-range forces between a monomer and the wall are important, where z is the distance from the wall, a^3 is the volume per monomer, and d is the layer thickness. In the central region ($d < z < \xi_b$)

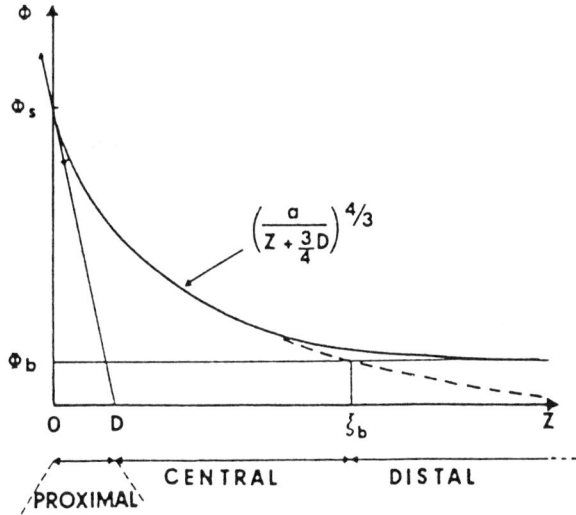

FIG. 68. Qualitative plot of a polymer volume fraction (Φ) vs. distance (z) from adsorbing wall. The extrapolation length (D) is fixed by the "free energy of sticking," γ_1, of monomers on the wall. Solid line: profile Φ (z) for a finite volume fraction Φ_b in the bulk solution. Dotted line: profile extrapolated to $\Phi_b = 0$. The two profiles coincide up to a distance ξ_b (the correlation length in bulk solution). (From Ref. 95, with permission of the American Chemical Society.)

the profile becomes independent of the bulk concentration of the polymer, and in the distal region ($z > \xi_b$) the concentration exponentially relaxes toward the bulk value. This description is of interest for the deposition of polymer films on electrodes, and in fact, such structure has been rendered likely in the case of electrochemical deposition of PVF [220] and poly(aniline) [158].

De Gennes also considered metastable adsorption, the subject of our present study, where the adsorbed polymer layer is in contact with a pure solvent. In this case the adsorption layer shrinks significantly, although the profile will strongly depend on the sample history. In the equilibrium case the layer thickness is scaled with the free energy of sticking, γ_1, as follows: $d \sim |\gamma_1|^{-3/2}$. For metastable adsorption the calculation of the profile is more complicated, but assuming a constant surface coverage, Γ, it is possible to minimize the free energy and estimate a density distribution. In fact, the behavior of several polymer film electrodes, e.g., PTCNQ [238], PVF [104, 220], and poly(aniline) [158,163], has been explained by the assumption of diminishing layer density from the metal surface. The reptation dynamics model [277] of de Gennes has been exploited in several works seeking to determine the diffusion of polymers at interfaces (see Ref. 282), most notably in recent works of Murray and co-workers [77,283] dealing with the determination of self-diffusion coefficient of a polymer in melt and solid state (vide infra).

De Gennes gave a scaling law for charge transport occurring in redox polymeric systems [280,281]. A relationship between the macroscopic conductance, Σ, and self-diffusion coefficient, D, was established, assuming intrachain conduction, σ_i, and jumping of electrons between two chains, σ_c:

$$\Sigma = \frac{\phi}{a^3} D = cD \tag{50}$$

where ϕ is the polymer volume fraction, a^3 is the unit volume, and $c = \phi/a^3$ is the concentrations of carriers.

Several cases were considered, taking into account the relative rates of σ_c and σ_i, which depend on the structure of the polymer layer described above. In the direct contact situation, the resistance, R_1, equals a^2/σ_1, where σ_1 describes transport between the metal and the monomers in direct contact with it. In the adsorbed layer $R_2 = a^2 \sigma^{-11/4}/\sigma_i$, and in the distal layer, when the σ_c becomes small, $R_3 = N\xi^2/\sigma_i$, where N is the number of monomer per chain and $\xi = a\phi^{-X}$ is the correlation length. Three major regimes were analyzed. The "free" regime, where the carrier jumps easily from one chain to the next,

$$\Sigma_f = \sigma_i a^{-1} \phi^2 \tag{51}$$

Charge Transport in Polymer-Modified Electrodes

the "captive" regime, when $\sigma_i >> \sigma_c$,

$$\Sigma_{capt} = a^{-1}N\phi^2\sigma_c \tag{52}$$

and the "semi-free" regime, which is an intermediate situation,

$$\Sigma_{sf} = (\sigma_i\sigma_c)^{1/2} a^{-1}\phi^{11/8} \tag{53}$$

De Gennes extended the theory for the case of gels, where $N = \infty$. Introducing a conductance between nodes, σ_n, and neglecting σ_c:

$$\Sigma = \frac{1}{\xi}\left(\frac{g}{\sigma_i} + \frac{2}{\sigma_n}\right)^{-1} \tag{54}$$

where g is the average distance between contacts (the average number of contacts per unit volume is $(\phi/g)a^{-3}$).

At present, we cannot use the results obtained by de Gennes because we have no information about σ_i, σ_c, or σ_n, and in many cases even ϕ is also unknown. In addition, it is valid only for a simple case, where charge carried by counterions can be neglected, i.e., this theory is inappropriate for ion-exchange polymers or for most of the fixed-site redox polymeric systems. However, we should not exclude the possibility that in the near future we will be able to apply this theory in the field of polymer film electrodes. For instance, the recent works of Murray and co-workers have already exploited some ideas of de Gennes.

B. Polymer Self-Diffusion

Recently, Murray and co-workers [283] took a large step in the understanding of the dynamics of polymeric motion. With the help of microelectrode-based solid-state voltammetry, the center-of-mass self-diffusion coefficient for ω-ferrocenecarboxamido-α-methoxy poly(ethylene oxide), Fc-MePEG (M_n = 2590), in its melt and solid state has been determined. The polymer chain diffusion coefficient $D_{Fc\text{-}MePEG}$ ranges from 3.3×10^{-8} cm^2/s at 89°C to 4×10^{-12} cm^2/s at room temperature, as shown in Fig. 69. The sharp change in D occurs at the melting temperature of the polymer ($T_m = 53.8$°C). Above T_m the polymer is in a highly viscous melt state, while below T_m the polymer is a partially crystalline wax. For the sake of comparison, the diffusion coefficients of ferrocene monomer, Fc and Fc-MePEG, in dimethyl poly(ethylene glycol), Me$_2$PEG, were also measured.

As Fig. 69 shows, the temperature dependence of D in all cases is similar qualitatively, and the Arrhenius plots give almost the same activation energy value, which is ca. 36 kJ/mol above T_m.

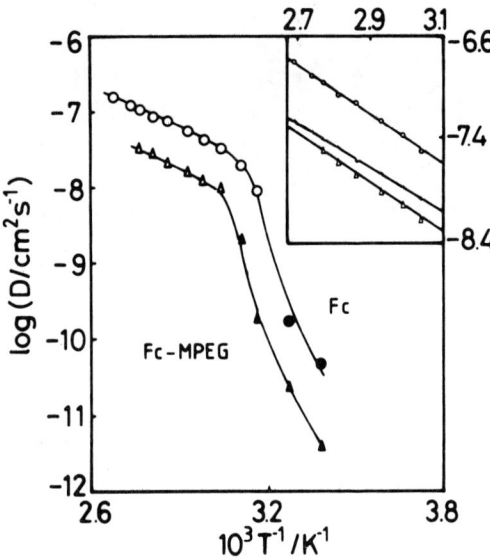

FIG. 69. Dependence of the apparent diffusion coefficient (D) on temperature for 53 mM ferrocene in Me$_2$PEG/LiClO$_4$ (circles) and bulk Fc-MePEG/LiClO$_4$ (triangles) mixtures. Filled symbols are for chronoamperometrically determined values; empty ones are for those obtained by using Eq. (37). The inset is an enlargement of the temperature plots above T_m and additional results for 42 mM Fc-MePEG diluted by Me$_2$PEG/LiClO$_4$ (+). (From Ref. 283, with permission of the American Chemical Society.)

The results shown in Fig. 69 indicate that the Fc-MePEG diffuses more slowly than the monomer ferrocene, 3.4 times when $T > T_m$ and 12 times when $T < T_m$. Later the authors corrected these values, taking into account the sublimation loss of ferrocene during the polymer solution preparation [284]. The monomer diffusion rate was redetermined with solutions of ferrocene carboxylic acid in unlabeled Me$_2$PEG. The monomer diffusion coefficient is, at $T > T_m$ and at 25°C, 14 and 19 times larger, respectively, than that of ferrocene-labeled polymer, as opposed to 3.4 and 12 times, as reported. The comparison between $D_{FcMePEG}$ and D_{Fc} provides evidence that $D_{FcMePEG}$ is the measure of self-diffusion of the polymer chain in FcMePEG.

The contribution of electron hopping to physical diffusion was estimated by using the Dahms-Ruff and Smoluchowski equations—Eqs. (5), (12), and (15). Assuming $\delta = 7.6 \times 10^{-8}$ cm and using $k_e = 7.8 \times 10^{10} \exp(-5.35$

Charge Transport in Polymer-Modified Electrodes 223

× 10^3/RT) [285], the calculation shows that the electron exchange reaction contributes less than 10% to the diffusion rates of Fc-MePEG and Fc dissolved at ca. 5 × 10^{-2} mol/dm^3 concentrations in unlabeled Me$_2$PEG at temperatures above T$_m$. (As discussed in Sec. III.B, the combination of Eqs. (5) or (6) with Eq. (12), and the use of Noyes expression, Eq. (13), is inappropriate. In addition, the estimation of the electron-hopping contribution is also affected by an error in k$_e$ value used for the calculation, because k$_e$ is probably less in viscous media [286,287] than in acetonitrile solution, and its dependence on temperature may also change.) D$_0$/D is, in contrast to the strong temperature dependence of D, virtually insensitive to temperature, as illustrated in Fig. 70. A further inspection reveals that D$_0$/D is actually not very sensitive to k$_e$, i.e., the electron-hopping contribution is controlled by the collision rate through D$_0$. In pure Fc-MePEG the electron exchange reaction might be appreciable, due to the 10-fold increase in the concentration of ferrocene sites. However, D values for the pure Fc-MePEG are lower than those for Fc-MePEG diluted by Me$_2$PEG, as seen in Fig. 70. This may be explained by a decrease of D$_0$, or, more likely, the actual value of k$_e$ is smaller than that used for the calculation.

The most important implication of this study is that during slow cyclic voltammetric or chronoamperometric experiments under favorable conditions, the polymer chains can diffuse to the metal surface within the time scale of the experiment, t$_{exp}$, i.e., the diffusional distance, d, is connected to D and t$_{exp}$ by the relation

$$d^2 = Dt_{exp} \tag{55}$$

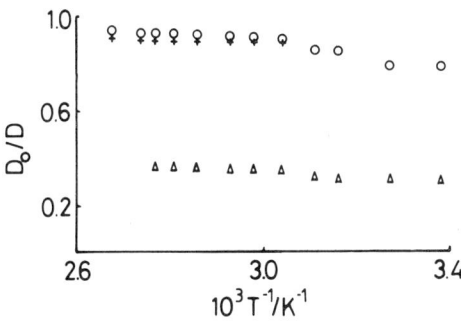

FIG. 70. Dependence of D$_0$/D on temperature for 53 mM ferrocene in Me$_2$PEG/LiClO$_4$ (○), pure Fc-MePEG/LiClO$_4$ (Δ), and 42 mM Fc-MePEG in Me$_2$PEG/LiClO$_4$ (+). (From Ref. 283, with permission of the American Chemical Society.)

Thus, when v = 5 mV/s and the potential interval of 500 mV, i.e., t_{exp} = 100 s, for D = 10^{-8} cm^2/s, d = 10^4 nm, and even for D = 10^{-12} cm^2/s, d_2 = 100 nm. At room temperature where $D_{Fc\text{-}MePEG}$ = 3.9 × 10^{-12} cm^2/s, the currents followed the Cottrell chronoamperometric relation over the experimental times of 30–100 s, so d ≅ 110–220 nm. (These calculations show that the condition d > > R_g is also met, i.e., the diffusional distance is much larger than the polymer radius of gyration, R_g, which is the requirement for the proper measurement of polymer self-diffusion.) Therefore, the possibility of direct electron uptake by redox groups of the polymers from the metal should be considered, when the redox polymer film containing a not too high molecular mass chain is highly swollen and the network is kept together by physical forces [31, 111]. It may also explain the incomplete film electroactivity observed for thick films and/or fast sweep rate in the case of several films.

Murray and co-workers [113–115,287] have carried out further solid-state voltammetric experiments on redox active molecules dissolved or attached to polymer electrolytes in order to investigate the coupling of physical diffusion and electron self-exchange reaction. Among other systems, they investigated a copolymer consisting of vinylferrocene and methoxy-nona (ethylene oxide)-methacrylate in which LiClO$_4$ was dissolved [114]. Reversible redox reaction in the bulk polymeric phase was observed for ferrocene sites in this copolymer. The amount of the oxidized ferrocene sites was ca. 800 times larger than the possible amount of ferrocene sites on the electrode surface. Because ferrocene sites are covalently fixed to the polymer backbone in the copolymers and their large displacement was assumed to be unlikely, the authors suggested that the redox reaction is brought about by the electron-hopping mechanism. Electron diffusivity at 40°C in the copolymer was estimated at ca. 10^{-10} cm^2/s. However, when LiTCNQ dissolved in poly(ethylene oxide) (PEO) containing LiClO$_4$ was studied [113,114], the results were interpreted on the basis of a mixed ion and electron conduction mechanism. Oxidation of TCNQ$^-$ in the PEO networks occurs via both physical diffusion of TCNQ$^-$ toward the electrode and the electron self-exchange between TCNQ$^-$ and TCNQ^0 in the diffusion layer. The electron diffusivity—calculated from the difference of the measured D values for TCNQ^0/TCNQ$^-$ and TCNQ$^-$/TCNQ^{2-} reactions—increases linearly with increasing LiTCNQ concentration up to ca. 0.1 mol/dm^3, i.e., it obeys Ruff's theory and surpasses the physical diffusivity at c_{LiTCNQ} = 0.05 mol/dm^3.

Finally, we mention the new results on the solvent dynamic effects on electron transfers [113,287], which clearly indicate that the dipolar relaxation times in polymer solvents are slower than those typical of the common monomeric fluid media. Consequently, the rate of electron exchange reactions in these systems may diminish.

C. Mechanical and Electrochemical Equilibria in Polymer Layers

As seen in Sec. V.A., the polymer film in contact with solution not containing the same polymer is in a metastable state. Assuming a constant Γ, a constrained equilibrium can be calculated, but we do not know whether even this equilibrium can be reached in a reasonable time scale [95]. It has also been recognized [64–70,218,219,236] that the different mobile species (ions, solvent molecules) do not reach equilibrium populations in the film and/or the redox transformation of the sites is not completed within the time scale of slow sweep rate experiments. Nevertheless, it seems to be useful to survey the equilibrium situation briefly.

The equilibrium condition involves an electrochemical equilibrium at the metal/film interface:

$$\tilde{\mu}_{ox} + \tilde{\mu}_e^M = \tilde{\mu}_{red} \tag{56}$$

where $\tilde{\mu}_{ox}$ and $\tilde{\mu}_{red}$ are the electrochemical potentials of the redox sites in the film and $\tilde{\mu}_e^M$ is that for the electrons in the metal. We should also state the equilibrium conditions for all mobile species,

$$\tilde{\mu}_i = \tilde{\mu}_i^s \tag{57}$$

where $\tilde{\mu}_i$ and $\tilde{\mu}_i^s$ are the electrochemical potentials of the i species in the film and solution, respectively. For any species i, the electrochemical potential is given by

$$\tilde{\mu}_i = \mu_i^0 + RT\ln \gamma_i c_i + z_i F\Phi + Pv_i \tag{58}$$

where μ_i^0 is the standard chemical potential, γ_i is the activity coefficient, c_i is the concentration, z_i is the ionic charge in signed units of the elementary charge, Φ is the electrostatic potential, P is the pressure relative to p^0, and v_i is the partial molar volume.

Bowden, Dautartas, and Evans [288–290] developed a model to describe the coupling of the mechanical and electrochemical thermodynamics of redox polymer films. They used the PVF electrode as a model system where the redox reaction is $ox^+A^- + e^- = red + A^-$. In this case, the potential difference between the metal and the solution can be expressed as follows:

$$\Delta \Phi = \Phi^M - \Phi = \frac{\mu_{ox}^0 - \mu_{red}^0 + \mu_e^{0M}}{F} + \frac{RT}{F} \ln(a_{ox}/a_{red})$$
$$+ \frac{\mu_A^0 - \mu_A^{OS}}{F} + \frac{RT}{F} \ln(a_A/a_A^s)$$
$$+ \frac{P(V_{ox} - V_{red})}{F} + \frac{V_A(P - P^S)}{F} \qquad (59)$$

Equation (59) is suitable for analyzing the effects of different contributions to the electrode potential. In most cases, the fourth term on the right-hand side of Eq. (59) is considered in order to find the Donnan-type equilibrium distribution of mobile ionic species between the film and the solution as a function of concentration or to establish the selectivity properties of the coating when more than one mobile counterions are present in the system [64, 65,148,149,236,291]. The fifth term in Eq. (59) is a mechanical work term, which is important when the partial molar volumes of the oxidized and reduced species in the film (V_{ox} and V_{red}) differ substantially from each other under conditions where the internal pressure of the film is significant. The last term in Eq. (59) is a mechanical work term, which depends on the pressure drop across the film/solution interface ($P - P^S$) and on the partial molar volume of the incorporated counterion (V_A). In a more general case, this work required for expanding and contracting the polymer network is in conjunction not only with the incorporation/expulsion of counterions but also with the swelling-deswelling caused by the sorption/desorption of solvent molecules. If so, Eq. (59) should be supplemented by additional terms expressing this contribution. In many cases, this contribution is coupled to the incorporation of ionic species through the hydration of counter- and co-ions, but it may also be uncoupled off the sorption of ions when the solvation of polymer chains and/or of their functional groups is more important. This model is oversimplified in other respects, as well. For instance, the Donnan-type contribution is also neglected, the activity coefficients are assumed to be independent of the redox state of the film, and the condition of the homogeneous distribution of sites may not valid, either. Nevertheless, the calculation carried out by Evans and co-workers [288–290] has so far been the only one dealing with the effect of the forced swelling of the surface polymer film. From a mechanical point of view, only the last term is considered important in the chosen model system, because the molar volume change between ferricenium and ferrocene is less than 2%. Also, only elastic deformations of the polymer network is discussed, which are reversibly coupled to the redox reaction. The plastic deformation of the film, which accompanies the break-in process, is not analyzed.

According to this mechanical/electrochemical model, the following relationship for the electrode potential as a function of fraction of oxidized sites, f, is valid:

$$E = E^{0\prime} + \frac{RT}{nF} \ln \frac{f}{(1-f)} + \frac{nV_A^2 cKf}{Fz^2} \tag{60}$$

where the third term accounts for the mechanical contribution containing the modulus of elasticity, K. The usefulness of the model was nicely demonstrated for plasma-polymerized [289] and spin-coated [290] PVF films.

Concerning the relationships between the measured potential and concentrations of redox centers, two important papers have appeared recently [292,293]. Fritsch-Faules and Faulkner demonstrated that potential measurements at a microelectrode array can give information on the spatial distribution of electrochemically generated redox species within a polymer film [292]. The technique developed is a very powerful one for determining concentration profiles, and unlike other methods, it does not require an interpretation of refractive indices, thus, one can determine the spatial distributions in highly complex systems. They also investigated the relationship between concentration and potential in QPVP films containing the $Fe(CN)_6^{3-/4-}$ redox couple [293]. They found a nearly nernstian behavior, which is somewhat surprising, considering the complex dynamics and thermodynamics of the system studied. Concentration profiles in these films were linear in spite of changes in film fluidity with the state of oxidation.

Rydzewski [294] examined the swelling and shrinking of a polyelectrolyte gel from a thermodynamical aspect. Three contributions to the free energy of the gel were considered: mixing of the constituents, network deformation, and electrostatic interactions. The most important results concerning our subject are related to the swelling of the gel as a function of the salt concentration in the bulk solution and of temperature. The increase of c_s or the decrease of T makes the gel shrink. Usually, the shrinking process occurs smoothly, but under certain conditions the process becomes discontinuous, as illustrated in Fig. 71. A tiny addition of salt leads to the collapse of the gel, i.e., a drastic decrease of volume to a fraction of its original value. The onset of shrinking and swelling strongly depends on temperature. This phenomenon is akin to thermodynamic phase transitions in other branches of physical chemistry.

The explanation given for the concentration and temperature dependence of the charge transport rate in PTCNQ and PVF films in Sec. IV.B is consistent with the predictions of this theory. The abrupt deterioration of the

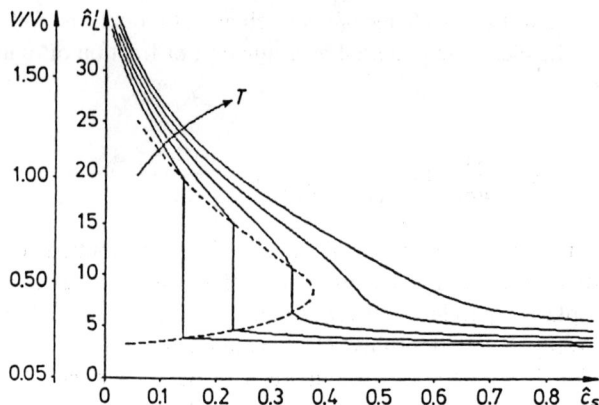

FIG. 71. Gel volume and number of solvent moles in the gel versus salt concentration of the bath. (From Ref. 294, with permission of Springer-Verlag.)

charge transport rate at high electrolyte concentrations, $c_s = 10$ mol/dm^3 (LiCl) or $c_s = 5$ mol/dm^3 (CaCl$_2$), now can be interpreted on the basis of Rydzewski's perceptions and of the free volume theory of diffusion. In the free volume theory, which is generally used to explain the permeability of polymer membranes, the diffusion coefficient can be expressed by

$$D_0 \sim \exp \frac{-v^*}{v_f} \tag{61}$$

where v^* is a characteristic volume required to accommodate the diffusing molecules in the sample and v_f is the free volume of solvent-filled cavities available for the mobile species in the film. Thus, an abrupt change in v_f causes a sharp decrease in D_0.

VI. CONCLUSION

Substantial developments have taken place in the field of polymer film electrodes over the last decade. New and interesting data have been collected about these systems, which, together with the new theories, provide a deeper insight into the nature of charge transport processes occurring in these electroactive surface films. Novel techniques, such as the in situ quartz crystal microbalance, radiotracer, probe beam deflection, luminescent indicator, optical methods, solid-state and fast sweep rate voltammetries, as well as impedance spectroscopy, have been especially useful in supplying new and reliable information. With these methods it became possible, for example, to determine the amount of sorbed ions and solvent molecules in the film, to separate—more or less successfully—the ion and electron transport processes, to recognize the linear relationship between the charge transport diffusion coefficient and the heterogeneous rate constant, to extend the limits of the measurements, especially in regard to the fast processes, to demonstrate the close relationship between the film morphology, including its changes with the experimental conditions, and the rate of the charge transport processes. However, it also became evident that the directly observable facts are usually the result of a number of complex interactions, in which the effects of the individual factors cannot be easily separated unequivocally.

The recent theories explain several relationships in detail, but none of them can be regarded as satisfactory in all respects. Much research is still needed to achieve a detailed understanding of transport phenomena occurring in polymer film electrodes. Essentially, the difficulties arise from the fact that the transport processes are related to the dynamic and static properties of several interacting molecules, which are confined in a polymer network. These properties, however, depend on the nature of the species forming the film and on the forces exerted by them on their environment. Because of the uncertainty of our knowledge about the structure and properties of polyelectrolyte layers (gels) and about the changes in their morphology caused by the potential and potential-induced processes (e.g., the incorporation of solvent molecules and ions) and temperature, the separate analysis of the particular effects and their relationships and, consequently, a reliable reconstruction of the reality by a theoretical treatment are extremely difficult. The interpretation of the transport phenomena in polymer films cannot be based on a single unified theory. The situation is further complicated by the fact that different experiments concerning the same property can lead to different and contradictory results, and sometimes even the same observations are explained in different ways. The uncertainties in the interpretation of the results are also reflected in this work.

The current theories of charge transport are based on a very simplified picture of the polymer films, which contains arbitrary assumptions. Nevertheless, they proved to be adequate for determining fundamental relationships in spite of their hypothetical nature. In the future, an even closer cooperation of electrochemists, polymer chemists, and physicists will be essential to clarify the unsolved problems and to set up more reliable theories about the mechanism of charge transport in these systems. The direction of future research should include the following areas.

1. One major problem is that the theory of polyelectrolytes is not yet developed to a level that would predict the behavior of the real systems of high complexity, especially in terms of quantitative relationships. Therefore, our task is to follow the achievements of polymer physics and chemistry and to adjust the theory of charge transport in polymer films to the new results. On the other hand, observations on polymer film electrodes may supply valuable information on the properties of polyelectrolytes. For instance, it is a unique opportunity that the amount of charge on the polymer chain can be controlled easily, and even the transition from a neutral state to a highly charged state can be carried out with no difficulty.

2. The morphology of the surface layer, and hence the rate of charge transport, is strongly related to the chemical nature of the polymer and to environmental conditions. Consequently, the experimental observations in regard to the morphology and its changes during the electrochemical transformation of the film are of utmost importance. Thus, the application of the in situ EQCM, radiotracer, optical measurements, etc., and the development of new techniques, which can give information in this respect, represent a challenge of vital importance in the near future.

3. Electrochemical methods, e.g., impedance spectroscopy and normal pulse voltammetry, which allow the simultaneous determination of the heterogeneous rate constant and the charge transport diffusion coefficient are important, because the relationship between k_s and D reveals a crucial aspect of the mechanism of charge transport and charge transfer processes. The linearity of the k_s vs. D functions found so far [17,31,78,111,188,196,211,212,214, 253,254] is perhaps the strongest evidence concerning the determining role of the chain and segmental motions; these processes govern both the site-site collisions within the film and the "collisions" between the metal substrate and the redox groups of the polymer. The recent diffusion studies in polymeric systems [113–115,283,287] and the application of SECM technique [83] also opened new vistas in this respect.

4. A crucial question is the determination of the meaning of D values measured experimentally by different techniques. The application of methods that may give separate information on the electronic and ionic charge transport processes, for example, luminescent indicator, tracer experiments, sandwich arrangements, etc.; and the comparison of D values obtained by techniques of large- and small-voltage perturbations, respectively, may help clarify the nature of the rate-determining step. In addition, the technical problem of the in situ thickness measurements should be solved in order to obtain reliable information on the redox site concentration and its variation during the charging/discharging processes, so as to extract correct D values

from the $D^{1/2}c$ products. The successful application of scanning electrochemical microscope for the in situ determination of film thickness [83] is perhaps the most important achievement in this respect. The SECM determination of polymer film thickness, while the film is in contact with an electrolyte solution, and thus it is in swollen state, is more straightforward than ellipsometry, where refractive index must be available, or profilometry, which is not applicable to easily deformed films.

5. The phenomenological theories, which can already be regarded as classical, lead to reliable relationships for many purposes on the basis of experimental data. However, a more thorough understanding of the processes, i.e., their molecular mechanism is also required. The main task of the molecular theory of the transport processes in thin electroactive films is to interpret the transport coefficients defined in phenomenological theories and determined experimentally. The appearance of new models [117,118,137, 141,142,153] on the charge transport in polymer films and the application of the advanced theories of electron transfer reaction [287] or of the polymer systems [283] indicate encouraging progress in this respect.

ACKNOWLEDGMENTS

I not only express my gratitude to my colleagues, especially to Professors J. Q. Chambers and G. Horányi, as well as to Drs. J. Bácskai and G. Láng, for their valuable contribution to our common research projects on polymer film electrodes, but I honor them even more for their friendship and for the stimulating atmosphere they created while working together during the last decade.

I acknowledge partial support from the National Scientific Research Fund (OTKA), Research Fund for Higher Education, Hungary (FFA), and the Soros Foundation throughout the writing of this chapter and throughout working on much of our own research described therein.

REFERENCES

1. A. Merz and A. J. Bard, J. Am. Chem. Soc. *100*:3222 (1978).
2. L. L. Miller and M. R. Van De Mark, J. Am. Chem. Soc. *100*:3223 (1978).
3. M. S. Wrighton, R. G. Austin, A. B. Bocarsly, J. M. Bolts, O. Haas, K. D. Legg, L. Nadjo, and M. C. Palazzotto, J. Electroanal. Chem. *87*:429 (1978).
4. R. Nowak, F. A. Schultz, M. Umana, H. Abruna, and R. W. Murray, J. Electroanal. Chem. *94*:219 (1978).
5. R. W. Murray, in *Electroanalytical Chemistry*, Vol. 13, (A. J. Bard, ed.), Marcel Dekker, New York, 1984, p. 191.

6. C. Ho, I. D. Raistrick, and R. A. Huggins, J. Electrochem. Soc. *127*:343 (1980).
7. D. R. Franceschetti and J. R. Macdonald, J. Electrochem. Soc. *129*:1754 (1982).
8. O. Contamin, E. Levart, G. Magner, R. Parsons, and M. Savy, J. Electroanal. Chem. *179*:41 (1984).
9. R. D. Armstrong, J. Electroanal. Chem. *198*:177 (1986).
10. C. Gabrielli, O. Haas, and H. Takenouti, J. Appl. Electrochem. *17*:82 (1987).
11. R. A. Bull, F.-R. F. Fan, and A. J. Bard, J. Electrochem. Soc. *129*:1009 (1982).
12. I. Rubinstein, J. Risphon, and S. Gottesfeld, J. Electrochem. Soc. *133*:729 (1986).
13. I. Rubinstein, E. Sabatini, and J. Risphon, J. Electrochem. Soc. *134*:3078 (1987).
14. S. H. Glarum and J. H. Marshall, J. Electrochem. Soc. *134*:142 (1987).
15. J. Tanguy, N. Mermilliod, and M. Hoclet, J. Electrochem. Soc. *134*:795 (1987).
16. R. D. Armstrong, B. Lindholm, and M. Sharp, J. Electroanal. Chem. *202*:69 (1986).
17. B. Lindholm, M. Sharp, and R. D. Armstrong, J. Electroanal. Chem. *235*:169 (1987).
18. T. B. Hunter, P. S. Tyler, W. H. Smyrl, and H. S. White, J. Electrochem. Soc. *134*:2198 (1987).
19. P. Fiordiponti and G. Pistoia, Electrochim. Acta *34*:215 (1989).
20. C. Deslouis, M. M. Musiani, and B. Tribollet, J. Electroanal. Chem. *264*:37, 57 (1989).
21. S. Panero, P. Prospieri, S. Passerini, B. Scrosati, and D. D. Perlmutter, J. Electrochem. Soc. *136*:3729 (1989).
22. R. Penner and C. R. Martin, J. Phys. Chem. *93*:984 (1989).
23. A. M. Waller and R. G. Compton, J. Chem. Soc. Faraday Trans. *85*:977 (1989).
24. W. J. Albery, Z. Chen, B. R. Horrocks, A. R. Mount, P. J. Wilson, D. Bloor, A. T. Monkman, and C. M. Elliot, Faraday Discuss. Chem. Soc. *88*:247 (1989).
25. W. J. Albery, C. M. Elliot, and A. R. Mount, J. Electroanal. Chem. *288*:15 (1990).
26. W. J. Albery and A. R. Mount, J. Electroanal. Chem. *305*:3 (1991).
27. F. Beck and P. Hülser, J. Electroanal. Chem. *280*:159 (1990).
28. K. Naoi, K. Ueyama, T. Osaka, and W. H. Smyrl, J. Electrochem. Soc. *137*:494 (1990).
29. G. Pourcelly, A. Oikonomou, C. Gavach, and H. D. Hurwitz, J. Electroanal. Chem. *287*:43 (1990).
30. G. Láng and G. Inzelt, Electrochim. Acta *36*:847 (1991).
31. G. Inzelt and G. Láng, Electrochim. Acta *36*:1355 (1991).

Charge Transport in Polymer-Modified Electrodes 233

32. C. Gabrielli, H. Takenouti, O. Haas, and A. Tsukada, J. Electroanal. Chem. *302*:59 (1991).
33. T. Osaka, T. Nakajima, K. Shiota, and T. Momma, J. Electrochem. Soc. *138*:2853 (1991).
34. M. Kalaji and L. M. Peter, J. Chem. Soc. Faraday Trans. *87*:853 (1991).
35. B. Lindholm, J. Electroanal. Chem. *289*:85 (1990).
36. M. M. Musiani, Electrochim. Acta *35*:1665 (1990).
37. D. A. Buttry, *Electroanalytical Chemistry* (A. J. Bard, ed.), Vol. 17, Marcel Dekker, New York, 1991, p. 1.
38. M. R. Deakin and D. A. Buttry, Anal. Chem. *61*:1147A (1989).
39. R. Schumacher, Angew. Chem. Int. Ed. Engl. *29*:329 (1990).
40. J. H. Kaufman, K. K. Kanazawa, and J. B. Street, Phys. Rev. Lett. *53*:2461 (1984).
41. D. Orata and D. A. Buttry, J. Am. Chem. Soc. *109*:3574 (1987).
42. P. T. Varineau and D. A. Buttry, J. Phys. Chem. *91*:1292 (1987).
43. D. Orata and D. A. Buttry, J. Electroanal. Chem. *257*:71 (1988).
44. S. Bruckenstein, C. P. Wilde, M. Shay, A. R. Hillman, and D. C. Loveday, J. Electroanal. Chem. *258*:457 (1989).
45. A. R. Hillman, D. C. Loveday, and S. Bruckenstein, J. Electroanal. Chem. *274*:157 (1989).
46. A. R. Hillman, D. C. Loveday, M. J. Swann, R. M. Eales, A. Hamnett, S. J. Higgins, S. Bruckenstein, and C. P. Wilde, Faraday Discuss. Chem. Soc. *88*:151 (1989).
47. S. Bruckenstein, C. P. Wilde, M. Shay, and A. R. Hillman, J. Phys. Chem. *94*:787 (1989).
48. S. Bruckenstein, A. R. Hillman, and M. J. Swann, J. Electrochem. Soc. *137*:1323 (1990).
49. A. R. Hillman, M. J. Swann, and S. Bruckenstein, J. Electroanal. Chem. *291*:147 (1990).
50. S. Servagent and E. Vieil, J. Electroanal. Chem. *280*:227 (1990).
51. G. Inzelt, J. Electroanal. Chem. *287*:171 (1990).
52. G. Inzelt and J. Bácskai, J. Electroanal. Chem. *308*:255 (1991).
53. J. Bácskai and G. Inzelt, J. Electroanal. Chem. *310*:379 (1991).
54. G. Inzelt and J. Bácskai, Electrochim. Acta *37*:647 (1992).
55. M. Daifuku, T. Kawagoe, N. Yamamoto, T. Ohsaka, and N. Oyama, J. Electroanal. Chem. *274*:313 (1989).
56. A. J. Kelly, T. Ohsaka, N. Oyama, R. J. Forster, and J. G. Vos, J. Electroanal. Chem. *287*:185 (1990).
57. A. J. Kelly and N. Oyama, J. Phys. Chem. *95*:9579 (1991).
58. J. Risphon, A. Redondo, C. Derouin, and S. Gottesfeld, J. Electroanal. Chem. *294*:73 (1990).
59. S. Cordoba-Torresi, C. Gabrielli, M. Keddam, H. Takenouti, and R. Torresi, J. Electroanal. Chem. *290*:269 (1990).
60. K. Naoi, M. Lien, and W. H. Smyrl, J. Electrochem. Soc. *138*:440 (1991).

61. C. K. Baker, Y.-J. Qiu, and J. R. Reynolds, J. Phys. Chem. 95:4446 (1991).
62. C. Barbero, M. C. Miras, O. Haas, and R. Kötz, J. Electroanal. Chem. 310:437 (1991).
63. G. Inzelt and G. Horányi, J. Electroanal. Chem. 200:405 (1986).
64. G. Inzelt, G. Horányi, J. Q. Chambers, and R. W. Day, J. Electroanal. Chem. 218:297 (1987).
65. G. Inzelt, G. Horányi, and J. Q. Chambers, Electrochim. Acta 32:757 (1987).
66. G. Inzelt and G. Horányi, J. Electroanal. Chem. 230:257 (1987).
67. G. Horányi and G. Inzelt, Electrochim. Acta 33:947 (1988).
68. G. Horányi and G. Inzelt, Electroanal. Chem. 257:311 (1988).
69. G. Inzelt and G. Horányi, J. Electrochem. Soc. 136:1747 (1989).
70. G. Horányi and G. Inzelt, J. Electroanal. Chem. 264:259 (1989).
71. A. V. Shlepakov, G. Horányi, G. Inzelt, and V. N. Andreev, Elektrokhimiya 25:1280 (1989).
72. V. E. Kazarinov, V. N. Andreev, M. A. Spytsin, and A. V. Shlepakov, Electrochim. Acta 35:899 (1990).
73. M. Kalaji, L. M. Peter, L. M. Abrantes, and J. C. Mesquita, J. Electroanal. Chem. 274:289 (1989).
74. M. Kalaji, L. Nyholm, and L. M. Peter, J. Electroanal. Chem. 313:271 (1991).
75. C. P. Andrieux, P. Audebert, P. Hapiot, M. Nechtschein, and C. Odin, J. Electroanal. Chem. 305:153 (1991).
76. P. G. Pickup, W. Kutner, C. R. Leider, and R. W. Murray, J. Am. Chem. Soc. 106:1991 (1984).
77. N. A. Surridge, J. C. Jernigan, E. F. Dalton, R. P. Buck, M. Watanabe, H. Zhang, M. Pinkerton, T. T. Wooster, M. L. Longmire, J. S. Facci, and R. W. Murray, Faraday Discuss. 88:1 (1989).
78. E. F. Dalton and R. W. Murray, J. Phys. Chem. 95:6383 (1991).
79. A. J. Bard, F.-R. F. Fan, D. T. Pierce, P. R. Unwin, D. O. Wipf, and F. Zhou, Science 254:68 (1991).
80. C. Lee and A. J. Bard, Anal. Chem. 62:1906 (1990).
81. J. Kwak and F. C. Anson, Anal. Chem. 64:250 (1992).
82. C. Lee and F. C. Anson, Anal. Chem. 64:528 (1992).
83. M. V. Mirkin, F.-R. F. Fan and A. J. Bard, Science 257:364 (1992).
84. M. Fujihira, *Topics in Organic Electrochemistry* (A. J. Fry and W. E. Britton, eds.), Plenum Press, New York, 1986, p. 255.
85. A. R. Hillman, *Electrochemical Science and Technology of Polymers-1* (R. G. Linford, ed.), Elsevier Applied Science, London, 1987, pp. 103, 241.
86. H. D. Abruna, Coord. Chem. Rev. 86:135 (1988).
87. A. F. Diaz, J. F. Rubinson, and H. B. Mark, Adv. Polym. Sci. 84:113 (1988).
88. M. Kaneko and D. Wöhrle, Adv. Polym. Sci. 84:143 (1988).
89. G. P. Evans, *Advances in Electrochemical Science and Engineering* (H. Gerischer and C. W. Tobias, eds.) Vol. 1, VCH Press, Weinheim, 1990, p. 1.
90. R. W. Murray (ed.), *Molecular Design of Electrode Surfaces*, J. Wiley, New York, 1992.

Charge Transport in Polymer-Modified Electrodes 235

91. R. W. Murray, private communication.
92. E. Laviron, *Electroanalytical Chemistry* (A. J. Bard, ed.), Vol. 12, Marcel Dekker, New York, 1982, p. 53.
93. L. R. Faulkner, Electrochim. Acta *34*:1699 (1989).
94. A. J. Bard and L. R. Faulkner, *Electrochemical Methods*, John Wiley, New York, 1980.
95. P. G. de Gennes, Macromolecules *14*:1637 (1981).
96. V. G. Levich, *Advances of Electrochemistry* (P. Delahay, ed.), Vol. 4, 1966, p. 314.
97. H. Dahms, J. Phys. Chem. *72*:362 (1968).
98. I. Ruff and V. J. Friedrich, J. Phys. Chem. *75*:3297 (1971).
99. I. Ruff, V. J. Friedrich, K. Demeter, and K. Csillag, J. Phys. Chem. *75*:3303 (1971).
100. I. Ruff and V. J. Friedrich, J. Phys. Chem. *76*:162 (1972).
101. I. Ruff and V. J. Friedrich, J. Phys. Chem. *76*:2957 (1972).
102. C. P. Andrieux and J. M. Savéant, J. Electroanal. Chem. *111*:377 (1980).
103. E. Laviron, J. Electroanal. Chem. *112*:1 (1980).
104. P. J. Peerce and A. J. Bard, J. Electroanal. Chem. *114*:89 (1980).
105. F. B. Kaufman and E. M. Engler, J. Am. Chem. Soc. *101*:547 (1979).
106. L. Botár and I. Ruff, Chem. Phys. Lett. *126*:348 (1986).
107. I. Ruff and L. Botár, J. Chem. Phys. *83*:1292 (1985).
108. R. P. Buck, J. Electroanal. Chem. *243*:279 (1988).
109. R. P. Buck, J. Phys. Chem. *92*:4196 (1988).
110. D. A. Buttry and F. C. Anson, J. Electroanal. Chem. *130*:333 (1981).
111. G. Inzelt and L. Szabó, Acta Chim. Hung. *126*:67 (1989).
112. G. Inzelt, Electrochim. Acta *34*:83 (1989).
113. M. Watanabe, T. T. Wooster, and R. W. Murray, J. Phys. Chem. *95*:4573 (1991).
114. M. Watanabe, H. Nagasaka, K. Sanui, N. Ogata, and R. W. Murray, Electrochim. Acta *37*:1521 (1992).
115. T. T. Wooster, M. L. Longmire, H. Zhang, M. Watanabe, and R. W. Murray, Anal. Chem. *64*:1132 (1992).
116. N. A. Surridge, M. E. Zvanut, F. R. Keene, C. S. Sosnoff, M. Silver, and R. W. Murray, J. Phys. Chem. *96*:962 (1992).
117. F. C. Anson, D. N. Blauch, J. M. Savéant, and C.-F. Shu, J. Am. Chem. Soc. *113*:1922 (1991).
118. D. N. Blauch and J. M. Savéant, J. Am. Chem. Soc. *114*:3323 (1992).
119. D. A. Buttry and F. C. Anson, J. Am. Chem. Soc. *105*:685 (1983).
120. K. Aoki, K. Tokuda, H. Matsuda, and N. Oyama, J. Electroanal. Chem. *176*:139 (1984).
121. N. Oyama, S. Yamaguchi, Y. Nishiki, K. Tokuda, and F. C. Anson, J. Electroanal. Chem. *139*:371 (1982).
122. K. Shigehara, N. Oyama, and F. C. Anson, J. Am. Chem. Soc. *103*:2552 (1981).

123. S. Nakahama and R. W. Murray, J. Electroanal. Chem. *158*:303 (1983).
124. L. Roullier and E. Laviron, J. Electroanal. Chem. *157*:193 (1983).
125. G. Inzelt, J. Farkas, and L. Dobos, Acta Chim. Hung. *120*:73 (1985).
126. G. Inzelt, E. Fekete, and L. Szabó, Acta Chim. Hung. *125*:435 (1988).
127. W. T. Yap and R. A. Durst, J. Electroanal. Chem. *216*:11 (1987).
128. W. Richtering and K. Doblhofer, Electrochim. Acta *34*:1685 (1989).
129. C. D. Paulse and P. G. Pickup, J. Phys. Chem. *92*:7002 (1988).
130. S. W. Feldberg, J. Electroanal. Chem. *198*:1 (1986).
131. J. R. Miller and J. Beitz, J. Chem. Phys. *78*:6746 (1981).
132. J. R. Miller, K. W. Hartman, and S. Abrash, J. Am. Chem. Soc. *104*:4296 (1982).
133. R. K. Huddleston and J. R. Miller, J. Chem. Phys. *79*:5337 (1982).
134. C. A. Naleway, L. A. Curtiss, and J. R. Miller, J. Phys. Chem. *95*:8434 (1991).
135. M. Smoluchowksi, Phys. Z. *17*:557 (1916).
136. I. Rubinstein, J. Electroanal. Chem. *188*:227 (1985).
137. P. He and X. Chen, J. Electroanal. Chem. *256*:353 (1988).
138. X. Chen, P. He, and L. R. Faulkner, J. Electroanal. Chem. *222*:223 (1987).
139. M. Sharp, B. Lindholm, and E.-L. Lind, J. Electroanal. Chem. *274*:35 (1989).
140. W. J. Vining and T. J. Meyer, J. Electroanal. Chem. *237*:191 (1987).
141. C. P. Andrieux and J. M. Savéant, J. Phys. Chem. *92*:6761 (1988).
142. I. Fritsch-Faules and L. R. Faulkner, J. Electroanal. Chem. *263*:237 (1989).
143. S.-M. Oh and L. R. Faulkner, J. Am. Chem. Soc. *111*:5613 (1989).
144. E. Leiva, P. Meyer, and W. Schmickler, J. Electrochem. Soc. *135*:1993 (1988).
145. J. M. Savéant, J. Electroanal. Chem. *242*:1 (1988).
146. J. S. Facci, R. H. Schmehl, and R. W. Murray, J. Am. Chem. Soc. *104*:4959 (1982).
147. R. P. Buck, J. Electroanal. Chem. *210*:1 (1986).
148. R. P. Buck, J. Electroanal. Chem. *219*:23 (1987).
149. R. P. Buck, J. Phys. Chem. *92*:6445 (1988).
150. R. P. Buck, J. Electroanal. Chem. *258*:1 (1989).
151. R. P. Buck, J. Electroanal. Chem. *271*:1 (1989).
152. C. J. Baldy, C. M. Elliot, and S. W. Feldberg, J. Electroanal. Chem. *283*:53 (1990).
153. L. Srinivasa Mohan and M. V. Sangaranarayanan, J. Electroanal. Chem. *323*:375 (1992).
154. C. E. D. Chidsey and R. W. Murray, J. Phys. Chem. *90*:1479 (1986).
155. S. W. Feldberg, J. Am. Chem. Soc. *106*:4671 (1984).
156. E. M. Genies, J. F. Penneau, and E. Vieil, J. Electroanal. Chem. *283*:205 (1990).
157. T. Kobayashi, H. Yoneyama, and H. Tamura, J. Electroanal. Chem. *177*:281 (1984).

158. C. M. Carlin, L. J. Kepley, and A. J. Bard, J. Electrochem. Soc. *132*:353 (1986).
159. R. C. M. Jakobs, L. J. J. Janssen, and E. Barendrecht, Recl. Trav. Chim. Pays-Bas *103*:275 (1984).
160. N. Mermilliod, J. Tanguy, M. Hoclet, and A. A. Syed, Synth. Met. *18*:359 (1987).
161. J. C. Lacroix, K. K. Kanazawa, and A. Diaz, J. Electrochem. Soc. *136*:1308 (1989).
162. G. Inzelt, J. Electroanal. Chem. *279*:169 (1990).
163. S. H. Glarum and J. H. Marshall, J. Electrochem. Soc. *134*:2160 (1987).
164. M. Kaya, A. Kitani, and K. Sasaki, Chem. Lett. Chem. Soc. Japan 147 (1986).
165. J. C. Scott, P. Pfluger, M. T. Krounbi, and G. B. Street, Phys. Rev. B *28*:2140 (1983).
166. M. E. Jozefowicz, R. Laversanne, H. H. S. Javadi, A. J. Epstein, J. P. Pouget, X. Tang, and A. G. MacDiarmid, Phys. Rev. B *39*:12958 (1989).
167. J. M. Ginden, A. J. Epstein, and A. G. MacDiarmid, Solid State Commun. *72*:987 (1989).
168. A. G. MacDiarmid and A. J. Epstein, Faraday Discuss. Chem. Soc. *88*:317 (1989).
169. T. A. Skotheim (ed.), *Handbook of Conducting Polymers*, Vols. 1 and 2, Marcel Dekker, New York, 1986.
170. J.-L. Bredas and G. B. Street, Acc. Chem. Res. *18*:309 (1985).
171. E. M. Geniés and M. Lapkowski, J. Electroanal. Chem. *236*:199 (1987).
172. E. M. Geniés and M. Lapkowski, Synth. Met. *24*:61 (1988).
173. W.-S. Huang, B. D. Humphrey, and A. G. MacDiarmid, J. Chem. Soc. Faraday Trans. *82*:2385 (1986).
174. B. Villeret and M. Nechstein, Phys. Rev. Lett. *63*:1285 (1989).
175. M. Lapkowski and E. M. Geniés, J. Electroanal. Chem. *279*:157 (1990).
176. F. Genoud, M. Guglielmi, M. Nechstein, E. M. Geniés, and M. Salmon, Phys. Rev. Lett. *55*:118 (1985).
177. M. Nechstein, F. Devreux, F. Genoud, E. Vieil, J. M. Pernaut, and E. M. Geniés, Synth. Met. *15*:59 (1986).
178. F. Li and W. J. Albery, J. Chem. Soc. Faraday Trans. *87*:2949 (1991).
179. M. E. G. Lyons, Faraday Discuss. Chem. Soc. *88*:291 (1989).
180. F. T. A. Vork, B. C. A. M. Schuermans, and E. Barendrecht, Electrochim. Acta *35*:567 (1990).
181. H. Mao, J. Ochmanska, C. D. Paulse, and P. G. Pickup, Faraday Discuss. Chem. Soc. *88*:165 (1989).
182. J.-C. LaCroix and A. F. Diaz, J. Electrochem. Soc. *135*:1457 (1988).
183. K. Chiba, T. Ohsaka, and N. Oyama, J. Electroanal. Chem. *217*:239 (1987).
184. S. Gottesfeld, A. Redondo, I. Rubinstein, and S. W. Feldberg, J. Electroanal. Chem. *265*:15 (1989).

185. K. Aoki, J. Electroanal. Chem. *310*:1 (1991).
186. C. R. Martin, I. Rubinstein, and A. J. Bard, J. Am. Chem. Soc. *104*:4817 (1982).
187. H. S. White, J. Leddy, and A. J. Bard, J. Am. Chem. Soc. *104*:4811 (1982).
188. J. Leddy and A. J. Bard, J. Electroanal. Chem. *189*:203 (1985).
189. J. G. Gaudiello, P. K. Ghosh, and A. J. Bard, J. Am. Chem. Soc. *107*:3027 (1985).
190. A. M. Hodges, O. Johansen, J. W. Loder, A. W.-H. Mau, J. Rabani, and W. H. F. Sasse, J. Phys. Chem. *95*:5966 (1991).
191. G. Inzelt, R. W. Day, J. F. Kinstle, and J. Q. Chambers, J. Phys. Chem. *87*:4592 (1983).
192. Y.-M. Tsou and F. C. Anson, J. Phys. Chem. *89*:3818 (1985).
193. M. Majda and L. R. Faulkner, J. Electroanal. Chem. *137*:149 (1982).
194. M. Majda and L. R. Faulkner, J. Electroanal. Chem. *169*:77, 97 (1984).
195. E. T. T. Jones and L. R. Faulkner, J. Electroanal. Chem. *222*:201 (1987).
196. T. Ohsaka, N. Oyama, K. Sato, and H. Matsuda, J. Electrochem. Soc. *132*:1871 (1985).
197. R. Lange, K. Doblhofer, and W. Storck, Electrochim. Acta *33*:385 (1988).
198. N. Oyama and F. C. Anson, J. Electrochem. Soc. *127*:640 (1980).
199. K. Shigehara, N. Oyama, and F. C. Anson, Inorg. Chem. *20*:518 (1981).
200. T. Inoue and F. C. Anson, J. Phys. Chem. *91*:1519 (187).
201. K. Doblhofer and R. Lange, J. Electroanal. Chem. *216*:241 (1987).
202. K. Doblhofer and R. Lange, J. Electroanal. Chem. *229*:239 (1987).
203. K. Niwa and K. Doblhofer, Electrochim. Acta *31*:549 (1986).
204. K. Doblhofer, H. Braun, and R. Lange, J. Electroanal. Chem. *206*:93 (1986).
205. S.-M. Oh and L. R. Faulkner, J. Electroanal. Chem. *269*:77 (1989).
206. N. Oyama, T. Ohsaka, M. Kaneko, K. Sato, and H. Matsuda, J. Am. Chem. Soc. *105*:6003 (1983).
207. F. C. Anson, J.-M. Savéant, and K. Shigehara, J. Am. Chem. Soc. *105*:1096 (1983).
208. F. C. Anson, T. Ohsaka, and J.-M. Savéant, J. Phys. Chem. *87*:640 (1983).
209. M. E. G. Lyons, H. G. Fay, J. G. Vos, and A. J. Kelly, J. Electroanal. Chem. *250*:207 (1988).
210. O. Haas, J. Rudnicki, F. R. McLarnon, and E. J. Cairns, J. Chem. Soc. Faraday Trans. *87*:939 (1991).
211. R. J. Forster and J. G. Vos, J. Electroanal. Chem. *314*:135 (1991).
212. R. J. Forster and J. G. Vos, Electrochim. Acta *37*:159 (1992).
213. P. Daum, J. R. Lenhard, D. R. Rolison, and R. W. Murray, J. Am. Chem. Soc. *102*:4649 (1980).
214. R. J. Forster and J. G. Vos, J. Electrochem. Soc.*139*:1503 (1992).
215. P. Daum and R. W. Murray, J. Phys. Chem. *85*:389 (1981).
216. J. Q. Chambers and G. Inzelt, Anal. Chem. *57*:1117 (1985).
217. G. Inzelt and L. Szabó, Electrochim. Acta *31*:1381 (1986).

218. A. R. Hillman, D. C. Loveday, and S. Bruckenstein, J. Electroanal. Chem. *300*:67 (1991).
219. A. R. Hillman, D. C. Loveday, M. J. Swann, S. Bruckenstein, and C. P. Wilde, J. Chem. Soc. Faraday Trans. *87*:2047 (1991).
220. A. R. Hillman, D. C. Loveday, and S. Bruckenstein, Langmuir *7*:191 (1991).
221. T. Kawai, C. Iwakura, and H. Yoneyama, J. Electrochem. Soc. *137*:2667 (1990).
222. A. H. Schroeder and F. B. Kaufman, J. Electroanal. Chem. *113*:209 (1980).
223. F. B. Kaufman, A. H. Schroeder, E. M. Engler, S. R. Kramer, and J. Q. Chambers, J. Am. Chem. Soc. *102*:483 (1980).
224. J. Q. Chambers, F. B Kaufman, and K. H. Nichols, J. Electroanal. Chem. *142*:277 (1982).
225. G. Inzelt, J. Q. Chambers, and F. B. Kaufman, J. Electroanal. Chem. *159*:443 (1983).
226. C. Degrand and L. L. Miller, J. Electroanal. Chem. *132*:136 (1982).
227. M. Fukui, A. Kitani, C. Degrand, and L. L. Miller, J. Am. Chem. Soc. *104*:28 (1982).
228. B. L. Funt and P. M. Hoang, J. Electroanal. Chem. *154*:229 (1983).
229. P. M. Hoang, S. Holdcroft, and B. L. Funt, J. Electrochem. Soc. *132*:2129 (1985).
230. R. W. Day, G. Inzelt, J. F. Kinstle, and J. Q. Chambers, J. Am. Chem. Soc. *104*:6804 (1982).
231. G. Inzelt, R. W. Day, J. F. Kinstle, and J. Q. Chambers, J. Electroanal. Chem. *161*:147 (1984).
232. G. Inzelt, J. Q. Chambers, J. F. Kinstle, and R. W. Day, J. Am. Chem. Soc. *106*:3396 (1984).
233. G. Inzelt, J. Q. Chambers, J. F. Kinstle, R. W. Day, and M. A. Lange, Anal. Chem. *56*:301 (1984).
234. G. Inzelt, J. Bácskai, J. Q. Chambers, and R. W. Day, J. Electroanal. Chem. *201*:301 (1986).
235. G. Inzelt, L. Szabó, J. Q. Chambers, and R. W. Day, J. Electroanal. Chem. *242*:265 (1988).
236. G. Inzelt and J. Q. Chambers, J. Electroanal. Chem. *266*:265 (1989).
237. P. Joo and J. Q. Chambers, J. Electrochem. Soc. *132*:1345 (1985).
238. H. Karimi and J. Q. Chambers, J. Electroanal. Chem. *217*:313 (1987).
239. C. V. Francis, P. Joo, and J. Q. Chambers, J. Phys. Chem. *91*:6315 (1987).
240. R. W. Day, H. Karimi, C. V. Francis, and J. Q. Chambers, J. Polym. Sci. Polym. Chem. Ed. *24*:645 (1986).
241. C. P. Andrieux, L. Nadjo, and J. M. Savéant, J. Electroanal. Chem. *26*:147 (1970).
242. E. Laviron, J. Electroanal. Chem. *146*:15 (1983).
243. K. M. O'Connell, E. Waldner, L. Roullier, and E. Laviron, J. Electroanal. Chem. *162*:77 (1984).

244. E. Laviron, J. Electroanal. Chem. *169*:29 (1984).
245. G. Inzelt, Acta Chim. Hung. *126*:611 (1989).
246. M. Nagy, Colloid Polym. Sci. *263*:245 (1985).
247. K. Dusek and W. Prins, Adv. Polym. Sci. *6*:1 (1969).
248. F. Castaneda and V. Plichon, J. Electroanal. Chem. *233*:77 (1987).
249. F. Castaneda and V. Plichon, J. Electroanal. Chem. *236*:163 (1987).
250. G. C. S. Collins and D. J. Schiffrin, J. Electrochem. Soc. *132*:1835 (1985).
251. D. C. Bookbinder and M. S. Wrighton, J. Am. Chem. Soc. *102*:5123 (1980).
252. D. C. Bookbinder and M. S. Wrighton, J. Electrochem. Soc. *130*:1080 (1983).
253. N. Oyama, T. Ohsaka, H. Yamamoto, and M. Kaneko, J. Phys. Chem. *90*:3850 (1986).
254. T. Ohsaka, H. Yamamoto, M. Kanako, A. Yamada, M. Nakamura, S. Nakamura, and N. Oyama, Bull. Chem. Soc. Jpn. *57*:1844 (1984).
255. Y.-M. Tsou, H.-Y. Liu, and A. J. Bard, J. Electrochem. Soc. *135*:1669 (1988).
256. A. F. Diaz, J. I. Castillo, J. A. Logan, and W.-Y. Lee, J. Electroanal. Chem. *129*:115 (1981).
257. J. H. Kaufman, N. Colaneri, J. C. Scott, and G. B. Street, Phys. Rev. Lett. *53*:1005 (1984).
258. M. J. Nowak, D. Spiegel, F. Hoppa, A. J. Heeger, and P. A. Pincus, Macromolecules *22*:2917 (1989).
259. A. Hamnett, Faraday Discuss. Chem. Soc. *88*:291 (1989).
260. E. M. Geniés, G. Bidan, and A. F. Diaz, J. Electroanal. Chem. *149*:101 (1983).
261. P. Burgmayer and R. W. Murray, J. Am. Chem. Soc. *88*:2515 (1984).
262. B. J. Feldman, P. Burgmayer, and R. W. Murray, J. Am. Chem. Soc. *107*:872 (1985).
263. J. B. Schlenoff and J. C. W. Chien, J. Am. Chem. Soc. *109*:6269 (1987).
264. R. J. Waltman and J. Bargon, Can. J. Chem. *64*:76 (1986).
265. P. Marque, J. Roncali, and F. Garnier, J. Electroanal. Chem. *218*:107 (1987).
266. A. G. MacDiarmid, J.-C. Chiang, M. Halpern, W.-S. Huang, S.-L. Mu, N. L. D. Somasiri, W. Wu, and S. T. Yaniger, Mol. Cryst. Liq. Cryst. *121*:173 (1985).
267. T. Kobayashi, H. Yoneyama, and H. Tamura, J. Electroanal. Chem. *177*:293 (1984).
268. P. M. McManus, R. J. Cushman, and S. C. Yang, J. Phys. Chem. *91*:744 (1987).
269. S. W. Feldberg and I. Rubinstein, J. Electroanal. Chem. *240*:1 (1988).
270. G. Inzelt and G. Horányi, Electrochim. Acta *35*:27 (1990).
271. H. Yang and A. J. Bard, J. Electroanal. Chem. *339*:423 (1992).
272. B. Pfeiffer, A. Thyssen, and J. W. Schultze, J. Electroanal. Chem. *260*:393 (1989).
273. K. Shimazu, K. Murakoshi, and H. Kita, J. Electroanal. Chem. *277*:347 (1990).

274. J. Bácskai, V. Kertész, and G. Inzelt, Electrochim. Acta *38*:393 (1993).
275. Y.-T. Kim, H. Yang, and A. J. Bard, J. Electrochem. Soc. *138*:L71 (1991).
276. C. Barbero, M. C. Miras, O. Haas, and R. Kötz, J. Electrochem. Soc. *138*:669 (1991).
277. P. G. de Gennes, *Scaling Concepts in Polymer Physics*, Cornell University Press, Ithaca, NY, 1979.
278. P. G. de Gennes, Nature *282*:367 (1979).
279. P. G. de Gennes, C. R. Acad. Sci. Paris *301*:1399 (1985).
280. P. G. de Gennes, C. R. Acad. Sci. Paris *302*:1 (1986).
281. P. G. de Gennes, Physica *138A*:206 (1986).
282. S. J. Whitlow and R. P. Wool, Macromolecules *24*:5926 (1991).
283. M. J. Pinkerton, Y. LeMest, H. Zhang, M. Watanabe and R. W. Murray, J. Am. Chem. Soc. *112*:3730 (1990).
284. M. J. Pinkerton, Y. LeMest, H. Zhang, M. Watanabe and R. W. Murray, J. Am. Chem. Soc. *112*:8217 (1990).
285. R. M. Nielson, G. E. McManis, L. K. Sanford, and M. J. Weaver, J. Phys. Chem. *93*:2152 (1989).
286. X. Zhang, H. Yand, and A. J. Bard, J. Am. Chem. Soc. *109*:1916 (1987).
287. H. Zhang and R. W. Murray, J. Am. Chem. Soc. *113*:5183 (1991).
288. E. F. Bowden, M. F. Dautartas, and J. F. Evans, J. Electroanal. Chem. *219*:49 (1987).
289. M. F. Dautartas, E. F. Bowden, and J. F. Evans, J. Electroanal. Chem. *219*:71 (1987).
290. E. F. Bowden, M. F. Dautartas, and J. F. Evans, J. Electroanal. Chem. *219*:91 (1987).
291. K. Doblhofer and R. D. Armstrong, Electrochim. Acta *33*:453 (1988).
292. I. Fritsch-Faules and L. R. Faulkner, Anal. Chem. *64*:1118 (1992).
293. I. Fritsch-Faules and L. R. Faulkner, Anal. Chem. *64*:1127 (1992).
294. R. Rydzewski, Continuum Mech. Thermodyn. *2*:77 (1990).

SCANNING ELECTROCHEMICAL MICROSCOPY

Allen J. Bard, Fu-Ren F. Fan, and Michael V. Mirkin

University of Texas at Austin
Austin, Texas

I. Introduction 244
   A. Principles 244
   B. Modes of application 247
II. Instrumentation 251
   A. Basic apparatus 251
   B. Tip position modulation 253
   C. Tip preparation 255
III. Theory 268
   A. SECM with a disk-shaped tip in the feedback mode of operation 268
   B. Generation/collection mode of SECM operation 287
   C. Processes with first- and second-order homogeneous reactions in the gap 293
   D. Adsorption/desorption processes 298
   E. SECM with a nondisk tip and tip shape characterization 303
IV. Applications 310
   A. SECM images 310
   B. Heterogeneous electron transfer and reaction rate imaging 319
   C. Studies of homogeneous chemical reactions and adsorption/desorption processes 331
   D. Fabrication 339
   E. Potentiometric and other tips 348
   F. Characterization of thin films and membranes 349
V. Conclusions 365
   Abbreviations 365
   List of Symbols 366
   References 370

I. INTRODUCTION

This chapter is devoted to a rather complete discussion of scanning electrochemical microscopy (SECM; this abbreviation is also used for the device, i.e., the microscope)—its principles, theory, instrumentation, and applications (electrochemical measurements, imaging, fabrication). SECM is one of the scanning probe microscopies (SPM), in which three-dimensional images of surfaces are obtained by scanning a small tip across a surface and recording an appropriate response [1]. The best known types of SPM are scanning tunneling microscopy (STM), in which the tunneling current between an atomically sharp tip and a conductive or semiconductive substrate is measured, and atomic force microscopy (AFM), in which the force between a sharp tip and a surface is monitored. SECM is also related to a variety of other techniques, for example, those in which a small reference electrode probe is used to study potential distributions near surfaces, e.g., in studies of corrosion [2,3], or in which small metal tips probe biological structures, e.g., to measure nerve action potentials or detect neurotransmitters [4]. As we will show, however, SECM differs from these in its high spatial resolution, its ability to image surfaces of various kinds, and its applicability to quantitative measurements (e.g., kinetic ones) through well-developed and fairly rigorous theory.

A. Principles

To understand the operation and response of the SECM, it is necessary to review briefly the behavior of a very small electrode (e.g., a disk with a radius, a, below about 10–20 μm), sometimes called an ultramicroelectrode (UME), in an electrochemical cell because the scanning tip in SECM is an UME. More detailed discussions of UMEs are available [5–7]. Consider an UME immersed in a solution containing an electrolyte and a reducible species, O, as well as auxiliary and reference electrodes (Fig. 1). When a potential sufficiently negative of the standard potential for the reaction $O + ne \rightleftarrows R$ is applied to the UME, the reduction of O occurs at the UME at a diffusion-controlled rate and a cathodic current passes. The current decays as a diffusion layer of O builds up around the electrode and rather quickly (i.e., in a time of the order of tens of a^2/D) attains a steady-state value, $i_{T,\infty}$, that depends upon the concentration of O, c, and its diffusion coefficient, D, and is given by

$$i_{T,\infty} = 4nFDca \qquad (1)$$

where F is the Faraday. The steady-state current results from the constant flux of O to the electrode surface driven by an expanding, essentially hemispheri-

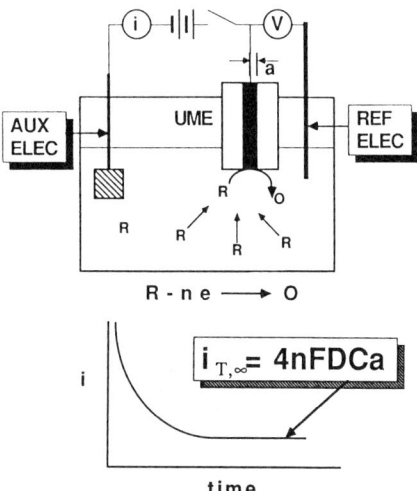

FIG. 1. (Top) Schematic of cell for ultramicroelectrode (UME) voltammetry. (Bottom) Current-time transient for potential step with UME showing attainment of steady-state current, $i_{T,\infty}$.

cal, diffusion layer around the electrode. Note that the steady-state diffusive flux of O to the UME (in mol/sec/cm^2), of the order Dc/a, can be rather large at small electrodes and increases as a decreases. Indeed, at UMEs, this flux is frequently larger than the convective flux to a rotating disk electrode or an electrode in a stirred solution. This means that measurements at a UME, and by extension at an SECM tip, are relatively immune to stirring or other convective effects. In SECM, it is the perturbation of the UME current when the UME is brought to within a few tip diameters of a surface (usually termed "the substrate") that constitutes the SECM response [8].

Consider first when the UME (or "tip") is brought close to an insulating substrate (Fig. 2C) [9]. The steady-state current flowing through the tip, i_T, will now be smaller than $i_{T,\infty}$ because the insulating substrate partially blocks the diffusion of O to the tip. Clearly, the closer the tip is to the insulator substrate, the smaller i_T will be, with $i_T \rightarrow 0$ as the distance between tip and substrate, d, approaches zero. This effect is sometimes termed "negative feedback." However, when the tip is close to a conductive substrate at which the reaction R → O + ne can occur, then a flux of O from substrate to tip occurs, in addition to some flux of O to the tip from the bulk solution (Fig. 2B). This results in $i_T > i_{T,\infty}$. This flux of O from substrate to tip causes an increase of i_T as d decreases; this is termed "positive feedback." In this case,

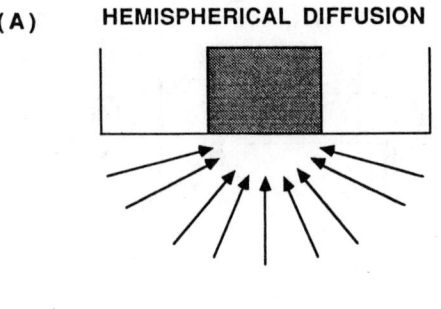

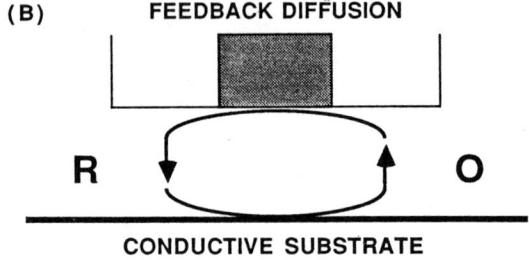

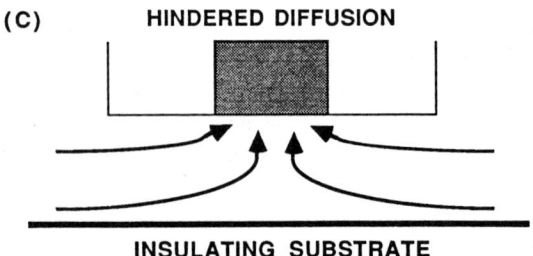

FIG. 2. Basic principles of scanning electrochemical microscopy (SECM): (A) far from the substrate, diffusion leads to a steady-state current, $i_{T,\infty}$; (B) near a conductive substrate, feedback diffusion leads to $i_T > i_{T,\infty}$; (C) near an insulating substrate, hindered diffusion leads to $i_T < i_{T,\infty}$. (Reprinted with permission from Ref. 9. Copyright 1990 American Chemical Society.)

the tip can be seen as both the generator of the signal sensing the substrate (the flux of the reduced species, R) and the detector (of the flux of O from the substrate), i.e., roughly a kind of "electrochemical radar." Thus, the magnitude of i_T compared to $i_{T,\infty}$ is governed by the nature of the substrate and by the tip-substrate spacing. The actual general situation can be somewhat more

Scanning Electrochemical Microscopy 247

complicated than the limiting cases described above, when the rate of the R → O reaction on the substrate is governed by the rate of heterogeneous electron-transfer kinetics rather than the rate of mass transfer (diffusion) of R to the substrate. This situation will be dealt with in Sec. IV.B.

In addition to the amperometric feedback mode described above, other SECM modes of operation are possible. For example, in the substrate generator/tip collector mode, the tip current, i_T, is used to monitor the flux of electroactive species from the substrate. If the substrate is an electrode, this flux is generated by the substrate current, i_s, and i_T/i_s represents the collection efficiency of the tip. Indeed, several early applications of UMEs as probes above an electrode surface, e.g., to determine concentration profiles above a substrate electrode, involved this configuration [10,11]. Alternatively, the substrate can monitor (collect) the flux from the tip, e.g., in studies of homogeneous reactions that occur in the tip-substrate gap (see Sec. IV.C). Finally, other types of tips, e.g., potentiometric probes or enzyme electrodes, are possible.

B. Modes of Application

The SECM can be used in a variety of ways: (1) as an electrochemical tool to study electrode processes and coupled homogeneous reactions, (2) as an imaging device (microscope), and (3) for fabrication at high resolution.

1. The SECM as an Electrochemical Tool

In applying the SECM to the study of electrode reactions, the x-y scanning feature is often not used. Measurements with the SECM in this mode combine many of the features of ultramicroelectrodes and thin layer electrochemistry [12] with a number of advantages. For example, as described in more detail below, the "characteristic flux" to an UME spaced a distance, d, from a conductive substrate is of the order of Dc/d, independent of the tip radius, a, when d < a. This means that very high fluxes, and thus high currents, can be obtained, even for tips of rather large (i.e., 1–10 μm) radii. Applications based on this feature to the determination of heterogeneous electron-transfer reaction rates are discussed below.

The SECM can also serve, in the generator/collector mode described above, in the same manner as the rotating ring disk electrode (RRDE) [13]. The SECM approach has the advantage that different substrates can be examined easily, i.e., without the need to construct rather difficult to fabricate RRDEs, and that higher interelectrode fluxes are available without the need to rotate the electrode or otherwise cause convection in the solution. Moreover, in the tip generation/substrate collection mode, the "collection efficiency" (i_s/i_T) in the absence of perturbing homogeneous chemical reactions is near

100%, compared to significantly lower values in practical RRDEs. Finally, although transient SECM measurements are possible, as discussed below, most applications have involved steady-state currents, which are easier to measure and are not perturbed by factors like double-layer charging and faradaic processes of adsorbed materials.

It is useful to consider the tiny electrode area and gap volume probed in SECM. Consider a tip of radius a = d/2 spaced at a distance d from a conductive substrate. Table 1 shows the approximate areas, volumes, and quantities of reactant in the interelectrode gap for different values of a and c [14]. Clearly, the SECM allows one to make quantitative electrochemical measurements in very small domains with small amounts of material. The measurement is made possible by the positive feedback, which acts as an amplifier of the tip signal because of repeated cycling of electroactive molecules between tip and substrate. This factor also suggests that the SECM may be useful in accelerated testing of stability and for analysis in very small volumes [15].

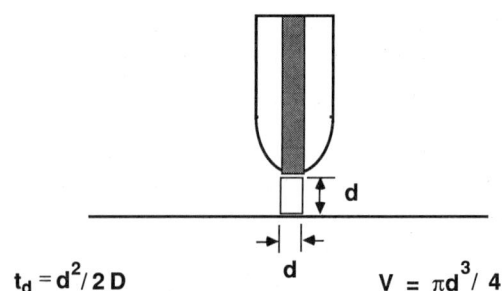

$t_d = d^2/2D$ d $V = \pi d^3/4$

TABLE 1

Approximate Volume (V) and Transit Time (t_d) Between Tip and Substrate in Solution Volume Below Tip at Different Values of d and c

d	V (cm^3)	Contents (molecules)		Transit time
		c = 0.01 M	c = 0.1 M	
10 μm	10^{-9}	4.5×10^9	4.5×10^{10}	0.1 sec
1 μm	10^{-12}	4.5×10^6	4.5×10^7	1 msec
0.1 μm	10^{-15}	4,500	4.5×10^4	10 μsec
10 nm	10^{-18}	4.5	45	0.1 μsec

Scanning Electrochemical Microscopy

2. The SECM as an Imaging Tool

As described in Sec. IV.A, the SECM can be employed to image the surfaces of different types of substrates, both insulators and conductors, immersed in solutions. In this application, the solution redox probe (mediator) and electrolyte must be chosen to be compatible with the type of surface and the surface processes to be probed. For example, if a particular surface species or substrate S_R is to be oxidized, the redox potential of the mediator pair, O/R, must be more positive than that of the pair S_O/S_R, so that the reaction $S_R + O \rightarrow S_O + R$ occurs spontaneously and rapidly. Generally, the mediator pairs are chosen to be those that show rapid, usually one-electron, heterogeneous reactions at the tip. A wide range of suitable redox couples, which are often also used in other applications (e.g., enzyme redox processes), are available for both aqueous and nonaqueous media and span a wide range of potential. Some typical examples are given in Table 2.

A particular advantage of SECM in imaging applications, compared to other types of scanning probe microscopy, is that the response observed is based on fairly rigorous theory (as discussed in Sec. III), and hence the measured current can be employed to estimate the tip-substrate distance. Thus, even when images are made in the "constant height" mode, i.e., where the tip is lowered to a fixed distance from the substrate and it is then scanned in the x-y plane with variations in tip currents used to image the substrate, the tip currents can be readily converted to a height or distance scale.

The resolution attainable with SECM depends upon the tip radius, a. The highest resolution reported to date is of the order of 30–50 nm. It is unlikely that SECM could ever attain the atomic resolution found with STM and AFM because of the difficulty in fabricating suitably insulated Å-size SECM tips and because the tunneling current will dominate when small tips are brought sufficiently close to a conductive substrate to observe an SECM response.

3. The SECM as a Fabrication Tool

As discussed in Sec. IV.D, the SECM can be employed to carry out high-resolution fabrication on surfaces using variants of conventional electrochemical approaches, i.e., electrodeposition, electroplating, and etching. For example, the tip can be lowered into a polymer film containing a metal ion (e.g., Ag^+) and metal structures can be formed by electrodeposition of metal while the tip is moved in a desired pattern. Substrates, e.g., metal or semiconductors, can be etched by generation of a suitable oxidant (such as Br_2) while the tip is moved near the substrate surface. SECM etching can also be carried out by making the substrate the anode and moving the cathode tip near it in a kind of ultramicroelectrochemical machining. As with imaging, the resolution attainable in fabrication is of the order of the tip diameter.

TABLE 2
Selected Oxidation Reduction Mediators in Aqueous Solution (pH 7)[a]

Mediator	Potential (V vs. NHE)
$Ru(bpy)_3^{3+/2+}$	1.27
$Ru(phen)_3^{3+/2+}$	1.22
Br_2/Br^-	1.09
$Fe(phen)_3^{3+/2+}$	1.07
$Fe(bpy)_3^{3+/2+}$	1.07
$IrCl_6^{2-/3-}$	1.00
$Ru(bpy)_2(NH_3)_2^{3+/2+}$	0.88
$Ru(CN)_6^{3-/4-}$	0.86
$Os(bpy)_3^{3+/2+}$	0.84
$Mo(CN)_8^{3-/4-}$	0.77
1,1-Dicarboxylic acid ferrocene	0.64
$Co(oxalate)_3^{3-/4-}$	0.57
$W(CN)_8^{3-/4-}$	0.49
Ferrocene$^{+/0}$	0.44
$Co(phen)_3^{3+/2+}$	0.38
CoEDTA$^{-/2-}$	0.38
$Fe(CN)_6^{3-/4-}$	0.36
$Co(bpy)_3^{3+/2+}$	0.32
1,4-Benzoquinone/hydroquinone	0.28
N,N,N',N'-Tetramethyl-p-Phenylenediamine (TMPD)	0.27
$Ru(en)_3^{3+/2+}$	0.18
1,2-Naphthoquinine/hydroquinone	0.14
FeEDTA$^{-/2-}$	0.12
$Ru(NH_3)_6^{3+/2+}$	0.05
Methylene blue	0.01
$Co(en)_3^{3+/2+}$	−0.22
Anthraquinone-2-sulfonate/hydroquinone	−0.22
Methyl viologen (+2/+)	−0.45
$Co(sepalchrate)^{3+/2+}$	−0.54
4,4′-Dimethyl-1,1′-trimethylene-2,2′-bipyridyl	−0.69

[a]More extensive listings of mediator compounds, especially for use with biological systems in aqueous solutions are given in W. Clark, *Oxidation-Reduction Potentials of Organic Systems*, Williams and Wilkins Co., Baltimore, 1960; M. L. Fultz and R. A. Durst, Anal. Chim. Acta *140*:1 (1982); R. Szentrimay, P. Yeh, and T. Kuwana, in *Electrochemical Studies of Biological Systems*, D. Sawyer, ed., ACS, Washington, DC, 1977, p. 143; P. N. Bartlett, P. Tebbutt, and R. G. Whitaker, Prog. Reaction Kinetics, *16*:55 (1991); and A. J. Bard, R. Parsons, and J. Jordan, *Standard Potentials in Aqueous Solution*, Marcel Dekker, New York, 1985.

II. INSTRUMENTATION

A. Basic Apparatus

The SECM instrument basically consists of a combination of familiar electrochemical components (cell, potentiostat) and those used in STM for manipulating a tip at high resolution (piezoelectrics, drivers) and acquiring the data (interface, computer). A block diagram of the SECM instrument is shown in Figure 3. It consists of four major components: tip movement and position controller, electrochemical cell (including tip, substrate, counterelectrode, and reference electrode), bipotentiostat/potential programmer, and computer/interface/display system [16].

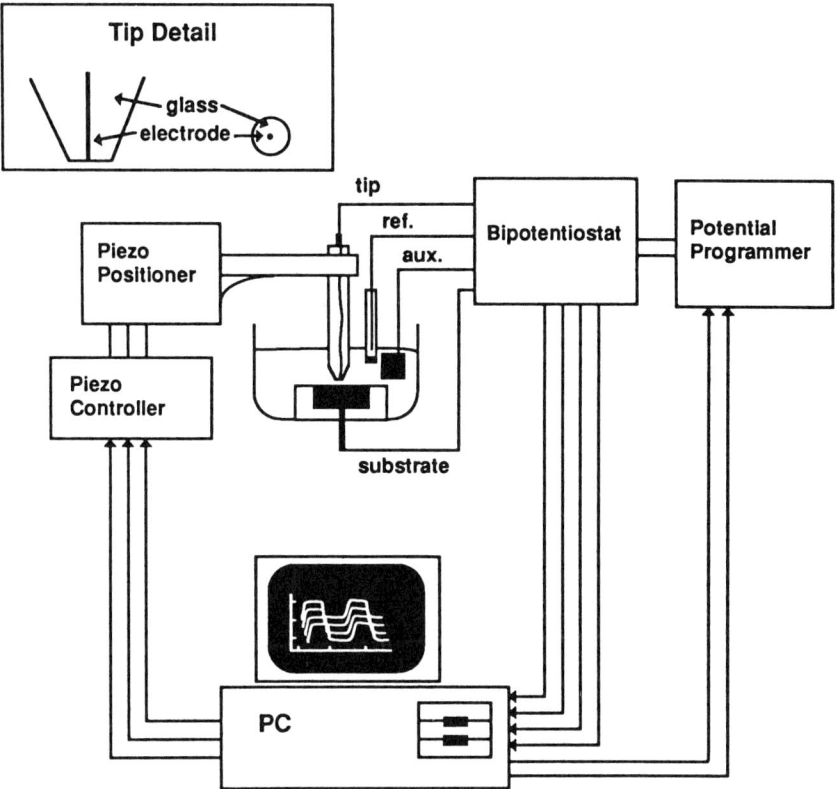

FIG. 3. Block diagram of the SECM apparatus. (Reprinted with permission from Ref. 9. Copyright 1990 American Chemical Society.)

In the apparatus currently used in this laboratory, the tip electrode is held on a piezoelectric pusher (PZT-44, Burleigh Instruments, Inc., Fishers, NY), which is mounted on an inchworm translator–driven x-y-z three-axis stage system (Burleigh). With this assembly, the position of the tip electrode above the substrate is controlled both by the digital movement of the inchworm translators (Type IW-502-2, Burleigh) and by the analog signal sent from a digital-to-analog converter to the pusher via a high-voltage amplifier (PZ-70, Burleigh). The substrate with or without a support can be attached to the electrochemical cell with an O-ring seal. The electrochemical cell is mounted on a steady platform. The movements of all three inchworms are controlled by a programmable controller (type CE-2000-3A00, Burleigh). The resolution of the digital movement of the inchworm translators can be adjusted from 1/128 of a "click distance" to two "click distances," depending on the selection of digital signals from the CE-2000-3A00 controller. A click distance represents a movement of 1.1–1.2 μm, depending on the piezoelectric material, age, and load on the inchworm translator. To calibrate an inchworm after mounting on the SECM, it is moved a relatively long distance (e.g., 200–400 "click distances"), and that distance is measured with a micrometer. The resolution of the piezoelectric pusher is better than 1 Å. To calibrate the piezoelectric pusher, the SECM feedback mode is used. The tip current is monitored while the electrode (a 5-μm-radius Pt disk) is moved toward a planar conductive substrate (a 2.5-mm-radius Pt disk) immersed in acetonitrile containing 5 mM ferrocene and 0.1 M tetrabutylammonium tetrafluoroborate (TBABF$_4$) by the z-axis inchworm. The inchworm is stopped when the tip current is about 50% higher than $i_{T,\infty}$. After recording the tip current, i_T, the tip is moved two more clicks toward the substrate with the z-axis inchworm. The tip is then moved backward from this location by changing the applied voltage of the PZT pusher until the tip current is equal to i_T. From this experiment, the sensitivity of the PZT pusher (i.e., displacement per volt) can be obtained from the known traveling distance of the z-axis inchworm during two clicks (~2.2–2.4 μm) and the voltage difference applied to the PZT pusher required to compensate for the tip current change. The tip and substrate potentials are controlled vs. a reference electrode with a bipotentiostat of conventional design [17,18] controlled with a potential programmer (Model 175, Princeton Applied Research, Princeton, NJ). The SECM instrument is controlled with a Deskpro 286 or 386 microcomputer (Compaq Computer Corp., Houston, TX) equipped with a data acquisition board (DT 2821-G-8DI; Data Translation, Inc., Marlboro, MA), which has 8 differential analog input channels (12-bit resolution, 250 kHz throughput) and 16 digital I/O lines. The DT2821-G-8DI board is used to acquire the electrochemical signals via the bipotentiostat, to supply the voltage for the PZT pusher via a PZ-70 high voltage operational amplifier, and to send the signals for the control of the x-y-z micropositioning devices via the Burleigh inchworm movement controller. Programs for these purposes are written in C language, and a commercially available Surfer program (Golden Software, Inc., Golden, CO) is adapted for image processing. The Surfer program has many features for data manipulation and presentation, e.g., data smoothing, three-dimensional plotting, rotation, and contour line plotting. The SECM image can be further improved (deblurred) by image-processing techniques that remove the diffusional broadening by use of a linear combination of Laplacian and Gaussian filtering [19,20].

The resolution attainable with the SECM is largely governed by the tip size and the distance between tip and sample. With a very small diameter tip (e.g., diameter < 100 nm), scanning the tip in close proximity to the substrate surface (e.g., 10 nm above the surface) and measuring the tip current become difficult because stray vibrations or irregularities in the sample surface can cause a "tip crash." Thus, for high resolution, the SECM must be operated in the constant current mode, as is often used with the STM, where the distance is adjusted by a feedback loop to the z-piezo to maintain i_T constant. This is straightforward when the sample is either all conductive or all insulating, since the piezo-feedback can be set to counter a decrease in current by either moving the tip closer (conductor) or farther away (insulator). However, for samples that contain both types of regions, a method of recognizing the nature of the substrate must be available. One approach is to modulate the motion of the tip normal to the sample surface and record di_T/dz as described in the next section.

B. Tip Position Modulation

In this section we describe another operation mode of the SECM, tip position modulation SECM (TPM SECM), in which the tip position (i.e., the tip-substrate distance) is modulated with a small-amplitude sinusoidal motion normal to the sample surface. The modulated current is then used as the imaging signal [21].

The principle of TPM is shown schematically in Figure 4A. The tip is moved sinusoidally with a frequency of f_m and an amplitude of $\delta/2$ in the z direction so that its distance from the substrate is modulated by $(\delta/2)\sin(2\pi f_m t)$ about its average distance, d. This modulation causes a modulation in the tip current at the same frequency, as shown in Figure 4B. Since a positive change in z (i.e., movement of the tip away from the substrate) causes a decrease in the tip current over a conductor and an increase over an insulator,

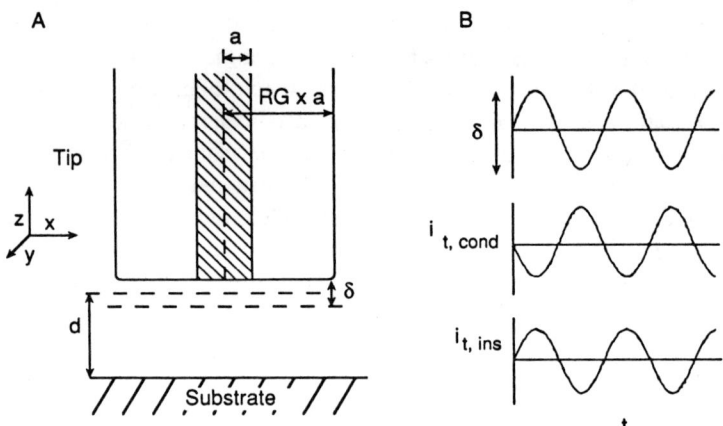

FIG. 4. (A) Schematic diagram of the tip-position modulation experiment. (B) Idealized representation of the tip position and currents observed over insulating and conducting substrates with time. (Reprinted with permission from Ref. 21. Copyright 1992 American Chemical Society.)

$i_{T,cond}$ and $i_{T,ins}$ are 180° out-of-phase. Thus, by detecting the in-phase component of the modulated current with a phase-sensitive technique, one can identify the conductive nature of the substrate as well as take advantage of the noise reduction.

The apparatus for the TPM SECM experiment is similar to that for the conventional SECM experiment, with some modification. Several additional components have been added to perform the modulation experiment. As shown in Figure 5, modulation of the tip position is achieved by mounting the tip onto a spring-loaded linear translation stage (Model 421-OMA, Newport Corp., Fountain Valley, CA) that is driven by a piezoelectric pusher (PZT-30, Burleigh) with a nominal displacement of 5 nm/V. The modulation voltage, V_m, for the pusher is derived from the sine-wave reference oscillator output of a lock-in amplifier (Model 5206 or 5210, EG&G PAR, Princeton, NJ) and is amplified to the desired value by a high-voltage operational amplifier (PZ-70, Burleigh). The dc response is measured after current-to-voltage conversion and filtering at 15 Hz to remove the ac component. The modulation signal is measured with the lock-in amplifier to generate the phase-resolved rms ac response. The dc and ac signals are acquired simultaneously through the A/D conversion card.

Tip position modulation provides several advantages. As mentioned above, by noting the phase angle, one can distinguish between conductive and insulating regions of the substrate under examination. Moreover, the lock-in

Scanning Electrochemical Microscopy

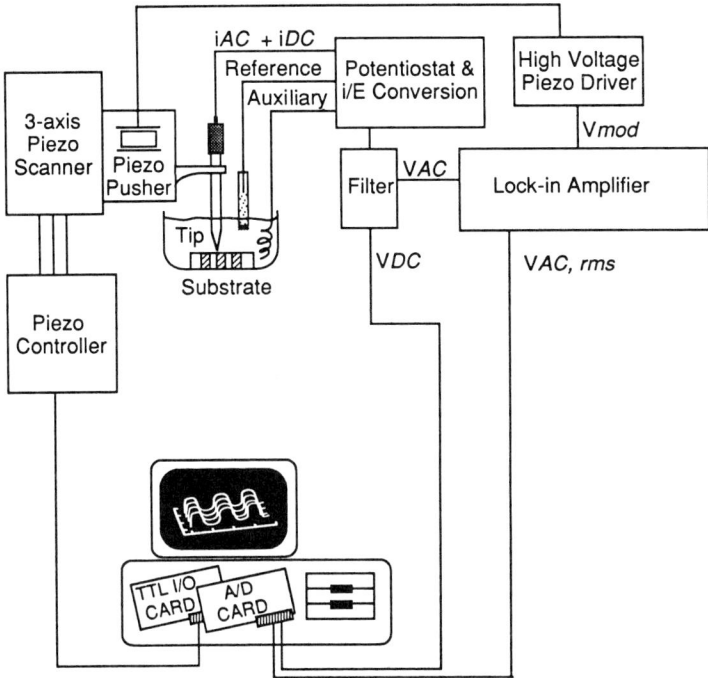

FIG. 5. Block diagram of the tip-position modulation SECM. (Reprinted with permission from Ref. 21. Copyright 1992 American Chemical Society.)

technique provides filtering of the SECM signal, and hence, improves signal-to-noise ratio and sensitivity. Finally, the modulated tip current for both insulator and conductor is measured from a zero baseline, since $i_{T,m} = 0$ when the tip is far from a substrate. This leads to better sensitivity in imaging, especially for insulators, where the dc response, $i_T/i_{T,\infty}$, only varies between 0 and 1.

C. Tip Preparation

In this section, several methods of fabrication of different types of tip electrodes suitable for SECM are described.

1. Disk-in-Glass Microelectrodes

This is the most popular type of tip currently used in this laboratory for SECM experiments (Fig. 6). Construction of this kind of tip is largely based on the fabrication techniques for disk-shaped microelectrodes, which have been described in the literature [5,6,18].

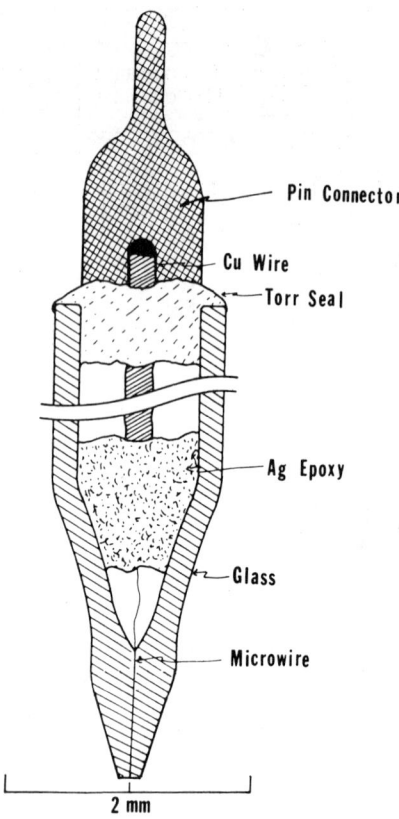

FIG. 6. Schematic diagram of a disk-in-glass microelectrode used in SECM experiments.

A fine wire of Pt or Au or a carbon fiber of the desired radius is placed in a 10-cm-long, 1-mm i.d. Pyrex (or soft glass for Au) tube sealed at one end. The open end of the tube is connected to a vacuum line and heated with a helix heating coil for one hour to desorb any impurities on the wire. One end of the wire is then sealed in the glass at the closed end of the tube by increasing the heating coil temperature. The glass should melt around the wire for 1–2 mm at the tip of the assembly. After the glass has cooled, the electrode is inspected under a microscope to determine that the wire is completely sealed at the tip and that there are no trapped air bubbles. The sealed end is polished with sandpaper, until the wire cross section is exposed, and then successively with

Scanning Electrochemical Microscopy

15-, 6-, 3-, 1-, and 0.25-μm diamond paste (Buehler, Lake Bluff, IL). Electrical connection to the unsealed end of the wire is made with silver paint (Ekote, No. 3030, Acme Chemicals and Insulation Co., New Haven, CT) or silver epoxy (Epotek H20E, Epoxy Technology, Inc., Billerica, MA) to a Cu wire. The glass wall surrounding the Pt or Au disk is conically sharpened with emery paper (Grit 600, Buehler) and 6-μm diamond paste, with frequent checking with an optical microscope until the diameter of the flat glass section surrounding the Pt or Au disk is less than 100 μm. This decreases the possibility of contact between glass and substrate because of any slight deviation in the axial alignment of the tip as the tip is moved close to the substrate.

The preparation of disk-shaped Pt tips of diameter less than 2 μm is similar to that described above but employs Wollaston wire (Goodfellow Metals, Cambridge, UK), i.e., Pt with a 50- to 100-μm silver coating. The silver is etched away with 30% nitric acid until 1–2 mm of the Pt wire is exposed. The etched wire is dipped into distilled water to remove nitric acid, and it is then rinsed with acetone and left to dry. Care must be taken during the etching procedure because the exposed Pt wire is very fragile and easily broken. The wire is sealed into the glass by the same procedures as stated above for the larger electrode. It is important, however, that the location where the Pt joins the silver is sealed in glass to prevent the breakage of the Pt during the polishing and wiring steps.

An alternative method for the preparation of submicrometer disks in glass was developed by Lee et al. [22]. A 1- to 2-cm-long Pt wire (25 μm diameter) is sharpened by electrochemical etching in a solution containing saturated $CaCl_2$ (60% by volume), H_2O (36%), and HCl (4%) at 2 V rms ac applied with a Variac transformer [23]. A carbon rod or plate serves as the counterelectrode in the two-electrode etching cell. A Pyrex capillary with a tip diameter greater than 100 μm is drawn from a piece of Pyrex tubing (2 mm o.d. and 1 mm i.d.) by using a microelectrode puller (Cat. No. 51217, Stoelting Co., Chicago, IL) with heat setting at 65 and pull setting at 100. The tip of the capillary is sealed by heating for 1 sec in a gas flame. The sharpened wire is transferred inside the glass capillary, and the open end of the tube is connected to a vacuum line. The capillary tip is then heated with four loops (1/4 in. i.d.) of nichrome wire (1.3 mm thick) at an ac voltage of 3.8 V (temperature ca. 800°C). Further manipulations involved in the fabrication of these tips are illustrated in Figure 7. Excess glass near the tip of the Pt wire (Fig. 7A) is removed by heating the electrode in the resistive heating coil, which pulls the glass toward the Pt wire. Heating is halted just before the Pt wire is exposed (Fig. 7B). The diameter of the exposed Pt disk could be controlled from 0.2 to 25 μm by controlling the extent of the final polishing.

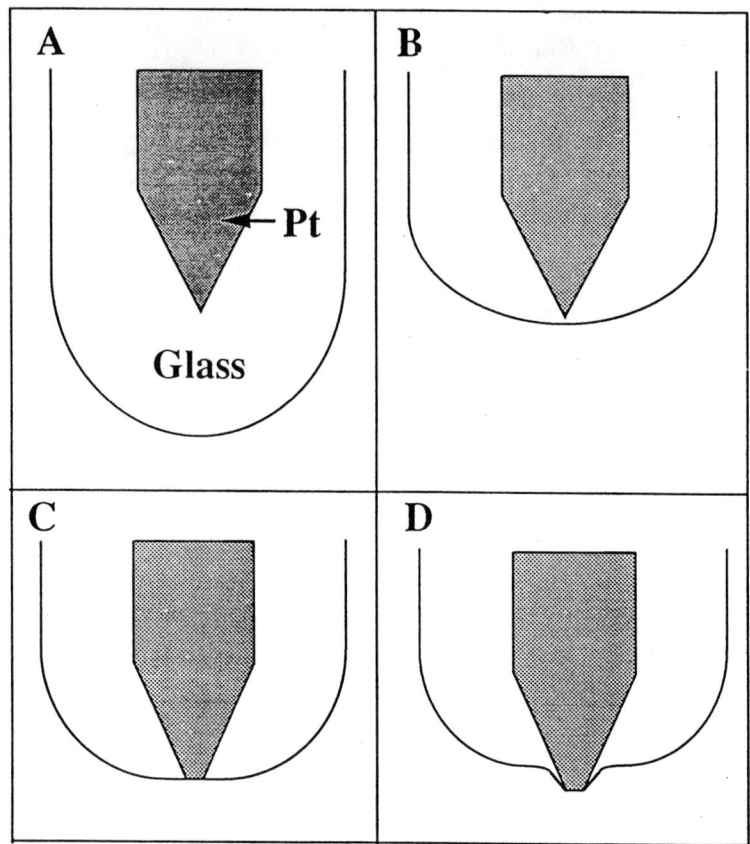

FIG. 7. Schematic procedure for producing a sharp-end microdisk electrode sealed in glass. (A) Sealing the etched Pt wire in glass under vacuum. (B) Allowing the excess glass at the electrode tip to flow by heating until the tip of the Pt wire is almost exposed. (C) Polishing to expose microdisk portion of the wire. (D) Further heating, which results in the Pt disk protruding from the glass insulation. (Reprinted with permission from Ref. 22. Copyright 1991 American Chemical Society.)

Further heating for 0.5–1 min causes the glass to pull away from the Pt wire so that the small exposed Pt disk protrudes from the sealed glass (Fig. 7D). Electrical connection to the unsealed end of the wire is made with silver paint to a Cu wire, which is then soldered to a pin socket for the connection to the SECM. After this, a small amount of Torr Seal Epoxy (Varian Associates, Palo Alto, CA) is packed into the open end of the glass capillary. This seals the barrel and provides strain relief for the contact wire.

2. Construction of Submicrometer Electrode with Electron Beam Lithography

This technique was developed in this laboratory by Lee et al. [22] and employed a commercial scanning electron microscope (SEM) to produce submicrometer features at the tip of an electron beam resist-coated polished Pt wire.

A 5-mm-long Pt wire (25-µm diameter) is bonded to a copper wire (1 mm diameter) with Ag epoxy and then sealed in Torr Seal. Figure 8 is a schematic representation of the electrode fabrication procedure. The insulated wire is polished with 600-grit sandpaper to expose the 25-µm cross section of the Pt wire. The Torr Seal insulation is polished down to a ~200-µm diameter centered around the exposed Pt disk. Further polishing of the electrode using a 1-µm-diameter alumina slurry results in the Pt wire protruding slightly from the epoxy insulation, as illustrated in Figure 8A. This is followed by a final polishing with 0.05-µm alumina. The electrode is then dip-coated with a 5.5% solution of poly(methyl methacrylate) (PMMA) in chlorobenzene, dried in air, and heated in an oven at 50°C for 1 hr. The PMMA (ca. 0.1–0.5 µm thick)-coated electrode (as shown in Fig. 8C) is mounted vertically onto an aluminum stub and placed into the SEM. After focusing and correcting for astigmatism, the electron beam is halted at the center of the Pt disk for 0.5 sec at a beam current of 5–10 pA. The exposed polymer is removed by dropping 1 ml of a 1:3 mixture of methyl isobutyl ketone and isopropyl alcohol on the substrate, followed by a brief rinse with ethanol (Fig. 8D). Pt is electroplated at -0.2 V vs. Ag/AgCl from a 1 mM H_2PtCl_6 solution in 0.1 M HCl. The electrolysis is halted once a predetermined amount of charge has passed (Fig. 8E).

3. Apiezon Wax Coated Ultramicroelectrode Tip

This type of tip was originally developed and used by Nagahara et al. [24] in electrochemical scanning tunneling microscopy and was later modified by us for use as an SECM tip [25]. A 125-µm-diameter Pt-Ir (80–20%) rod is sharpened by electrochemical etching in a solution consisting of saturated $CaCl_2$ (60% by volume), H_2O (36%), and HCl (4%) at ca. 20 V rms ac applied with a Variac transformer. A carbon rod or plate serves as the counterelectrode in a two-electrode cell. After etching, the ultramicrotip is washed with Millipore reagent water and ethanol and dried in air prior to insulation. Insulation of the tip is done with molten Apiezon wax. The sharply etched tip, after soldering in a pin socket, is mounted vertically on a manipulator. A copper plate (1.5 mm thick), held horizontally as shown in Figure 9, is heated using a soldering iron element and is used to melt the Apiezon wax. A 1-mm-wide rectangular slit extends from one side to the center of the

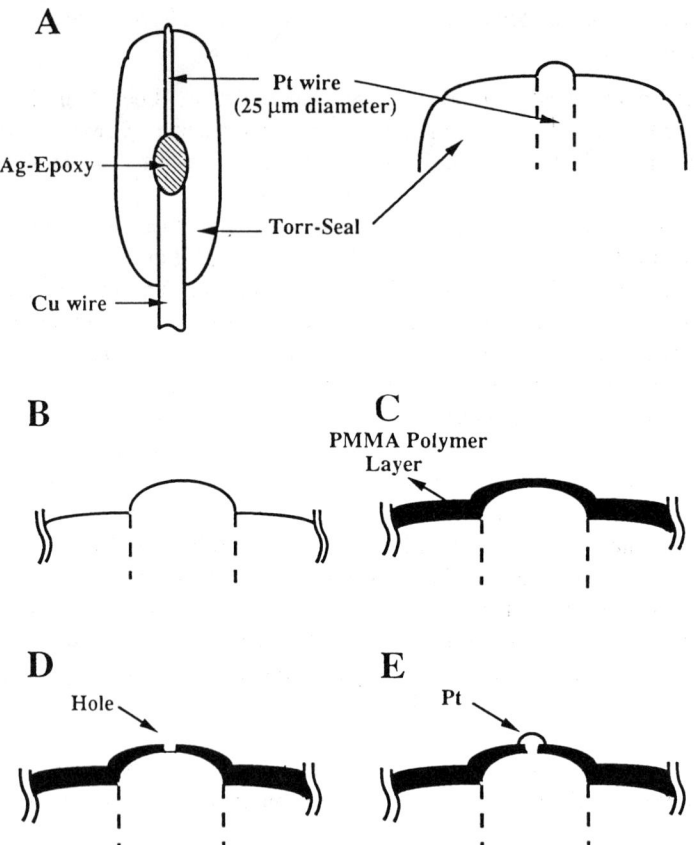

FIG. 8. Schematic procedure for producing a microdisk electrode by electron beam lithography. (A) A cross section of the Pt electrode design and an enlarged view of the tip of the electrode after polishing. (B–E) Schematic representation of the Pt tip after polishing (B), coating with PMMA (C), developing the exposed area of the polymer to produce a cylindrical hole (D), and electroplating Pt into the exposed hole (E) (see text). (Reprinted with permission from Ref. 22. Copyright 1991 American Chemical Society.)

copper plate, providing a temperature gradient (colder at the open end and hotter at the closed end) for the melted wax. The tip is raised through the molten wax and allowed to break the top surface of the melt. If the tip breaks the surface at a too hot region, it is mostly bare. If the tip breaks the surface at the colder region of the melt, it raises a bulb of wax above it and the tip must

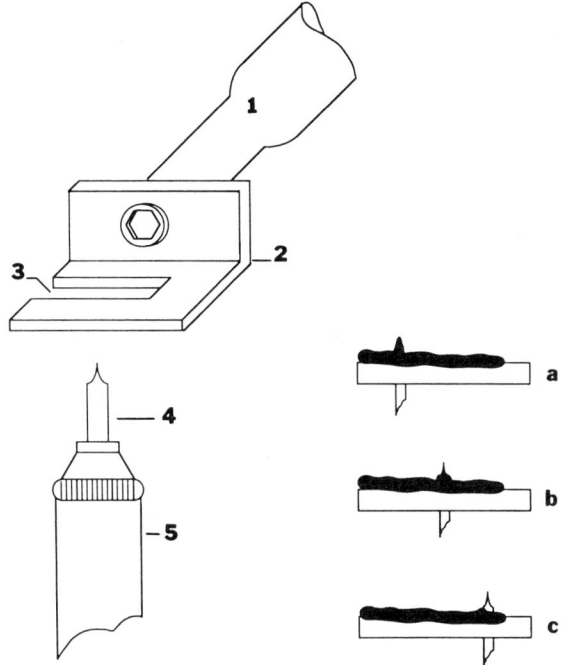

FIG. 9. Illustration of the apparatus/technique used for tip insulation. (1) soldering iron, (2) copper plate, (3) 1-mm-wide slit, (4) STM tip, and (5) holder for manipulator. Where the tip penetrates the Apiezon wax along the slit determines how much of the tip gets insulated. (a) At too cold a region, the tip is completely covered with Apiezon wax. (b) Optimum point allows only the extreme end of the tip to be exposed. (c) At too hot a region, the tip receives little insulation and is thus mostly bare. (Reprinted with permission from Ref. 24. Copyright 1989 American Institute of Physics.)

be reentered into a hot region to remove the excess wax. Between these two regions exists an optimum point where the wax coats the tip well. The insulated tip is moved sideways out of the groove so as to leave the very end of the tip unperturbed. Several coatings are usually required to insulate the tip completely or nearly completely. The degree of insulation of a tip can be checked by carrying out cyclic voltammetry in a solution containing a redox species. A well-insulated tip will not show detectable voltammetric waves. For a completely insulated tip, the very end of the tip can be exposed by placing it in a scanning tunneling microscope (STM) with a bias of a few volts between the tip and a conductive substrate (e.g., a Pt disk) and the STM set in the constant current mode. The onset of current flow (e.g., 0.5 nA) produces a hole in the tip insulation at the point of closest approach of tip to substrate, while leaving most of the tip still insulated. The amount of exposed area of the tip can be controlled by the bias voltage and the onset current flow. A further treatment of "micropolishing" these tips, by continuously scanning them over the substrate, can enlarge the exposed area relative to that of tips not receiving this treatment. Such tips have been used in a number of studies with aqueous solutions but are less useful when nonaqueous solvents, such as acetonitrile, are used because they tend to swell and dissolve in organic solvents. These electrodes are fragile and are usually discarded after a few days' use. For SECM experiments performed in nonaqueous solvents, the very small electrodes ("nanodes") reported by Penner et al. [26] are perhaps more stable. These electrodes were made by electrochemically etching a Pt-Ir wire to produce a tapered tip and then pushing this, at a controlled rate, through molten glass to insulate most of wire. Careful control is apparently necessary to produce very small tips without complete insulation. The tip geometry of these may make them unsuitable for SECM, e.g., if the tip is recessed into the glass insulator.

4. Characterization of the Tip

The microelectrode tips are first characterized by cyclic voltammetry. Figure 10A shows the cyclic voltammogram (CV) for a 0.1 M ferrocyanide solution using an exposed PMMA-coated electrode after developing. The radius of the microelectrode can be estimated by the diffusion-limited plateau current, $i_{T,\infty}$, by using the relation for a disk-shaped microelectrode [Eq. (1)]. The calculated diameter of the electrode used in Figure 10A is 0.25 μm, which is somewhat smaller than the true one because the electrode is recessed by the thickness of the polymer layer (ca. 0.25 μm). Figure 10B shows CVs obtained in a 0.2 M solution of methyl viologen dichloride and 2 M KCl using a glass-coated microelectrode fabricated by the procedure of Lee et al. [22] and a disk-in-glass microelectrode prepared from 0.6 μm-diameter Pt (Wollaston) wire. By comparison with the CV of the 0.6-μm-diameter electrode, the size of the smaller electrode was determined to be approximately 0.2 μm in diameter.

The geometry and size of micrometer-size tips have also been investigated by SEM. Figures 11A and B show the exposed Pt tips of two separate electrodes made by the electron beam lithographic technique. Figure 11A shows an electron beam lithographically produced hole in the PMMA polymer layer. The electrode shown in Figure 11B was produced by electroplating Pt into the lithographically generated hole. An electric charge of 2×10^{-8} C

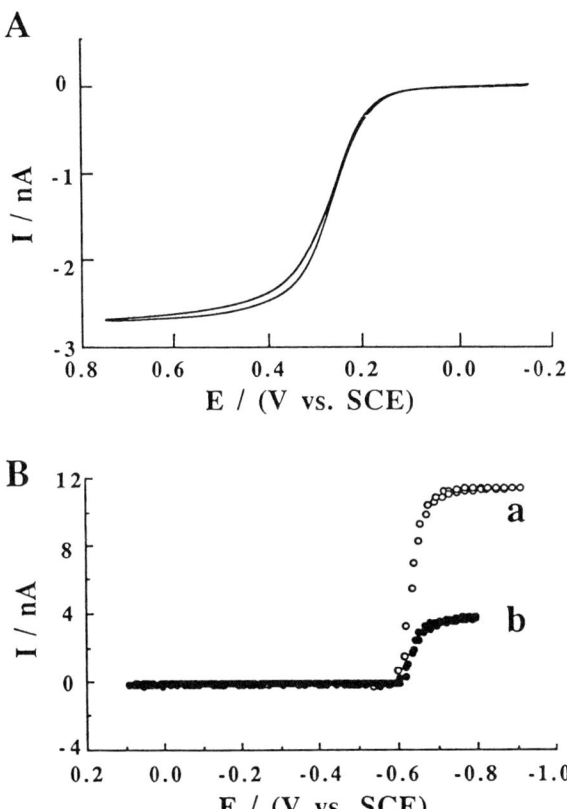

FIG. 10. Cyclic voltammograms of microelectrodes. (A) Electrode produced by electron beam lithography (procedure given in Fig. 8). Solution, 0.1 M $K_4Fe(CN)_6$ and 1 M KCl; scan rate (v) of 0.1 V/s. (B) Solution, 0.2 M $MVCl_2$ and 2 M KCl; (a) disk-in glass electrode using 0.6-μm-diameter commercial wire; v = 1 V/s; (b) small electrode produced by the procedure given in Fig. 7; v = 0.1 V/s. (Reprinted with permission from Ref. 22. Copyright 1991 American Chemical Society.)

was passed during the electrolysis corresponding to an approximately 0.5-μm thickness in the ca. 1-μm hole. Because the polymer layer was only 0.25 μm thick, the Pt electrode should extend beyond the polymer insulation. Figures 11C and 11D show a top view and a side view of a glass-coated electrode fabricated by Lee's technique (see Fig. 7), which has a disk diameter of 0.6 μm measured by cyclic voltammetry using Eq. (1).

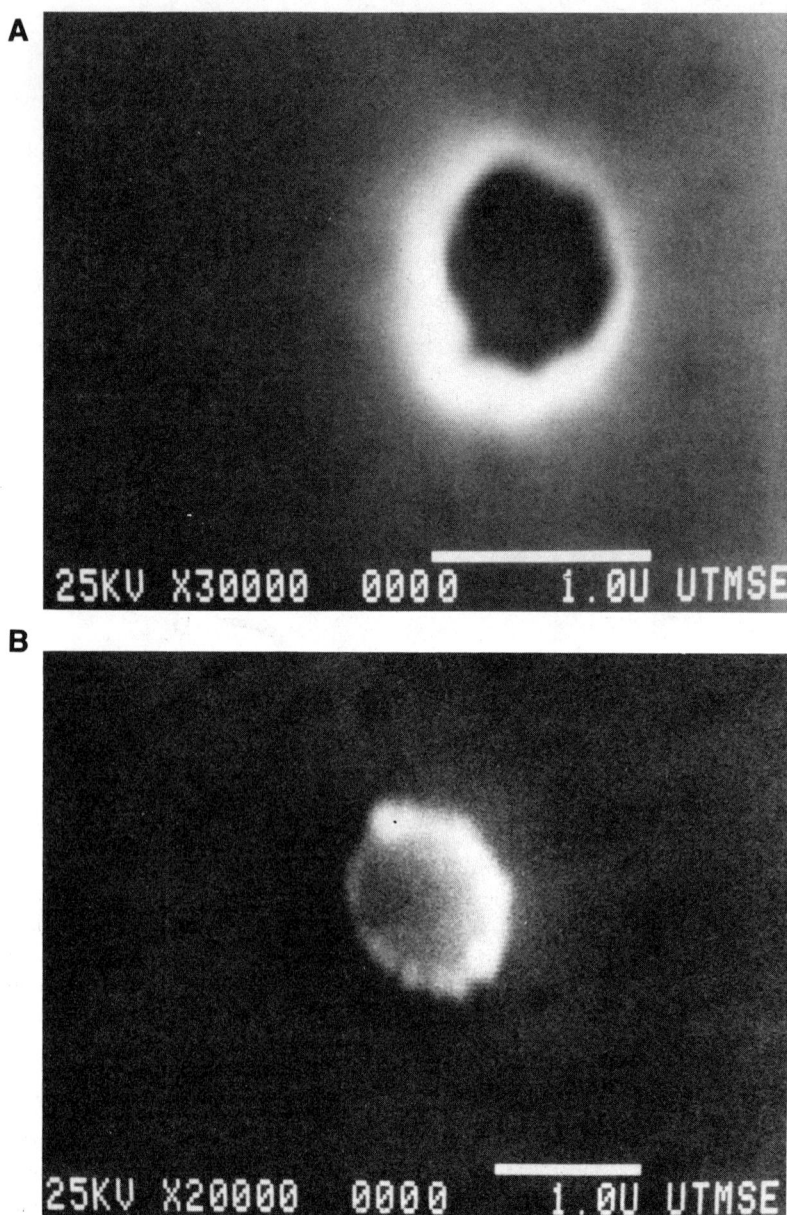

FIG. 11. Scanning electron micrographs of microelectrodes. (A) An electron beam lithographically produced hole in a PMMA polymer layer. (B) An electron beam lithographically produced electrode after the electrodeposition of Pt. (C) Top view and (D) side view of a glass-coated Pt electrode. (Reprinted with permission from Ref. 22. Copyright 1991 American Chemical Society.)

Scanning Electrochemical Microscopy

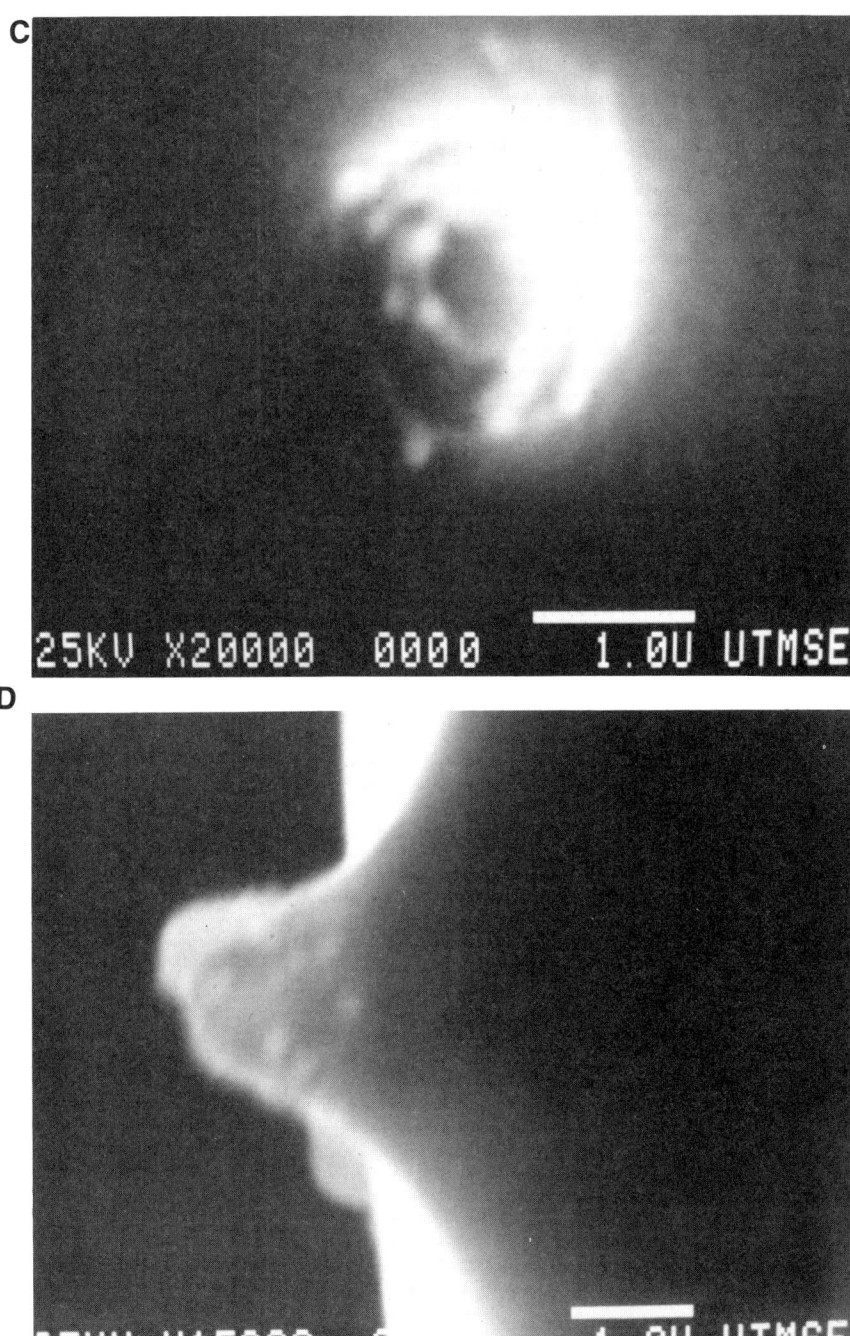

For very small electrodes ("nanodes"), one cannot obtain good images by electron microscopic techniques and SECM approach curves, i.e., the tip current vs. distance curves, rather than the voltammetric measurements, are used to evaluate ultramicroelectrode geometry. (For theory, see Sec. III.E) As an example, the normalized tip current (I) for an Apiezon wax coated Ir-Pt tip, as a function of the normalized tip-substrate distance (L), for a conductive substrate, e.g., Pt, is shown in Figure 12 [25]. The experiment was carried

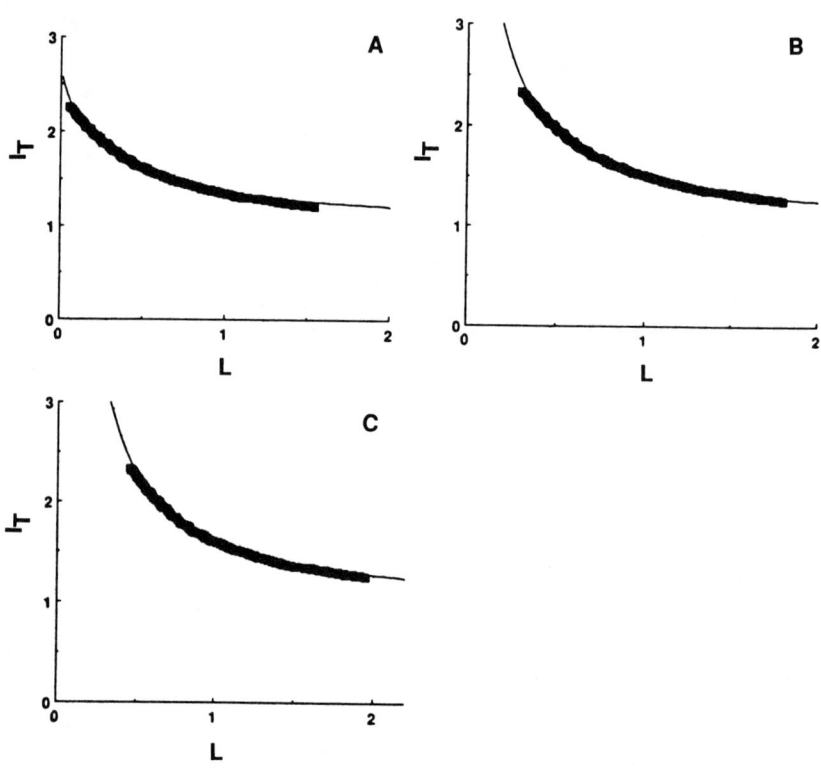

FIG. 12. Normalized tip current vs. normalized distance for a tip electrode over a conductive substrate (a Pt disk) in a solution containing 1.5 mM $Os(bpy)_3^{2+}$ and 0.5 M Na_2SO_4 as the supporting electrolyte (thick curve). The Pt substrate (5-mm diameter) was held at 0.2 V vs. SCE; $i_{T,\infty}$ = 0.059 nA. The thin curves show the theoretical curves computed for (A) conical 420-nm-radius tip, k = 0.8, and offset = 0; (B) spherical segment 420-nm-base radius tip, α_0 = 0.7069, and offset = 105 nm; and (C) 420-nm-radius disk tip and offset = 170 nm. (Reprinted with permission from Ref. 25. Copyright 1992 Elsevier.)

out with an aqueous solution containing 1.5 mM $Os(bpy)_3^{2+}$ and 0.5 M Na_2SO_4 as the supporting electrolyte. During the distance scan, the tip was held at 0.80 V vs. SCE, where a steady-state plateau was obtained in the CV, and the Pt substrate was biased at 0.20 V vs. SCE, where $Os(bpy)_3^{3+}$ generated at the tip was reduced back to $Os(bpy)_3^{2+}$. The steady-state tip current obtained at an infinite tip-substrate separation ($i_{T,\infty}$ = 0.059 nA suggests the radius value to be in the range of 200–500 nm (depending on the exact shape of the tip). Three ways of fitting these experimental data by using the families of theoretical current-distance curves computed for conical, spherical, and disk geometries were tried. Three parameters can be varied for conical or spherical geometries, i.e., the base radius (a_0), the height parameter (the ratio of the cone height to its radius, k, or α_0), and the offset value (see Fig. 36 for explanation). In the case of a disk, only two parameters, a_0 and the offset, are applicable. The offset value characterizes any error in the determination of zero distance. In an SECM experiment with a disk tip embedded in an insulating sheath, the point of zero tip-substrate separation cannot be determined exactly because small protrusions in the insulating sheath, or lack of precise alignment of the tip with respect to substrate, will often cause the insulator to touch the substrate before the actual d = 0 is reached. The distance scale is thus usually obtained by an arbitrary adjustment of the i_T vs. d curve to the theoretical response for a disk. For cone-shaped or spherical-segment-shaped tips, the actual zero distance can be evaluated within a few nm and any errors in the d = 0 value should be in this range.

The fitting procedure showed that the radius is not an adjustable parameter; a good fit can only be obtained for a certain radius value. This value appeared to be constant whether the simulation was based on a conical, spherical, or disk approximation. The experimental approach curve shown in Figure 12 can be fitted using either conical (Fig. 12A), spherical segment (Fig. 12B), or disk (Fig. 12C) approximations with the same radius value, a_0 = 420 nm, and k = 0.8 or α_0 = 0.7069. The most significant difference between these three cases is the offset value required for a best fit, which is equal to 0 nm, 105 nm, and 170 nm for cone, spherical segment, and disk, respectively. The equal radii of the disk, spherical segment, and cone base apparently lead to a fixed size projection on the substrate plane, and the value of this radius is essential for determination of the diffusion flux normal to this plane. The sharp top of the cone (as well as the top of spherical segment) prevents the substrate from approaching the main part of the tip surface. The differences in the heights of these three bodies determine the offset values required to fit the experimental curve using each of these approximations. From the offset values, it is concluded that the tip used in Figure 12 is shaped more like a cone than a disk or a spherical segment. Of course, most tips deviate more or less from the idealized geometric shapes taken in theoretical treatments. This procedure is also useful for electrodes with smaller radii [52].

III. THEORY

The availability of well-developed quantitative theory covering various regimes of measurements and electrochemical mechanisms is a significant advantage of the SECM. The different operating modes of the SECM, e.g., feedback and generation/collection modes, steady-state and transient measurements, require significantly different theoretical descriptions. Quantitative treatments are available for both diffusion-controlled processes and finite kinetics as well as for more complicated mechanisms involving adsorption and homogeneous reactions in the gap. Most of the reported theoretical results concern an SECM with a disk-shaped tip. Other tip geometries will be discussed at the end of this section.

A. SECM with a Disk-Shaped Tip in the Feedback Mode of Operation

Feedback theory has been the basis for most quantitative SECM applications reported to date. The first theoretical treatment of the feedback response was the finite-element simulation of a diffusion-controlled process by Kwak and Bard [27], but we will start from a more general formulation for a quasi-reversible process under non–steady-state conditions and then consider some important particular cases.

1. General Theory for an Uncomplicated Non–Steady-State Process in SECM

The considerable complexity of SECM theory is due to the combination of a cylindrical diffusion to the microtip electrode and a thin layer–type diffusion space. The geometry shown in Figure 2 suggests a time-dependent diffusion problem for a simple quasi-reversible mediator in cylindrical coordinates as follows [28,29]:

$$0 < z < d,\ 0 \leq r,\ 0 < t$$

$$\frac{\partial c_O}{\partial t} = D_O \left(\frac{\partial^2 c_O}{\partial z^2} + \frac{\partial^2 c_O}{\partial r^2} + \frac{1}{r} \frac{\partial c_O}{\partial r} \right); \qquad (2)$$

$$\frac{\partial c_R}{\partial t} = D_R \left(\frac{\partial^2 c_R}{\partial z^2} + \frac{\partial^2 c_R}{\partial r^2} + \frac{1}{r} \frac{\partial c_R}{\partial r} \right)$$

Scanning Electrochemical Microscopy

$$t = 0, \; 0 \leq r, \; 0 < z < d; \; c_O(t,r,z) = c_O^\circ; \; c_R(t,r,z) = 0 \quad (3)$$

$$0 < t, \; 0 \leq r, \; z = 0; \; j_T(t,r) = D_O\left[\frac{\partial c_O}{\partial z}\right]_{z=0} =$$
$$-D_R\left[\frac{\partial c_R}{\partial z}\right]_{z=0} \quad (4)$$

$$0 < t, \; 0 \leq r, \; z = d; \; j_S(t,r) = -D_O\left[\frac{\partial c_O}{\partial z}\right]_{z=d} = D_R\left[\frac{\partial c_R}{\partial z}\right]_{z=d} \quad (5)$$

$$i_T(t) = 2\pi nF \int_0^a j_T(t,r) r \, dr; \quad i_S(t) = 2\pi nF \int_0^{a_S} j_S(t,r) r \, dr \quad (6)$$

where r and z are spatial variables, t is time, a is the radius of the tip and a_S is the radius of the substrate, c(t,r,z) is the concentration of electroactive species, c° is its bulk value, c(t,r) is the surface concentration, D is the diffusion coefficient, j is the diffusion flux density, i is the total faradaic current, and d is the tip-substrate separation. The subscript O relates to the oxidized form and R to the reduced form; the subscripts T and S refer to the tip and substrate electrodes, respectively. The faradaic current density is

$$nFj(t,r) = nF[k_f c_O(t,r) - k_b c_R(t,r)] \quad (7)$$

and $j_T(t,r) = 0$ at $r > a$, and $j_S(t,r) = 0$ at $r > a_S$. If the substrate is an insulator, $j_S \equiv 0$. The rate constants for oxidation (k_b) and reduction (k_f) at the tip and substrate are given by the Butler-Volmer relations [17]:

$$k_f = k^\circ \exp[-\alpha nf(E - E^{\circ\prime})] \quad (8)$$
$$k_b = k^\circ \exp[(1 - \alpha)nf(E - E^{\circ\prime})] \quad (9)$$

where k° is the standard rate constant, E is the electrode potential, $E^{\circ\prime}$ is the formal potential, α is the transfer coefficient, n is the number of electrons transferred per redox event, and $f = F/RT$ (here F is the Faraday, R is the gas constant, and T is the temperature).

The above formulation implies that the overall process is confined to the thin layer of solution between the substrate and the tip with its insulating sheath. This is an approximation because the insulating sheath surrounding the tip is not infinitely thick and condition (10):

$$Rg < r; \; c_O(t,r,z) = c_O^\circ; \; c_R(t,r,z) = 0 \quad (10)$$

can be used to account for the finite value of the sheath radius, Rg, [30,31]. However, the calculations in Refs. 29 and 30 showed that the differences caused by the infinite sheath approximation are negligible in the case of a conductive substrate and can be easily taken into account for an insulating one. We prefer to use the formulation given by Eqs. (2) to (7) because it is the only one suitable for analytical treatment. Moreover, the actual geometry of the glass sheath is conical [8] rather than cylindrical [as assumed in Eq. (10)] and is always imperfect. Thus Rg is the least reliable parameter in SECM theory and one should avoid using it in calculations.

Although the above problem can be solved for the most general case of unequal diffusion coefficients, this would require a quite cumbersome numerical or analytical treatment. To our knowledge, all computational results reported for a non–steady-state process were obtained for $D_O = D_R$ (this leads to $c_O + c_R = c°$ and reduces system (2) to a single equation). It is advantageous to use the dimensionless variables:

$$R = \frac{r}{a} \tag{11}$$

$$Z = \frac{z}{a} \tag{12}$$

$$C = 1 - \frac{c}{c_O^°} \tag{13}$$

$$T = \frac{tD}{a^2} \tag{14}$$

$$J = \frac{ja}{Dc_O^°} \tag{15}$$

Thus the solution of the problem can be obtained in terms of the dimensionless currents $I_T(T)$ and $I_S(T)$:

$$I_T(T) = -\frac{\pi}{2} \int_0^1 J_T(T,R)R \, dR; \tag{16a}$$

$$I_S(T) = -\frac{\pi}{2} \int_0^\rho J_S(T,R)R \, dR \tag{16b}$$

where $\rho = a_S/a$, and I_T and I_S are equal to the physical currents normalized by the limiting diffusion tip current, $I = i/i_{T,\infty}$, where $i_{T,\infty}$ is given by Eq. (1).

The numerical solution of this problem is challenging because the combination of the high spatial resolution required at both the tip and the substrate and the relatively large thickness of the solution gap results in a large number of volume elements, and hundreds of time iterations are necessary to calculate the potentiostatic transient to a steady-state value [30–32]. Two advanced computational methods, i.e., the Krylov integrator [31] and the alternating-direction implicit (ADI [33]) finite difference method [30,32], were applied to compute potentiostatic transients for two limiting cases: a diffusion-controlled process and totally irreversible kinetics. The analysis of the simulation results [31] revealed several time regions typical for SECM transients (Fig. 13). In

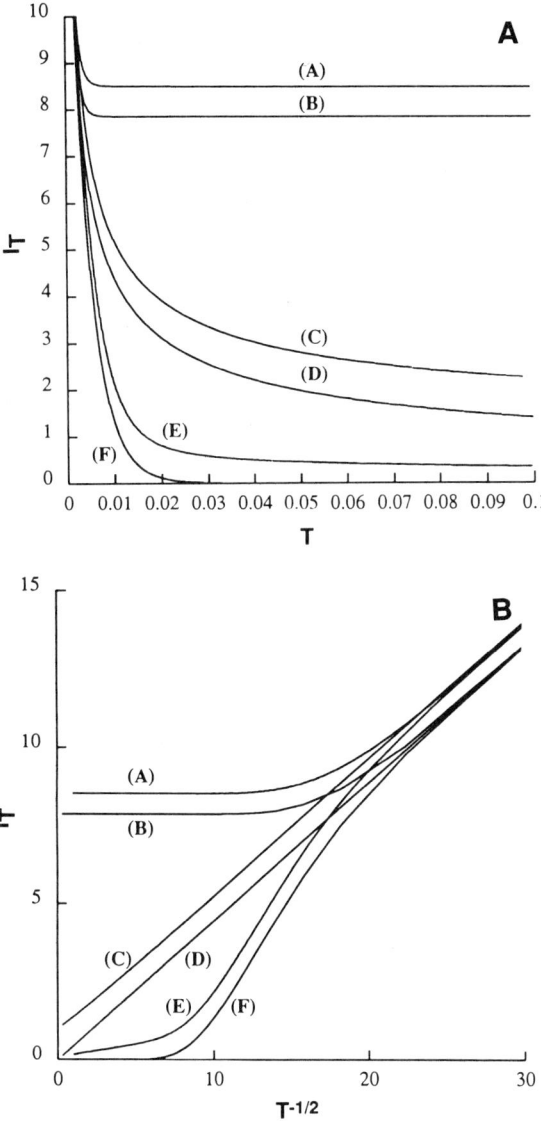

FIG. 13. Comparison of simulated SECM transients with transients corresponding to different electrode geometries (all processes are diffusion controlled). (A) SECM transient for a conductive substrate; (B) two-electrode thin-layer cell; (C) microdisk; (D) planar electrode; (E) SECM with an insulating substrate; (F) one-electrode thin layer cell. Curves A, B, E, and F were computed with $L = d/a = 0.1$. (Reprinted with permission from Ref. 31. Copyright 1991 American Chemical Society.)

the short time region (t ≤ 0.001) an SECM transient follows closely the microdisk transient, then it starts to deviate and finally levels at a constant value of i_T. For a conductive substrate, $i_T \geq i_{T,\infty}$, and it is always larger than the value calculated for the same time from thin layer cell (TLC) theory. For an insulating substrate, $i_T < i_{T,\infty}$. The smaller the tip/substrate separation, the earlier the deviations between the SECM and microdisk transients. The time when the SECM undergoes a transition from the microdisk regime to the TLC regime is related to the time needed by the species to diffuse across the gap between the tip and the substrate. This time can be determined experimentally and can be used to evaluate the diffusion coefficient [31].

Similarly, the transients computed for different rates of an irreversible heterogeneous reaction [30] showed a microdisk-type behavior at short times. The current magnitude at longer times and its eventual steady-state value are determined by the value of the dimensionless heterogeneous rate constant, $K_{b,S} = k_{b,S}a/D$ (Fig. 14). The substrate behaves as a conductor as $K_{b,S} \to \infty$ and as an insulator as $K_{b,S} \to 0$.

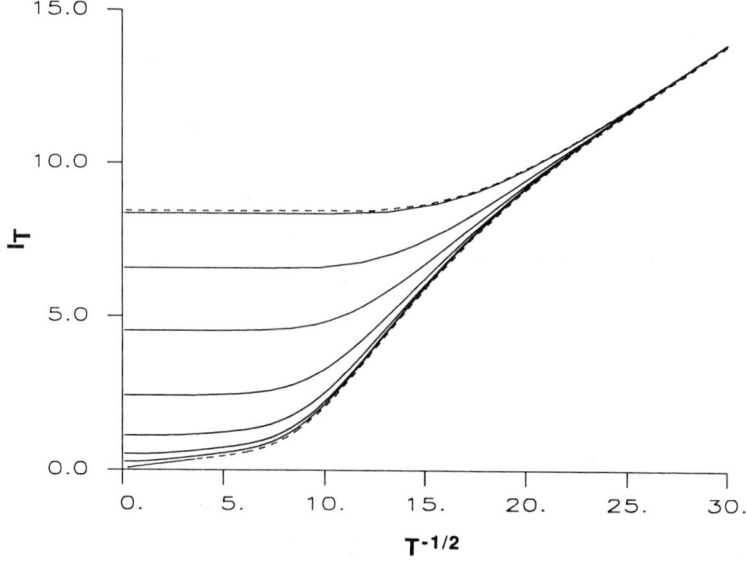

FIG. 14. Simulated feedback SECM transients with various rate constants for an irreversible heterogeneous process at the substrate. The upper and lower dashed curves correspond respectively to the limits $K_{b,S} \to \infty$ and $K_{b,S} = 0$. The solid curves (from top to bottom) represent log $K_{b,S}$ = 3.0, 1.5, 1.0, 0.5, 0, −0.5, and −1.0. RG = 10. (Reprinted with permission from Ref. 30. Copyright 1992 American Chemical Society.)

Scanning Electrochemical Microscopy

For many electrochemical systems, digital simulation is the only way of generating theoretical data. However, an analytical treatment of the SECM diffusion problem is possible. The solution of Eqs. (2) to (7) can be obtained in the form of two-dimensional integral equations [28,29] (the notation in Refs. 28 and 29 is slightly different than that used here) by applying a Hanckel transform in the spatial domain and a Laplace transform in the temporal one.

$$\frac{J_T(T,R)\exp(\alpha_T n_T f[E_T(T) - E°])/\Lambda_T + 1}{1 + \exp(n_T f[E_T(T) - E°])} = -\frac{1}{2L}\int_0^x u\,du \int_0^T \frac{\exp\left(-\frac{R^2 + u^2}{4(T - \tau)}\right)}{T - \tau}$$

$$\times I_0\left(\frac{Ru}{2(T - \tau)}\right)\left[J_S(\tau,u)\theta_4\left(0\left|\frac{i\pi(T - \tau)}{L^2}\right.\right) + J_T(\tau,u)\theta_3\left(0\left|\frac{i\pi(T - \tau)}{L^2}\right.\right)\right]d\tau \quad (17)$$

$$\frac{J_S(T,R)\exp(\alpha_S n_S f[E_S(T) - E°])/\Lambda_S + 1}{1 + \exp(n_S f[E_S(T) - E°])} = -\frac{1}{2L}\int_0^x u\,du \int_0^T \frac{\exp\left(-\frac{R^2 + u^2}{4(T - \tau)}\right)}{T - \tau}$$

$$\times I_0\left(\frac{Ru}{2(T - \tau)}\right)\left[J_S(\tau,u)\theta_3\left(0\left|\frac{i\pi(T - \tau)}{L^2}\right.\right) + J_T(\tau,u)\theta_4\left(0\left|\frac{i\pi(T - \tau)}{L^2}\right.\right)\right]d\tau \quad (18)$$

where $L = d/a$ is the dimensionless distance between the tip and the substrate, $x = \max(1,\rho)$, $\Lambda = ak°/D$, I_0 is a modified Bessel function of the first kind of order zero, and θ_3 and θ_4 are theta functions [34]. Equations (17) and (18) describe the SECM response for both conductive and insulating substrates. However, in the last case, $J_S \equiv 0$, and only Eq. (17) needs to be solved, resulting in a simpler problem. Equations (17) and (18) were solved numerically using the algorithms described in Refs. 29 and 30. The amount of computation time is much smaller than in the case of numerical solution of corresponding differential equations. This is because of the drastic decrease in the number of nodes in the space-time grid. The typical number of spatial points over both the tip and substrate surfaces is about 10–20 (vs. $\sim 10^4$ points in the two-dimensional spatial grid required for differential equations), and the number of T-points is also at least 10 times smaller.

The SECM description in the form of Eqs. (17) and (18) is somewhat overly general, since it includes the possibility of mixed diffusion/kinetic control of both the tip and substrate processes and two sets of kinetic parameters, Λ, α, and n, for both working electrodes. In actual experiments, at least one of those electrodes is held under diffusion control while the other is used for kinetic measurements. Even with this simplification, the SECM response depends on too many parameters to allow presentation of a complete

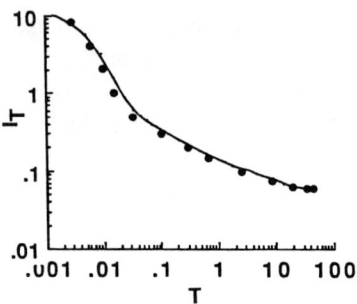

FIG. 15. Transient computed for SECM with an insulating substrate and an inlaid tip. L = d/a = 0.1; $nf(E_T - E°) = 5.87$. Filled circles were computed from Ref. 32 for RG = 10. (Reprinted with permission from Ref. 29. Copyright 1992 Elsevier.)

set of working curves which would cover all experimental possibilities. A few transients (Fig. 15) and non–steady-state CVs (Fig. 16) for a quasi-reversible process were computed [29] for both insulating and conducting substrates. The analysis of transients calculated for an insulating substrate showed that the influence of the RG (RG = Rg/a) parameter can be detected only at dimensionless times $T > RG^2$. This effect should be barely noticeable on the time scale of a typical SECM experiment [29].

The voltammograms in Figure 16 represent the current at a finite substrate electrode (a_S = a) produced by the linear sweep of the tip potential (E_S is constant and held where the tip-generated product is oxidized at the substrate). In this case, the tip behaves as a generator and the substrate behaves as a collector electrode. The near-steady-state collection efficiency (Fig. 16A) is near unity, even for a small substrate at small tip-substrate separation (e.g., L = 0.1). One can expect some delay in the substrate response as the tip potential is scanned. This delay is a function of the time for the tip-generated species to transit the gap, and hence, it depends on the L, D, and scan rate (v) values. This delay is not seen in Figure 16A because of the small L and v. When the sweep is fast (Fig. 16B), the delay between the cathodic peak of the tip voltammogram (1) and the substrate anodic peak (curve 2) is substantial. Pairs of such curves should contain information about the diffusional and electrochemical kinetics of the redox couple in the gap.

2. *Steady-State Conditions*

Although non–steady-state SECM measurements may be advantageous in some cases, most experimental results were obtained under steady-state conditions. However, specific SECM theory for kinetically controlled processes

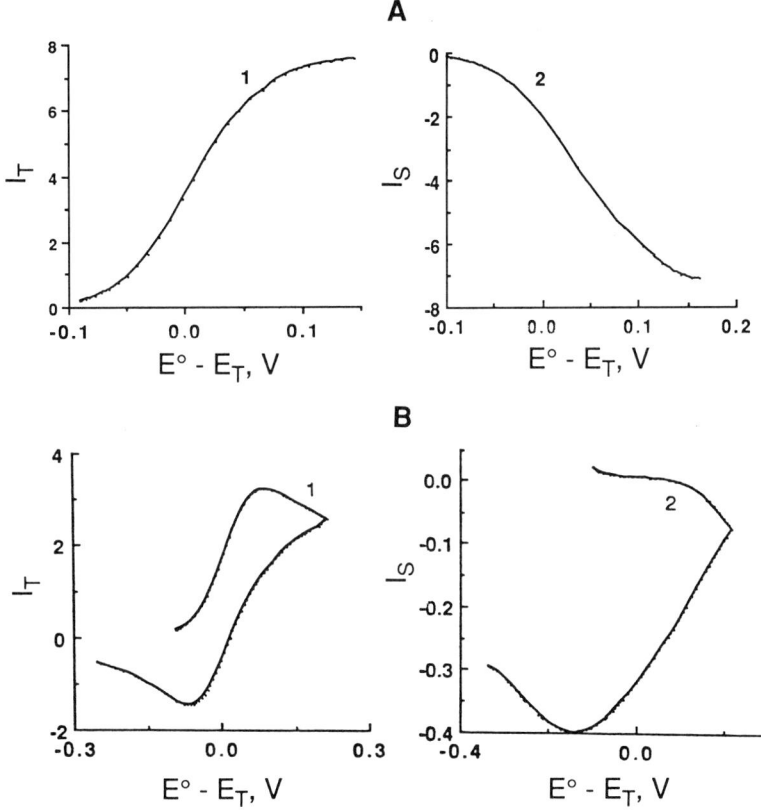

FIG. 16. Substrate voltammograms (curves 2) produced at a constant potential-biased substrate by linear (A) or triangular (B) sweep of the tip potential (curves 1). $\alpha = 0.5$, $nfa^2/D = 0.01$ (A) and 100 (B). $L = 0.1$ (A) and 1 (B). $nf(E_s - E°) = 5.87$. (Reprinted with permission from Ref. 29. Copyright 1992 Elsevier.)

at steady state has not yet been published. All computational results were obtained by solving non–steady-state equations and taking limiting values of current as $t \to \infty$ or $v \to 0$ (for CV). In this section, we will survey those results. Then the specific steady-state theory will be considered along with the practically important particular case of the diffusion-controlled process.

The special case of totally irreversible substrate kinetics was treated numerically [30]. The family of theoretical working curves I_T vs. $K_{b,s}$ calculated for various L (Fig. 17) can be used to determine the rate constant for the substrate reaction. The range of measurable rate constants as a function of the

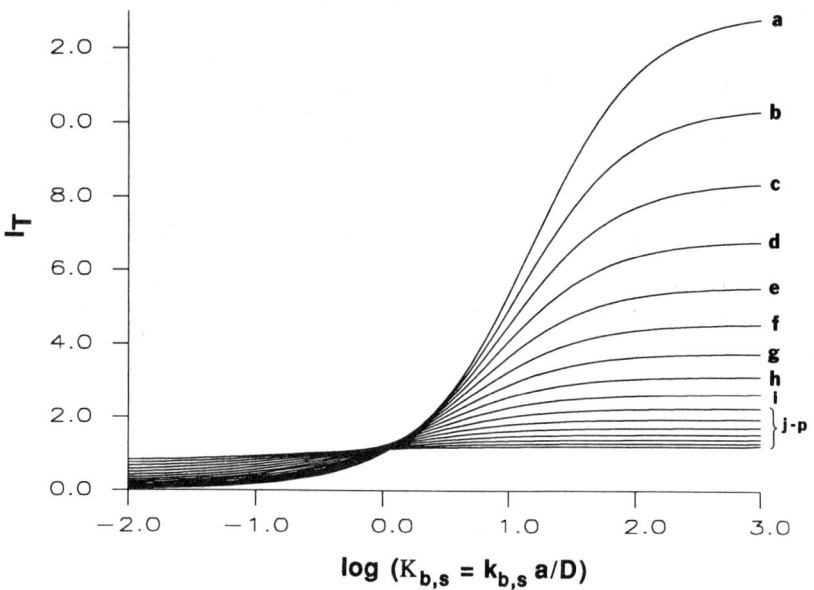

FIG. 17. Working curves of I_T vs. log $K_{b,s}$. log L = $-1.2, -1.1, -1.0, \ldots 0.2$ (a-p). (Reprinted with permission from Ref. 30. Copyright 1992 American Chemical Society.)

dimensionless tip-substrate separation can be determined from the kinetic zone diagram (Fig. 18).

The computational results in Ref. 30 illustrate the essential features of the feedback response in the qausi-reversible regime. Table 3, representing a range of kinetic situations, contains values of the dimensionless steady-state SECM current for various values of the dimensionless rate constant, Λ_S, and the dimensionless tip-substrate separation, L. An α value of 0.5 was assumed for all calculations. Each row in Table 3 corresponds to a given value of the normalized tip/substrate distance. Seven groups of columns (three columns in each group) correspond to various values of Λ_S spanning the range of quasi-reversibility from virtually reversible ($\Lambda_S = 25$) to essentially irreversible processes ($\Lambda_S = 0.001$). Three contrasting values of the dimensionless substrate potential, $E_1 = (E_S - E^{o\prime})nf$, were considered for each L; these correspond to the formal potential, a potential where the substrate feedback reaction is almost diffusion controlled, and a potential between the two. Table 3 indicates that an increase in the overpotential (while keeping the values of the other parameters constant) leads to a higher feedback current. At suffi-

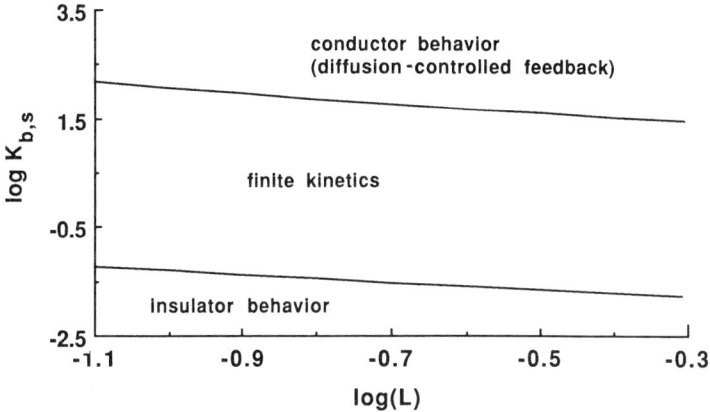

FIG. 18. Kinetic zone diagram illustrating the regions of finite irreversible kinetics, diffusion-controlled feedback, and insulating substrate behavior. (Reprinted with permission from Ref. 30. Copyright 1992 American Chemical Society.)

ciently high overpotentials, the substrate behaves like a conductor with diffusion-controlled feedback, i.e., with increasing L, I_T decreases monotonically to unity. At $E_1 = 0$ and small Λ_S, the rate of the feedback process at the substrate is negligible; the substrate shows insulator behavior, and the tip current increases with L. However, at $E_1 = 0$ and large Λ_S, a significant positive feedback is observed with a small L. Dependencies of I_T vs. Λ_S calculated for a few values of L and E_S (Fig. 19) showed [29] that a significant difference (about 10%) between I_T values, corresponding to $\Lambda_S = 25$ and $\Lambda_S = 50$, can be detected at close tip-substrate separations (L = 0.1). Thus rate constants up to 20 cm/sec should be accessible to steady-state measurement with the presently available 1-μm-diameter microtips.

The data in Table 3 were computed for an infinite substrate. This approximation is valid when a substrate is somewhat larger than the tip (see below). One should, however, be cautious when working with a large substrate ($a_S \gg a$) biased at a potential close to $E^{\circ\prime}$. In this situation, some reduction of O occurs at the large portion of the substrate surface distant from the tip. This non–steady-state process tends to gradually deplete the mediator in the solution surrounding the tip and may also result in a significant IR-drop.

The dependence of the near-steady-state voltammogram shape on Λ_T, as well as the effect of the tip-substrate spacing, are shown in Figure 20. The sequences of tip CV computed with different values of Λ_T are similar to those for a microdisk electrode [5]. However, at a small value of L (Fig. 20A) the

TABLE 3

SECM Steady-State Current Computed for Various Values of the Kinetic Parameter Λ_S, Normalized Tip/Substrate Distance L, and Dimensionless Substrate Potential E_1 [a]

L	$\Lambda_S = 25$		$\Lambda_S = 5$		$\Lambda_S = 1$		$\Lambda_S = 0.5$		$\Lambda_S = 0.1$		$\Lambda_S = 0.05$		$\Lambda_S = 0.001$	
	\multicolumn{2}{c}{E_1}	\multicolumn{2}{c}{E_1}	\multicolumn{2}{c}{E_1}	\multicolumn{2}{c}{E_1}	\multicolumn{2}{c}{E_1}	\multicolumn{2}{c}{E_1}	\multicolumn{2}{c}{E_1}
	0.0	0.585	2.926	0.0	1.171	4.682	0.0	2.341	7.803	0.0	3.902	9.754	0.0	7.803	11.71	0.0	7.803	13.66	0.0	15.61	21.46

L	$\Lambda_S=25$, $E_1=0.0$	$E_1=0.585$	$E_1=2.926$	$\Lambda_S=5$, $E_1=0.0$	$E_1=1.171$	$E_1=4.682$	$\Lambda_S=1$, $E_1=0.0$	$E_1=2.341$	$E_1=7.803$	$\Lambda_S=0.5$, $E_1=0.0$	$E_1=3.902$	$E_1=9.754$	$\Lambda_S=0.1$, $E_1=0.0$	$E_1=7.803$	$E_1=11.71$	$\Lambda_S=0.05$, $E_1=0.0$	$E_1=7.803$	$E_1=13.66$	$\Lambda_S=0.001$, $E_1=0.0$	$E_1=15.61$	$E_1=21.46$
0.1	3.71	4.79	7.66	2.40	3.84	7.38	1.00	2.42	7.39	0.68	2.62	7.67	0.21	3.25	6.95	0.17	2.07	7.31	0.11	2.04	7.30
0.2	2.12	2.74	4.20	1.66	2.62	4.22	0.93	2.05	4.24	0.72	2.22	4.32	0.24	2.59	4.10	0.19	1.89	4.22	0.14	1.86	4.21
0.5	1.16	1.48	2.20	1.07	1.62	2.26	0.84	1.61	2.28	0.69	1.72	2.29	0.49	1.87	2.25	0.43	1.60	2.27	0.36	1.60	2.27
0.8	0.89	1.14	1.68	0.85	1.30	1.74	0.78	1.39	1.76	0.68	1.49	1.76	0.60	1.58	1.75	0.57	1.44	1.76	0.50	1.44	1.76
1.0	0.81	1.03	1.51	0.79	1.18	1.57	0.72	1.31	1.58	0.67	1.40	1.58	0.61	1.47	1.57	0.60	1.37	1.58	0.55	1.37	1.58
1.5	0.70	0.89	1.32	0.69	1.03	1.35	0.68	1.18	1.36	0.67	1.27	1.36	0.66	1.31	1.35	0.66	1.26	1.36	0.67	1.26	1.36
2.0	0.66	0.82	1.21	0.65	0.96	1.24	0.66	1.12	1.25	0.66	1.19	1.25	0.70	1.22	1.25	0.73	1.20	1.25	0.78	1.20	1.25
5.0	0.63	0.78	1.05	0.64	0.90	1.08	0.65	1.01	1.08	0.66	1.06	1.08	0.74	1.07	1.08	0.79	1.07	1.08	0.92	1.07	1.08

[a] $E_1 = (E_S - E^{o\prime})nf$, $\Lambda_S = ak^o_S/D$, and $\alpha = 0.5$ for all cases.

Source: Ref. 30.

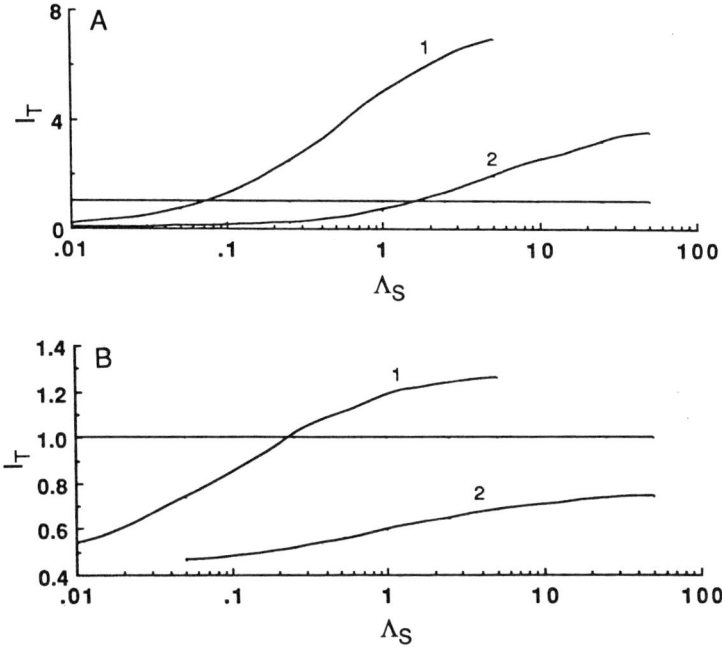

FIG. 19. Working curves of I_T vs. Λ_S. $\alpha = 0.5$, $nf(E_T - E°) = 5.87$, and L = 0.1 (A) and 1 (B). Curve 1, $nf(E_S - E°) = 5.87$ and curve 2, $nf(E_S - E°) = 0$. The horizontal line corresponds to zero feedback current. (Reprinted with permission from Ref. 29. Copyright 1992 Elsevier.)

half-wave potential is more sensitive to the charge-transfer rate than is $E_{1/2}$ of the microdisk voltammograms (with the same disk radius) because of the higher mass-transfer coefficient [35]. When L increases (Fig. 20B), the feedback current drops dramatically, and the voltammograms become more closely spaced. Thus the dimensionless parameter characterizing heterogeneous kinetics changes from $ak°/D$ to $dk°/D$ as a conductive substrate is approached by the tip UME. This suggests that the SECM should be useful for studying rapid heterogeneous electron transfer kinetics, since it should be easier to obtain very small (and variable) tip-substrate spacings than to produce microdisks with equally small radii.

Equations (16), (17), and (18) are suitable for any size substrate (infinite or finite), which is represented by the parameter ρ (if $R > \rho$, $J_S = 0$). The influence of a finite substrate size is important for interpretation of images of small irregular substrate features [36] as well as experiments with a micro-

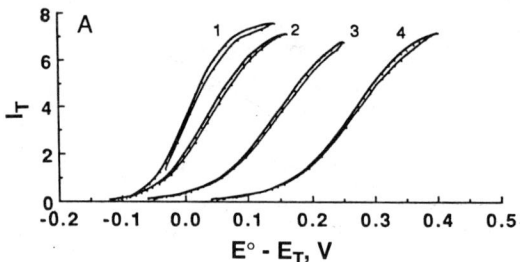

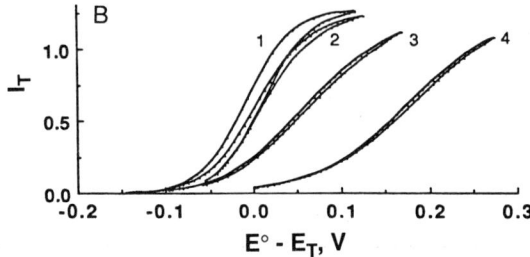

FIG. 20. Steady-state tip CV as a function of Λ_T and L. $\alpha = 0.5$, $nfa^2/D = 0.01$, $nf(E_S - E°) = 5.87$, and $\Lambda_S = 5$. $L = 0.1$ (A) and (B). $\Lambda_T = 25$ (1), 5 (2), 0.5 (3), and 0.05 (4). (Reprinted with permission from Ref. 29. Copyright 1992 Elsevier.)

electrode substrate [30,37]. The current-distance curves in Figure 21 were computed [30] for various values of ρ using both numerical solution of integral equations and the ADI simulations. Apparently, the size of the substrate is important when $ρ \leq 1$. At $ρ \geq ρ^∞ = 1 + 1.5L$, the substrate behaves essentially as an infinite one. The value of $ρ^∞$ defines the surface area, which is actually seen in an SECM feedback experiment and, thus, has important implications in terms of the resolution of SECM images. In particular, the highest resolution is obtained at close tip-substrate separations. On the other hand, the difference between the steady-state feedback current for a finite conductive substrate and that for an insulating one becomes small at $ρ < 0.05–0.1$. Thus it should be possible to identify particles (or other objects) 10–20 times smaller than the tip by SECM (if these particles are sufficiently well separated). With a tip 1 μm in diameter, this corresponds to particles 50–100 nm in size. Particular caution is necessary for interpretation of images containing small conductive spots ($0 < ρ < ρ^∞$); from Figure 21, one can see that such features may appear as conductors at smaller L and as insulators at larger L.

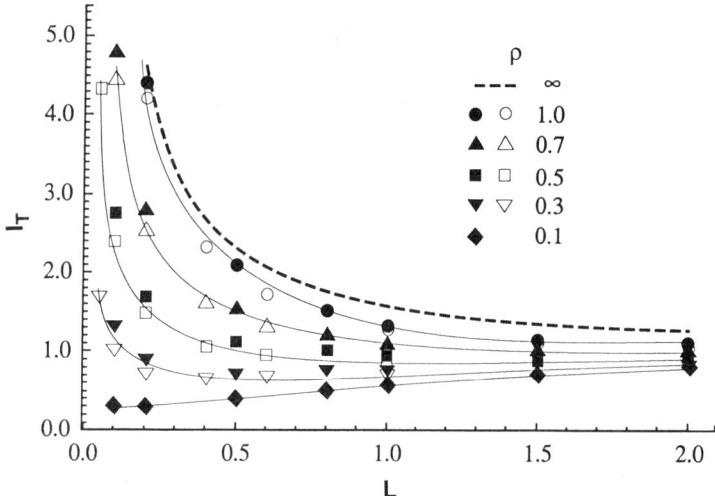

FIG. 21. I_T vs. L curves for finite disk-shaped substrates. Filled symbols calculated from Eqs. (17) and (18); open symbols are from ADI simulation. The ρ values are as indicated. Dashed line is simulation for $\rho = \rho^\infty$ from Ref. 27. The lines through the symbols are drawn as a guide. (Reprinted with permission from Ref. 30. Copyright 1992 American Chemical Society.)

All of the above steady-state results represent the long-time limit of the data obtained from time-dependent equations. The formulation of the true steady-state SECM problem is significantly simpler. It includes a single Laplace equation:

$$\frac{\partial^2 c_O}{\partial z^2} + \frac{\partial^2 c_O}{\partial r^2} + \frac{1}{r}\frac{\partial c_O}{\partial r} = 0 \qquad (19)$$

for oxidized (or reduced) form of the mediator with boundary condition (4) for the tip surface and Eq. (20) for the substrate surface:

$$z = d, \quad c_O(r) = c^\circ \qquad (20)$$

All variables in these equations are now time-independent. The assumption of the equality of the diffusion coefficients is now unnecessary, because the concentrations of O and R at steady-state are interrelated as

$$c_O + c_R D_R/D_O = c^\circ \qquad (21)$$

With the dimensionless variables given by Eqs. (11) to (13) and Eq. (15), the solution is

$$\frac{1 - \pi J_T(R)/4\kappa'}{\theta} = \int_0^1 uJ_T(u)du \int_0^\infty J_0(pR)J_0(pu)\tanh(pL)dp \quad (22)$$

where J_0 is the Bessel function of the first kind of order zero [34], $\kappa' = \pi a k° \exp[-\alpha n f (E - E°')]/(4D_O)$, and $\theta = 1 + \exp[nf(E - E°')]D_O/D_R$ [35]. Equations (22) and (15) can be used to compute the steady-state tip current when the substrate reaction is diffusion limited. The normalized steady-state current, $I_T(\kappa, \theta, L)$, is a trivariate function whose representation would require very extensive tabulation. Instead of this, we will present analytical approximations for diffusion-controlled, nernstian, and quasi-reversible tip electrochemistry.

a. Diffusion-Controlled and Nernstian Steady-State Processes in SECM

The knowledge of the shape of the I_T-L (the "approach") curve for a diffusion-controlled process is critical for both imaging and quantitative kinetic measurements because it allows one to establish the distance scale. The problem describing a diffusion-controlled steady-state process represents the limiting case of the above where the boundary conditions (4) and (5) are replaced by:

$Z = 0, 0 \leq R \leq 1, C_O = 1; 1 < R \leq RG, J(R) = 0;$

$RG < R, C_O = 0$ (23)

$Z = L, 0 \leq R, C_O = 0$ (conductive substrate)

or

$0 \leq R, J(R) = 0$ (insulating substrate) (24)

This problem was solved numerically [27]. The dimensionless current-distance curves were tabulated for both insulating and conductive substrates (Fig. 22) and several values of RG, assuming the equality of the diffusion coefficients and an infinitely large substrate. Later [25], analytical approximations were obtained for both working curves:

$$I_T(L) = 0.68 + \frac{0.78377}{L} + 0.3315\exp\left(\frac{-1.0672}{L}\right) \quad (25)$$

For a conductive substrate, Eq. (25) fits within 0.7% the simulated I_T-L curve over an L interval from 0.05 to 20 (Fig. 22A). For an insulating substrate, a similar equation:

$$I_T(L) = \frac{1}{0.292 + 1.5151/L + 0.6553\exp(-2.4035/L)} \quad (26)$$

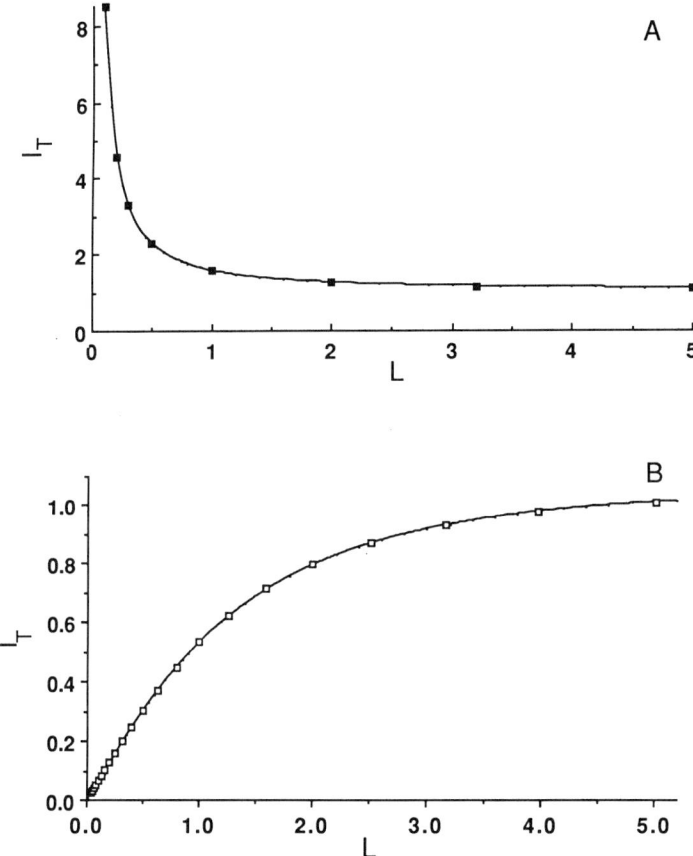

FIG. 22. Diffusion-controlled steady-state tip current as a function of tip-substrate separation. (A) Substrate is a conductor; (B) substrate is an insulator. Solid lines are computed from (A) Eq. (25) and (B) Eq. (27). Filled and open squares are from Ref. 27.

is slightly less accurate (within 1.2%); however, the longer expression:

$$I_T(L) = \frac{1}{0.15 + 1.5385/L + 0.58\exp(-1.14/L) + 0.0908\exp[(L - 6.3)/(1.017L)]} \quad (27)$$

is accurate to within 0.5% over the same L interval (Fig. 22B). Equations (25) to (27) were derived using the data tabulated in Ref. 27 for $RG = 10$. Since the errors associated with these equations are typically less than experimental errors, one can consider them as "exact."

Fitting an experimental current-distance curve to a theoretical one enables one to determine a zero tip-substrate separation point (L = 0), which in turn allows the L values essential for any quantitative SECM measurements. Note that the dimensionless $I_T(L)$ vs. L curves contain no adjustable parameters and do not depend on solution concentration or diffusion coefficients.

Equation (25) can be rewritten as

$$i_T(L) = \frac{3.14a^2nFC°D}{d} + i_{T,\infty}[0.68 + 0.3315\exp(-1.0672/L)] \quad (28)$$

The first term in this expression represents the current in a TLC with a working electrode surface area $A = \pi a^2$ and thickness d. The second term tends to $i_{T,\infty}$ as $L \to \infty$ (~1% of error is associated with the fitting uncertainties and RG effect). This term represents the contribution of the microdisk steady-state current to the total current. This contribution is diminished by the blocking effect of the substrate and becomes negligibly small as $L \to 0$. Thus,

$$i_T(L) = i_{TLC}(L) + i_{T,\infty}[0.68 + 0.3315\exp(-1.0672/L)] \quad (29)$$

It is easy to demonstrate (see Ref. 38 for extensive discussion) that the reversible (nernstian) steady-state voltammogram for any electrode geometry obeys the following equation:

$$i(E) = \frac{i_{dif}}{\theta} \quad (30)$$

where i_{dif} is the diffusion limiting current and θ was defined above. Thus, for the SECM with a conductive substrate,

$$\frac{i_T(E,L)}{i_{T,\infty}} = \frac{0.68 + 0.78377/L + 0.3315\exp(-1.0672/L)}{\theta} \quad (31)$$

A similar equation for SECM with an insulating substrate can be obtained by the combination of Eq. (30) with Eq. (26) or Eq. (27).

b. Analytical Approximations for a Quasi-Reversible Process

The simplest approximation for the SECM feedback current can be obtained assuming the uniform accessibility of the tip surface, e.g., a uniform surface concentration of the electroactive species [35]. With this assumption, the SECM is treated as a modified TLC with the diffusion-limiting current expressed by Eq. (25). The approximate equation for a quasi-reversible steady-state voltammogram is as follows [35]:

$$\frac{i_T(E,L)}{i_{T,\infty}} = \frac{0.68 + 0.78377/L + 0.3315\exp(-1.0672/L)}{\theta + 1/\kappa} \quad (32)$$

Scanning Electrochemical Microscopy

where the kinetic parameter,

$$\kappa = k^\circ \exp \frac{-\alpha n f(E - E^{\circ\prime})}{m_O} \tag{33}$$

and the effective mass-transfer coefficient for SECM is

$$m_O = 4D_O \frac{0.68 + 0.78377/L + 0.3315\exp(-1.0672/L)}{\pi a}$$

$$= \frac{i_T(L)}{\pi a^2 nFc^\circ} \tag{34}$$

The approximation given by Eq. (32) is in a way similar to the approximation for a microdisk voltammetric response by that of the equivalent size hemisphere [39]. In both cases, the assumption of the uniform surface concentration is exact for a reversible electrode reaction [as $\kappa \to \infty$, Eq. (32) reduces to Eq. (31)]. With a decrease in the reaction rate, Eq. (32) becomes less accurate; however, the errors are negligible at $L \ll 1$. In this case, the second term in Eq. (32) is much smaller than the first one, leading to TLC behavior (i.e., uniform surface concentration).

A more precise approximation can be obtained without the uniform accessibility assumption. According to Ref. 39, a quasi-reversible voltammogram at a microdisk can be calculated as:

$$\frac{i_{disk}(E)}{i_{T,\infty}} = \frac{1}{\theta\left(1 + \dfrac{\pi}{\kappa'\theta} \dfrac{2\kappa'\theta + 3\pi}{4\kappa'\theta + 3\pi^2}\right)} \tag{35}$$

The analogous equation for the TLC is [12,35]:

$$i_{TLC}(E,L) = \frac{i_{TLC}(L)}{\theta + 1/\kappa} \tag{36}$$

and m_O in this case is:

$$m_{TLC} = \frac{2D_O D_R}{(D_O + D_R)d} \tag{37}$$

The substitution of Eqs. (35) and (36) for a microdisk and TLC contributions to the SECM current leads to:

$$I_T = \frac{0.78377}{L(\theta + 1/\kappa)} + \frac{0.68 + 0.3315\exp(-1.0672/L)}{\theta\left[1 + \dfrac{\pi}{\kappa'\theta} \dfrac{2\kappa'\theta + 3\pi}{2\kappa'\theta + 3\pi^2}\right]} \tag{38}$$

One can see from Eq. (38) that at large L the SECM response is essentially that of a single microdisk electrode, and the apparent reversibility of the steady-state voltammogram is determined by parameter $\Lambda_T = ak°/D$. At smaller L, however, a TLC-type behavior is expected and another parameter, $\Lambda' = dk°/D$ becomes the measure of apparent reversibility. Voltammograms calculated from Eqs. (32) and (38) are shown in Figure 23. Voltammograms

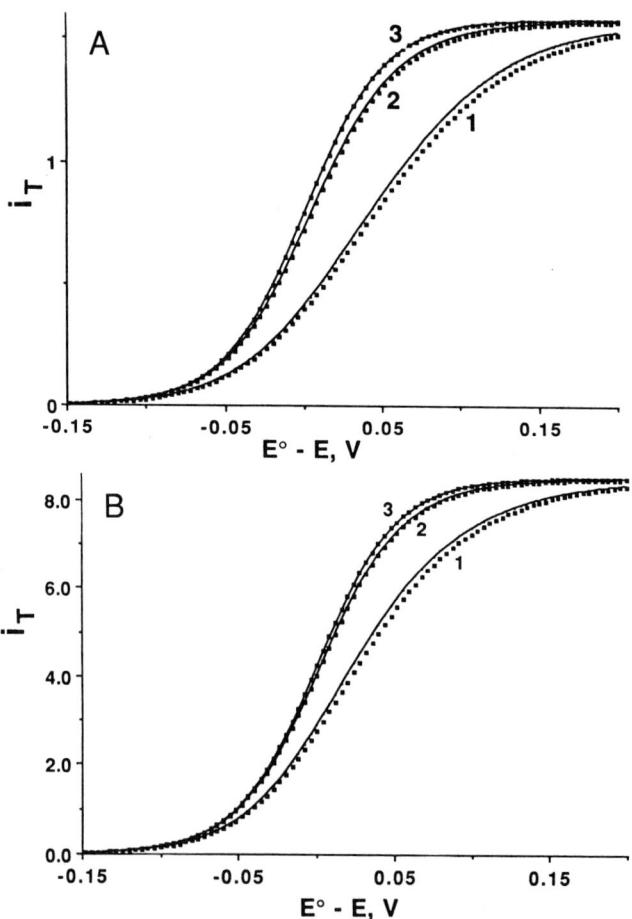

FIG. 23. Steady-state voltammograms at different values of tip-substrate separation and different rates of the tip electrode reaction. (A) L = 1, (B) L = 0.1. Squares were calculated using the uniform accessibility approximation, Eq. (32); solid curves were obtained from Eq. (38) for Λ' = 1 (1), 10 (2), and 100 (3). Nernstian voltammograms are indistinguishable from curves 3. (Reprinted with permission from Ref. 64. Copyright 1993 American Chemical Society.)

of an essentially reversible process ($\Lambda' = 100$) are indistinguishable from each other and from the nernstian one obtained from Eq. (31) (curves 3). As expected, there are significant differences between quasi-reversible curves computed from Eqs. (32) and (38) ($\Lambda' = 1$, curves 1). The difference becomes much smaller as either Λ' increases ($\Lambda' = 10$, curves 2) or L decreases (see curves corresponding to the same Λ' in Figs. 23A and 23B).

B. Generation/Collection Mode of SECM Operation

In the feedback mode of the SECM operation, the overall redox process is essentially confined to the thin layer between the tip and the substrate. The generation/collection (G/C) situation (when the substrate is a generator and the tip is a collector) is quite different; the tip travels within a thick diffusion layer generated by the substrate electrode. The rigorous theoretical description is problematic because (1) the moving tip stirs the substrate diffusion layer; disturbances are especially significant when the tip is an amperometric sensor and has its own diffusion layer; (2) when the substrate is large, no steady-state can be achieved; and (3) when the tip is close to the substrate it blocks the diffusion to its surface, and this screening effect is hard to take into account because of the imperfect geometry of the tip insulating sheath.

1. Potentiometric G/C Mode

The potentiometric tip is a passive sensor, i.e., it does not change the concentration profile of electroactive species generated (or consumed) chemically or electrochemically at the substrate. A consistent theoretical treatment can be proposed only for a steady-state situation when a small substrate (a microdisk or a spherical cap) generates stable species. The concentration of such species can be measured by an ion-selective microtip [37,40] as a function of the tip position. Two types of source geometry will be considered: a microdisk and a hemisphere. The solution of the time-independent diffusion problem is

$$c = c^\circ + \frac{r^{*2}j_S}{Dr'} \tag{39}$$

for a hemispherical source, and

$$C(R,Z) = \frac{2}{\pi} \int_0^1 \sqrt{u/R} J_S(u) p K(p)\, du \tag{40}$$

for a disk-shaped substrate [37], where r^* is the hemisphere radius, r' is the polar radius, $K(p)$ is the complete elliptic integral of the first kind [34], and $p = 2\sqrt{Ru}/\sqrt{Z^2 + (R + u)^2}$. Note that R and Z in Eq. (40) are normalized by the substrate radius a_S, rather than the tip radius, a. Now the flux, $J_S(R)$,

TABLE 4

Dimensionless Steady-State Concentration Profile $c(z,r)/J_{AV}$ for Electroactive Species Generated at a Disk-Shaped Substrate (The Flux Is Assumed to Be Uniform over the Disk Surface.)

$z\backslash r$	0.0	0.1	0.2	0.3	0.4	0.5	0.6	0.7	0.8	0.9	1.0
0.05	0.9512	1.0654	1.0581	1.0384	1.0118	0.9793	0.9405	0.8941	0.8381	0.7677	0.6747
0.10	0.9050	1.0143	1.0093	0.9907	0.9649	0.9331	0.8951	0.8499	0.7958	0.7302	0.6523
0.15	0.8612	0.9644	0.9618	0.9447	0.9201	0.8894	0.8526	0.8092	0.7580	0.6980	0.6310
0.20	0.8198	0.9166	0.9161	0.9007	0.8775	0.8481	0.8130	0.7718	0.7240	0.6695	0.6108
0.25	0.7808	0.8712	0.8723	0.8587	0.8370	0.8093	0.7761	0.7373	0.6930	0.6437	0.5916
0.30	0.7440	0.8283	0.8308	0.8188	0.7988	0.7729	0.7416	0.7054	0.6646	0.6198	0.5732
0.35	0.7095	0.7880	0.7915	0.7811	0.7627	0.7386	0.7095	0.6759	0.6383	0.5976	0.5556
0.40	0.6770	0.7501	0.7544	0.7454	0.7287	0.7064	0.6794	0.6483	0.6138	0.5768	0.5387
0.45	0.6466	0.7147	0.7196	0.7118	0.6967	0.6762	0.6513	0.6226	0.5910	0.5572	0.5226
0.50	0.6180	0.6815	0.6868	0.6803	0.6666	0.6478	0.6249	0.5986	0.5695	0.5387	0.5071
0.55	0.5913	0.6504	0.6561	0.6506	0.6383	0.6212	0.6001	0.5760	0.5494	0.5212	0.4922
0.60	0.5662	0.6214	0.6273	0.6227	0.6117	0.5961	0.5769	0.5547	0.5303	0.5045	0.4780
0.65	0.5427	0.5942	0.6002	0.5965	0.5867	0.5726	0.5550	0.5347	0.5123	0.4886	0.4643
0.70	0.5207	0.5688	0.5749	0.5719	0.5633	0.5504	0.5344	0.5158	0.4953	0.4735	0.4511
0.75	0.5000	0.5450	0.5512	0.5489	0.5412	0.5296	0.5149	0.4979	0.4791	0.4592	0.4385
0.80	0.4806	0.5228	0.5290	0.5272	0.5204	0.5099	0.4966	0.4810	0.4638	0.4454	0.4264
0.85	0.4624	0.5020	0.5082	0.5069	0.5009	0.4914	0.4793	0.4650	0.4492	0.4323	0.4147
0.90	0.4454	0.4825	0.4886	0.4877	0.4825	0.4740	0.4629	0.4499	0.4354	0.4198	0.4035
0.95	0.4293	0.4642	0.4702	0.4698	0.4651	0.4575	0.4474	0.4355	0.4222	0.4078	0.3927
1.00	0.4142	0.4471	0.4530	0.4528	0.4488	0.4419	0.4328	0.4219	0.4096	0.3963	0.3824
1.10	0.3866	0.4158	0.4215	0.4219	0.4188	0.4133	0.4057	0.3966	0.3862	0.3748	0.3629
1.20	0.3620	0.3881	0.3935	0.3943	0.3921	0.3876	0.3813	0.3736	0.3648	0.3551	0.3448
1.30	0.3401	0.3635	0.3686	0.3697	0.3681	0.3645	0.3593	0.3528	0.3453	0.3370	0.3281
1.40	0.3205	0.3415	0.3464	0.3476	0.3465	0.3437	0.3394	0.3339	0.3275	0.3204	0.3127
1.50	0.3028	0.3218	0.3264	0.3278	0.3271	0.3248	0.3212	0.3166	0.3112	0.3050	0.2983
1.60	0.2868	0.3041	0.3084	0.3098	0.3095	0.3077	0.3047	0.3008	0.2962	0.2908	0.2850
1.70	0.2723	0.2880	0.2921	0.2936	0.2935	0.2921	0.2896	0.2863	0.2823	0.2777	0.2726
1.80	0.2591	0.2735	0.2773	0.2789	0.2789	0.2778	0.2758	0.2730	0.2696	0.2656	0.2611
1.90	0.2471	0.2603	0.2639	0.2654	0.2656	0.2648	0.2631	0.2608	0.2578	0.2543	0.2504
2.00	0.2361	0.2482	0.2516	0.2531	0.2534	0.2528	0.2515	0.2495	0.2469	0.2439	0.2404
2.10	0.2259	0.2371	0.2403	0.2418	0.2422	0.2418	0.2407	0.2390	0.2368	0.2341	0.2311
2.20	0.2166	0.2270	0.2300	0.2315	0.2319	0.2316	0.2307	0.2293	0.2274	0.2250	0.2224
2.30	0.2080	0.2176	0.2204	0.2219	0.2224	0.2222	0.2215	0.2203	0.2186	0.2166	0.2142
2.40	0.2000	0.2089	0.2116	0.2130	0.2136	0.2135	0.2129	0.2119	0.2104	0.2086	0.2066
2.50	0.1926	0.2009	0.2035	0.2048	0.2054	0.2054	0.2049	0.2040	0.2028	0.2012	0.1994
2.60	0.1857	0.1934	0.1959	0.1972	0.1978	0.1978	0.1975	0.1967	0.1956	0.1943	0.1926
2.70	0.1792	0.1865	0.1888	0.1901	0.1907	0.1908	0.1905	0.1899	0.1889	0.1877	0.1863
2.80	0.1732	0.1800	0.1822	0.1834	0.1840	0.1842	0.1840	0.1835	0.1827	0.1816	0.1803
2.90	0.1676	0.1740	0.1760	0.1772	0.1778	0.1780	0.1779	0.1775	0.1767	0.1758	0.1746
3.00	0.1623	0.1683	0.1703	0.1714	0.1720	0.1723	0.1722	0.1718	0.1712	0.1703	0.1693
3.20	0.1526	0.1580	0.1598	0.1608	0.1614	0.1617	0.1617	0.1614	0.1610	0.1603	0.1595
3.40	0.1440	0.1488	0.1504	0.1514	0.1520	0.1523	0.1523	0.1522	0.1518	0.1513	0.1507
3.60	0.1363	0.1406	0.1421	0.1430	0.1436	0.1439	0.1440	0.1439	0.1437	0.1433	0.1427
3.80	0.1294	0.1333	0.1346	0.1355	0.1360	0.1363	0.1365	0.1364	0.1363	0.1359	0.1355
4.00	0.1231	0.1267	0.1279	0.1287	0.1292	0.1295	0.1297	0.1297	0.1296	0.1293	0.1290
4.20	0.1174	0.1207	0.1218	0.1225	0.1230	0.1233	0.1235	0.1235	0.1235	0.1233	0.1230
4.40	0.1122	0.1152	0.1162	0.1169	0.1174	0.1177	0.1179	0.1179	0.1179	0.1177	0.1175
4.60	0.1074	0.1102	0.1112	0.1118	0.1122	0.1125	0.1127	0.1128	0.1128	0.1127	0.1125
4.80	0.1031	0.1056	0.1065	0.1071	0.1075	0.1078	0.1080	0.1081	0.1081	0.1080	0.1079
5.00	0.0990	0.1014	0.1022	0.1028	0.1032	0.1035	0.1036	0.1037	0.1037	0.1037	0.1036
5.50	0.0902	0.0921	0.0928	0.0933	0.0937	0.0939	0.0941	0.0942	0.0943	0.0943	0.0942
6.00	0.0828	0.0844	0.0850	0.0854	0.0858	0.0860	0.0862	0.0863	0.0863	0.0864	0.0863
6.50	0.0765	0.0779	0.0784	0.0788	0.0791	0.0793	0.0794	0.0795	0.0796	0.0796	0.0796
7.00	0.0711	0.0723	0.0728	0.0731	0.0733	0.0735	0.0737	0.0738	0.0738	0.0739	0.0739
7.50	0.0664	0.0674	0.0679	0.0681	0.0684	0.0685	0.0687	0.0688	0.0688	0.0689	0.0689
8.00	0.0623	0.0632	0.0636	0.0638	0.0640	0.0642	0.0643	0.0644	0.0645	0.0645	0.0646
8.50	0.0586	0.0595	0.0598	0.0600	0.0602	0.0603	0.0605	0.0605	0.0606	0.0607	0.0607
9.00	0.0554	0.0561	0.0564	0.0566	0.0568	0.0569	0.0570	0.0571	0.0572	0.0572	0.0573
9.50	0.0525	0.0532	0.0534	0.0536	0.0538	0.0539	0.0540	0.0541	0.0541	0.0542	0.0542
10.0	0.0499	0.0505	0.0507	0.0509	0.0510	0.0511	0.0512	0.0513	0.0514	0.0514	0.0515

TABLE 4 (*continued*)

z/r	1.1	1.2	1.3	1.4	1.5	1.6	1.7	1.8	1.9	2.0
0.05	0.5848	0.5206	0.4716	0.4321	0.3993	0.3715	0.3475	0.3265	0.3080	0.2915
0.10	0.5768	0.5167	0.4693	0.4305	0.3982	0.3706	0.3468	0.3259	0.3075	0.2911
0.15	0.5660	0.5108	0.4655	0.4280	0.3963	0.3692	0.3457	0.3251	0.3068	0.2905
0.20	0.5537	0.5033	0.4606	0.4245	0.3938	0.3673	0.3442	0.3239	0.3058	0.2897
0.25	0.5407	0.4947	0.4548	0.4204	0.3907	0.3649	0.3423	0.3223	0.3046	0.2887
0.30	0.5275	0.4855	0.4482	0.4155	0.3870	0.3620	0.3400	0.3205	0.3031	0.2874
0.35	0.5143	0.4758	0.4410	0.4101	0.3829	0.3588	0.3374	0.3184	0.3013	0.2859
0.40	0.5012	0.4659	0.4335	0.4044	0.3784	0.3552	0.3345	0.3160	0.2993	0.2843
0.45	0.4884	0.4558	0.4257	0.3983	0.3735	0.3513	0.3314	0.3134	0.2972	0.2825
0.50	0.4758	0.4457	0.4177	0.3919	0.3684	0.3472	0.3280	0.3106	0.2948	0.2805
0.55	0.4635	0.4357	0.4096	0.3854	0.3631	0.3428	0.3244	0.3076	0.2923	0.2783
0.60	0.4515	0.4259	0.4015	0.3787	0.3577	0.3383	0.3206	0.3044	0.2896	0.2760
0.65	0.4399	0.4161	0.3934	0.3720	0.3521	0.3337	0.3167	0.3011	0.2868	0.2736
0.70	0.4286	0.4066	0.3854	0.3653	0.3464	0.3289	0.3127	0.2977	0.2838	0.2711
0.75	0.4177	0.3972	0.3774	0.3585	0.3407	0.3241	0.3085	0.2942	0.2808	0.2685
0.80	0.4071	0.3881	0.3696	0.3518	0.3350	0.3192	0.3043	0.2905	0.2777	0.2658
0.85	0.3969	0.3791	0.3618	0.3452	0.3293	0.3142	0.3001	0.2869	0.2745	0.2630
0.90	0.3870	0.3705	0.3543	0.3386	0.3236	0.3093	0.2958	0.2831	0.2713	0.2601
0.95	0.3774	0.3620	0.3468	0.3321	0.3179	0.3044	0.2915	0.2794	0.2680	0.2572
1.00	0.3681	0.3538	0.3396	0.3257	0.3123	0.2994	0.2872	0.2756	0.2646	0.2543
1.10	0.3505	0.3380	0.3255	0.3133	0.3013	0.2897	0.2786	0.2680	0.2579	0.2483
1.20	0.3341	0.3232	0.3122	0.3013	0.2906	0.2802	0.2701	0.2604	0.2512	0.2423
1.30	0.3188	0.3092	0.2995	0.2899	0.2803	0.2709	0.2618	0.2529	0.2444	0.2363
1.40	0.3045	0.2961	0.2876	0.2789	0.2704	0.2619	0.2536	0.2456	0.2378	0.2302
1.50	0.2912	0.2838	0.2762	0.2685	0.2609	0.2532	0.2457	0.2384	0.2312	0.2243
1.60	0.2788	0.2722	0.2655	0.2587	0.2518	0.2449	0.2381	0.2314	0.2248	0.2184
1.70	0.2672	0.2614	0.2554	0.2493	0.2431	0.2369	0.2307	0.2246	0.2186	0.2127
1.80	0.2563	0.2512	0.2459	0.2404	0.2348	0.2292	0.2236	0.2180	0.2125	0.2071
1.90	0.2462	0.2416	0.2369	0.2320	0.2269	0.2219	0.2167	0.2116	0.2066	0.2016
2.00	0.2367	0.2326	0.2284	0.2240	0.2194	0.2148	0.2102	0.2055	0.2009	0.1963
2.10	0.2278	0.2242	0.2204	0.2164	0.2123	0.2081	0.2039	0.1996	0.1954	0.1911
2.20	0.2194	0.2162	0.2128	0.2092	0.2055	0.2017	0.1979	0.1940	0.1900	0.1861
2.30	0.2116	0.2087	0.2056	0.2024	0.1991	0.1956	0.1921	0.1885	0.1849	0.1813
2.40	0.2042	0.2016	0.1989	0.1960	0.1929	0.1898	0.1866	0.1833	0.1800	0.1766
2.50	0.1973	0.1950	0.1925	0.1899	0.1871	0.1842	0.1813	0.1783	0.1752	0.1721
2.60	0.1908	0.1887	0.1864	0.1840	0.1815	0.1789	0.1762	0.1734	0.1706	0.1678
2.70	0.1846	0.1827	0.1807	0.1785	0.1762	0.1738	0.1714	0.1688	0.1662	0.1636
2.80	0.1788	0.1771	0.1753	0.1733	0.1712	0.1690	0.1667	0.1644	0.1620	0.1596
2.90	0.1733	0.1718	0.1701	0.1683	0.1664	0.1644	0.1623	0.1601	0.1579	0.1557
3.00	0.1681	0.1667	0.1652	0.1636	0.1618	0.1600	0.1581	0.1561	0.1540	0.1519
3.20	0.1585	0.1574	0.1561	0.1548	0.1533	0.1517	0.1501	0.1484	0.1467	0.1448
3.40	0.1499	0.1489	0.1479	0.1468	0.1455	0.1442	0.1428	0.1414	0.1398	0.1383
3.60	0.1421	0.1413	0.1404	0.1395	0.1384	0.1373	0.1361	0.1348	0.1335	0.1322
3.80	0.1350	0.1343	0.1336	0.1328	0.1319	0.1309	0.1299	0.1288	0.1277	0.1265
4.00	0.1285	0.1280	0.1274	0.1267	0.1259	0.1251	0.1242	0.1233	0.1223	0.1213
4.20	0.1226	0.1222	0.1217	0.1211	0.1204	0.1197	0.1189	0.1181	0.1173	0.1164
4.40	0.1172	0.1169	0.1164	0.1159	0.1153	0.1147	0.1141	0.1134	0.1126	0.1118
4.60	0.1123	0.1119	0.1116	0.1111	0.1106	0.1101	0.1095	0.1089	0.1083	0.1076
4.80	0.1077	0.1074	0.1071	0.1067	0.1063	0.1058	0.1053	0.1048	0.1042	0.1036
5.00	0.1034	0.1032	0.1029	0.1026	0.1022	0.1018	0.1014	0.1009	0.1004	0.0999
5.50	0.0941	0.0940	0.0938	0.0936	0.0933	0.0930	0.0927	0.0923	0.0920	0.0915
6.00	0.0863	0.0862	0.0861	0.0859	0.0857	0.0855	0.0853	0.0850	0.0847	0.0844
6.50	0.0796	0.0796	0.0795	0.0794	0.0792	0.0791	0.0789	0.0787	0.0785	0.0783
7.00	0.0739	0.0739	0.0738	0.0737	0.0736	0.0735	0.0734	0.0732	0.0731	0.0729
7.50	0.0689	0.0689	0.0689	0.0688	0.0688	0.0687	0.0686	0.0685	0.0683	0.0682
8.00	0.0646	0.0646	0.0645	0.0645	0.0645	0.0644	0.0643	0.0642	0.0641	0.0640
8.50	0.0607	0.0607	0.0607	0.0607	0.0607	0.0606	0.0606	0.0605	0.0604	0.0603
9.00	0.0573	0.0573	0.0573	0.0573	0.0573	0.0572	0.0572	0.0571	0.0571	0.0570
9.50	0.0542	0.0542	0.0543	0.0542	0.0542	0.0542	0.0542	0.0541	0.0541	0.0540
10.0	0.0515	0.0515	0.0515	0.0515	0.0515	0.0515	0.0515	0.0514	0.0514	0.0513

Source: Ref. 37.

can be defined using either a constant current or a constant concentration approximation [41]. In the first case, J_{AV} = const = $j^*/(\pi c^\circ D a_S)$, where j^* is the total flux at the substrate, and

$$C(R,Z) = \frac{2J_{AV}}{\pi} \int_0^1 \sqrt{u/R} pK(p)\, du \qquad (41)$$

The practically important value, $C(0,Z)$, corresponds to a small tip electrode positioned over the center of the substrate. In this case, $K(0) = \pi/2$, and

$$C(0,Z) = 2J_{AV}(\sqrt{Z^2 + 1} - z) \qquad (42)$$

in agreement with Ref. 42, where the same result was obtained in a different way.

The constant concentration assumption with the primary current distribution [43],

$$J_S(R) = 0.5 J_{AV}/\sqrt{1 - r^2} \qquad (43)$$

where the average dimensionless flux density, J_{AV}, is equal to the total flux divided by the disk area, is given above. With this assumption, the concentration profile is

$$C(R,Z) = \frac{J_{AV}}{\pi} \int_0^1 \sqrt{u/R(1 - u^2)}\, pK(p)\, du \qquad (44)$$

The comparison of the dimensionless concentration profiles, i.e., the function $C(R,Z)/J_{AV}$, computed from Eq. (41) (Table 4) and Eq. (44) showed that the differences are minor and can be detected only at quite small tip/substrate separations.

The potentiometric SECM experiment yields the potential of the tip electrode, E, as a function of the tip/substrate separation. To establish a correspondence between these data and the above theory, one needs to use a calibration curve, which is an E vs. c plot. A typical form of such a dependence is

$$E = E' + S \log c \qquad (45)$$

where E' is a constant and S is the experimentally determined slope equal to 0.059 V for an ideal one-electron nernstian reaction at 25°C. Using such a calibration, one can transform the experimental results to the c vs. (z,r) dependence and fit them to the theoretical values in Table 4 to find J_S and establish the distance scale. Similar procedures were described above for the feedback mode of the SECM.

When the tip electrode is not small compared to the substrate, its potential is determined by the average value of surface concentration. For a hemispherical substrate, the average distance from the source center to the surface is, when the tip and substrate centers lay on the same vertical line,

$$<r'> = \frac{2[(z^2 + a^2)^{3/2} - z^3]}{3a^2} \quad (46)$$

where z is the distance between the tip and substrate centers and a is the tip radius. The average value of the surface concentration can be obtained by substituting $<r'>$ in Eq. (39). When the substrate is a microdisk, the approximate average value of the surface concentration can be computed from Table 4 according to

$$<c> = \frac{2}{a^2} \int_0^a c(r) r \, dr \quad (47)$$

One should note that the above theory implies that the products generated on the substrate do not participate in any chemical reaction in solution. Otherwise (e.g., when the tip is a pH sensor used to monitor proton concentration in a buffered aqueous solution [37]), a more complicated treatment may be necessary.

2. *G/C Mode with Amperometric Tip*

Two significantly different G/C SECM arrangements are shown schematically in Figure 24. The substrate generation/tip collection (SG/TC) mode (Fig. 24A) was historically the first SECM-type measurement performed [10,11,44]. The aim of such experiments was to probe the diffusion layer generated by the large substrate electrode using a much smaller tip electrode as an amperometric sensor. A simple approximate theory [10,11] using the well-known c(z,t) function for a potentiostatic transient at a planar electrode [17] was developed to predict the evolution of the concentration profile

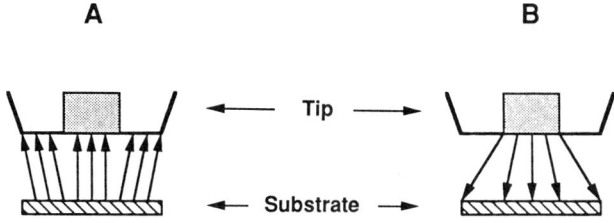

FIG. 24. Schematic representation of the (A) SG/TC and (B) TG/SC modes of the SECM operation. (Reprinted with permission from Ref. 46. Copyright 1992 American Chemical Society.)

following the substrate potential perturbation. A more complicated theory was based on the concept of the "impulse response function" [44]. While these theories have been successful in calculating concentration profiles, the prediction of the time-dependent tip current response is not straightforward because it is a complex function of the concentration distribution. Moreover, these theories do not account for distortions caused by interference of the tip and substrate diffusion layers and feedback effects.

A somewhat different approach to SG/TC theory [45] deals with the possibility of the detection of the current for a reactant flux diverging from a point source. Two features of this proposed model make it inapplicable to real experimental situations: (1) the point source is not inlaid, i.e., an object much smaller than the tip electrode is not supported by any macroscopic substrate; and (2) the tip is assumed to be a sphere (or a part of a sphere, also noninlaid). Thus, the results of Ref. 45 can be considered only as qualitatively useful in SECM.

The following shortcomings (in addition to the theoretical problems listed in the beginning of this section) limit the applicability of the SG/TC mode: (1) the process at a large substrate electrode is always non–steady state; (2) a large substrate current may cause significant IR-drop; and (3) the collection efficiency, i.e., the i_T to i_S ratio, is low. The tip-generation/substrate collection (TG/SC) mode (Fig. 24B) was found to be better [46,47]. The TG/SC experiment (unlike the feedback mode of the SECM) includes simultaneous measurements of both transient tip and substrate currents. For an uncomplicated process at short times, the substrate current is initially close to zero and grows as the tip current decreases (Fig. 25), until the difference between them vanishes gradually. At steady state, these quantities are almost identical, if L is not very large [46,47] (the collection efficiency, i_S/i_T, is more than 0.99 at $L \leq 2$, and $i_S/i_T \cong 0.8$ at $L = 5$ [46]). Under these conditions, the tip-generated species predominantly diffuses to the large substrate, rather than escape from the tip-substrate gap. For a process with a coupled chemical reaction, there are large differences between i_S and i_T, and both quantities provide important kinetic information. The TG/SC theory for a process with a second-order following chemical reaction [46] will be considered in the next section.

Concluding our discussion of the G/C mode of the SECM, the TG/SC arrangement is certainly advisable for kinetic measurements, while SG/TC (preferably, in potentiometric mode) can be used for monitoring corrosion, enzymatic reactions, and other heterogeneous processes at the substrate surface.

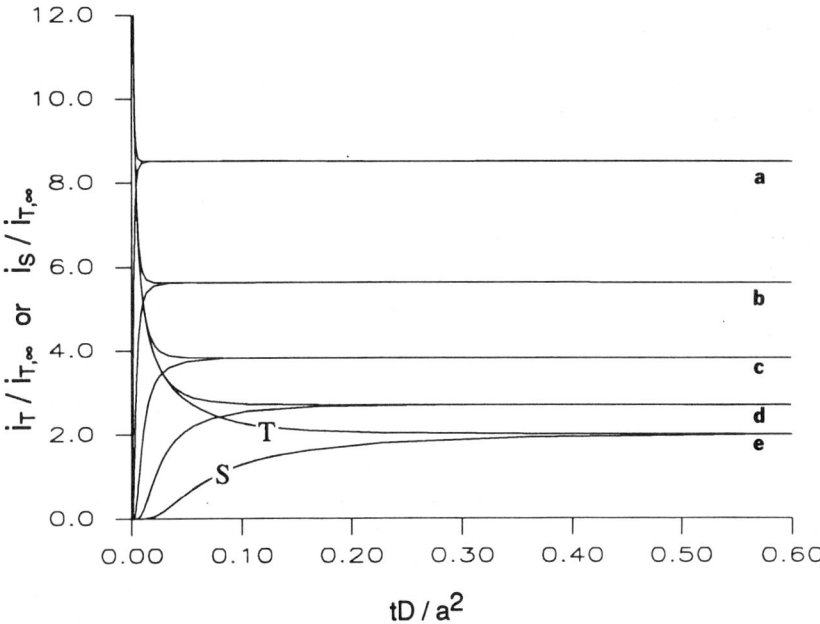

FIG. 25. Tip and substrate current transients for diffusion-controlled conditions. The data are for Log L = (a) −1, (b) −0.8, (c) −0.6, (d) −0.4, and (e) −0.2. The tip and substrate currents are denoted by T and S on curve e. (Reprinted with permission from Ref. 46. Copyright 1992 American Chemical Society.)

C. Processes with First- and Second-Order Homogeneous Reactions in the Gap

A homogeneous chemical reaction occurring in the gap between the tip and substrate electrodes causes a change in i_T. Thus, its rate can be determined from either feedback or G/C SECM measurements. If both heterogeneous processes at the tip and substrate electrodes are fast (at extreme potentials of both working electrodes) and the chemical reaction (rate constant, k_c) is irreversible, the SECM response is a function of a single kinetic parameter K = const × k_c/D, and its value can be extracted from I_T vs. L dependencies.

SECM theory has been reported for electrochemical processes coupled with either first- or second-order homogeneous chemical reactions. Two approaches to the evaluation of the kinetics of the first-order irreversible reaction, i.e., the E_rC_i mechanism

$$R - ne^- \rightarrow O; \quad O \xrightarrow{k_c} P \tag{48}$$

have been proposed [32]: (1) analysis of the chronoamperometric SECM response (Fig. 26) as a function of L and k_c, and (2) fitting of experimental steady-state $i_T - d$ curves to the theoretical dependencies calculated for various values of the dimensionless homogeneous rate constant $K = k_c a^2/D$ (Fig. 27). All theoretical data were generated by numerical solution of the time-dependent diffusion problem using the ADI finite-difference method. With an increase in K, the shape of transients in Figure 26 changes gradually from one corresponding to the kinetically uncomplicated conductive substrate behavior (upper dashed curve) to that of an insulating substrate (lower dashed curve). The intermediate curves correspond to the range of rate constants measurable in a transient experiment, which can be (semi-qualitatively) represented by a zone diagram (Fig. 28). The approach utilizing steady-state current-distance curves [or, equivalently, I_T vs. K working curves (Fig. 29)] is free of problems typical for transient measurements and is probably more reliable. First-order rate constants in excess of 2×10^4 sec^{-1} should be accessible to measurements under steady-state conditions.

An analytical solution for a somewhat more general E_qC_r mechanism was obtained in terms of multidimensional integral equations [29], but it has not been utilized in any calculations.

The detailed TG/SC theory developed for an electrode process with a following dimerization reaction (E_rC_{2i} mechanism, see also Fig. 30) [46]:

tip: $\quad O + ne^- \rightarrow R \tag{49}$

gap: $\quad 2R \xrightarrow{k_c'} \text{products} \tag{50}$

substrate: $\quad R - ne^- \rightarrow O \tag{51}$

dealt with both chronoamperometric and steady-state responses. Both tip and substrate transients were computed for various values of the chemical rate constant, k_c' (incorporated in the dimensionless parameter $K' = k_c' a^2 c_O^\circ/D$), and are compared with those for a diffusion-controlled process (Fig. 31). At short times, both the tip and substrate chronoamperograms follow the behavior predicted for a kinetically uncomplicated situation. At times comparable with the lifetime of R, a deviation from this behavior occurs because the following chemical reaction consumes this intermediate. The competition between the chemical reaction and the diffusion of the intermediate from the tip results in the appearance of the peak on I_S-T curves. The larger the K' value, the smaller the peak current, and the shorter the time corresponding to this peak. The steady-state values of both tip and substrate currents are clearly

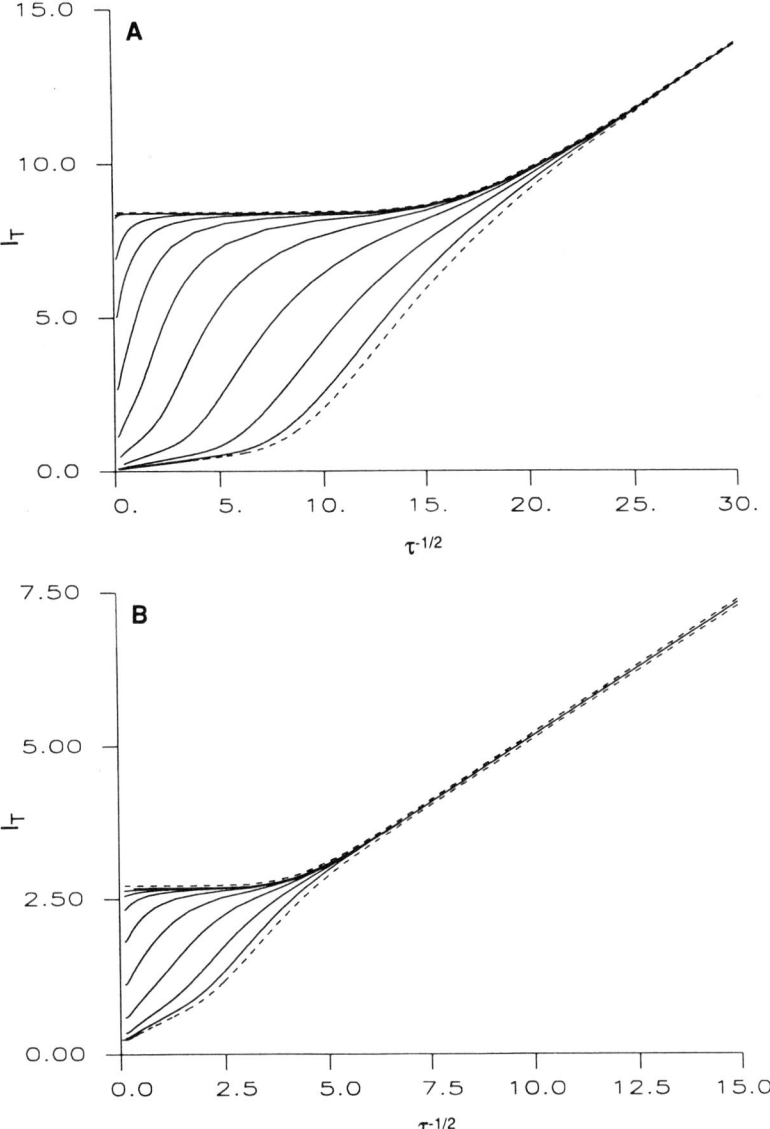

FIG. 26. Tip-current transients for the E_rC_i mechanism for log (d/a) = (A) −1 and (B) −0.4. The upper and lower dashed curves in each diagram correspond, respectively to diffusion-controlled processes with conductive and insulating substrates. The solid curves (from top to bottom) represent (A) log K = −2.0, −1.5, −0.5, 0, 0.5, 1.0, 1.5, 2.0, 2.5, and 3.0 and (B) log K = −2.2, −1.7, −1.2, −0.7, −0.2, 0.3, 0.8, 1.3, and 1.8. (Reprinted with permission from Ref. 32. Copyright 1991 American Chemical Society.)

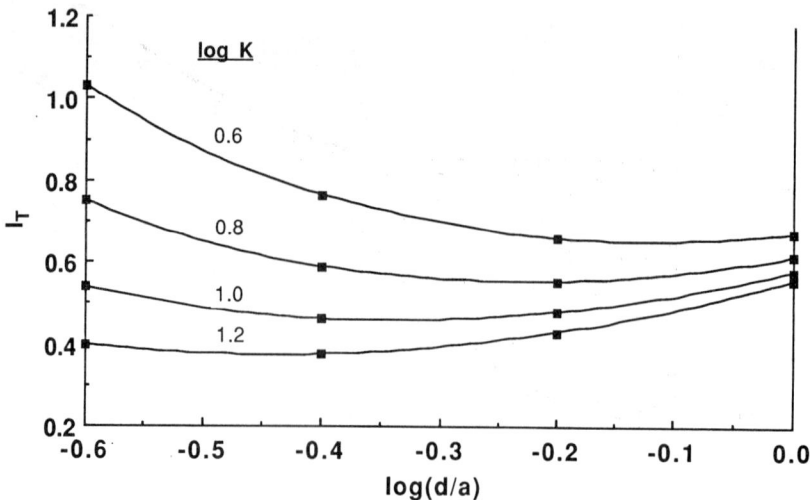

FIG. 27. Steady-state feedback current-distance curves for various values of log K for the E_rC_i mechanism. (Reprinted with permission from Ref. 32. Copyright 1991 American Chemical Society.)

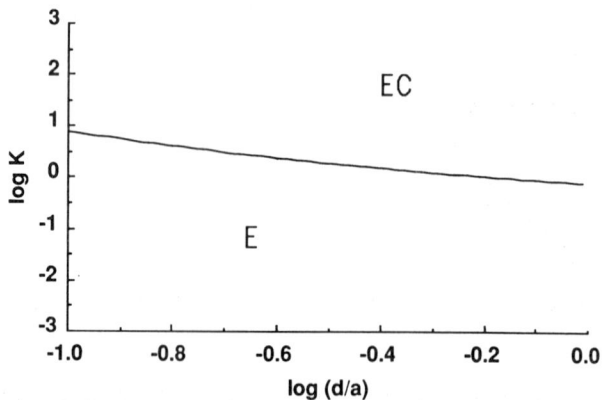

FIG. 28. Kinetic zone diagram for the I_T-T characteristics (up to T = 1.65 T_C, where T_C is the dimensionless critical diffusion time [31,32]) for the E_rC_i mechanism representing the regions of finite homogeneous kinetics influencing the transient behavior (EC) and simple positive feedback (E). $K = k_c a^2/D$. (Reprinted with permission from Ref. 32. Copyright 1991 American Chemical Society.)

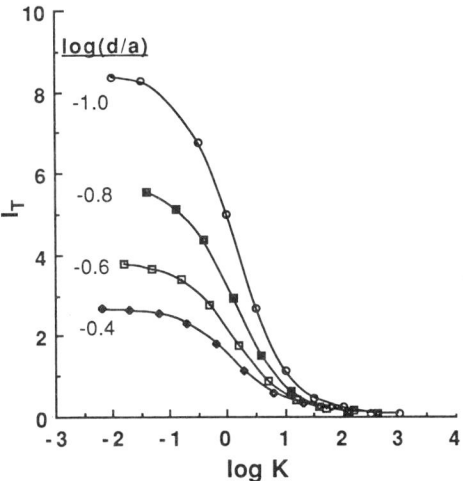

FIG. 29. Steady-state I_T-K working curves for various values of log L (L = d/a) for the E_rC_i mechanism. (Reprinted with permission from Ref. 32. Copyright 1991 American Chemical Society.)

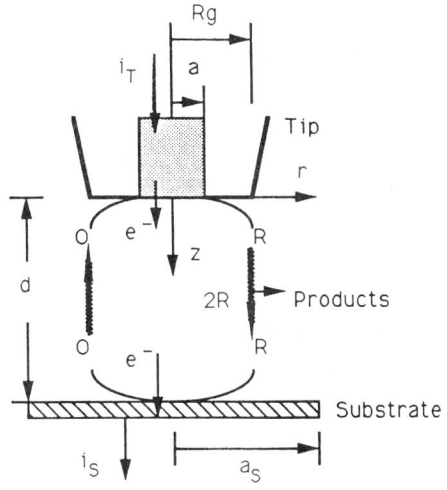

FIG. 30. Principles of the TG/SC mode in studying dimerization kinetics of electrogenerated species. The electrode and chemical and diffusion processes are shown along with the relevant coordinate system and notation. (Reprinted with permission from Ref. 46. Copyright 1992 American Chemical Society.)

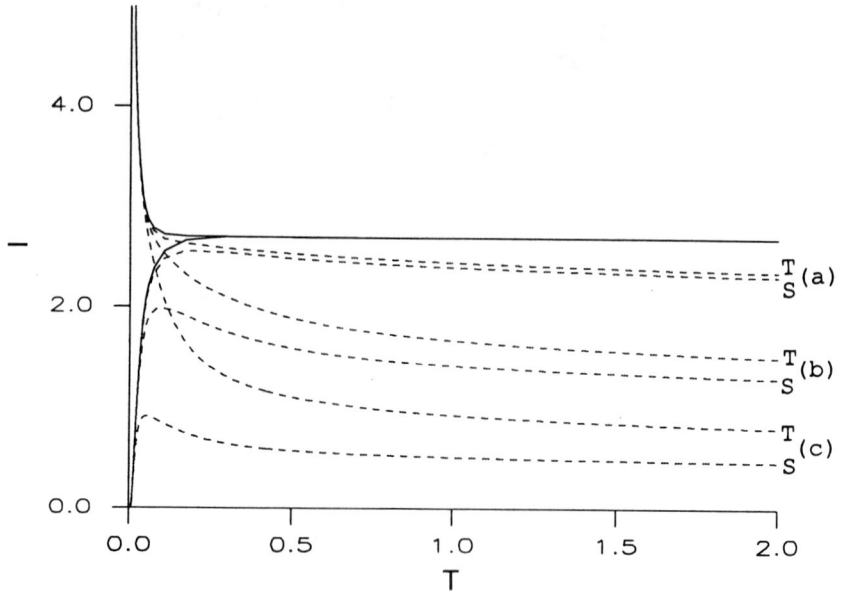

FIG. 31. SECM tip and substrate chronoamperometric characteristics for the E_rC_{2i} mechanism with log L = −0.4. The solid line shows the behavior in the absence of a following homogeneous reaction, and the labeled dashed lines are for K′ = (a) 1, (b) 10, and (c) 100. The labels T and S denote tip and substrate, respectively. (Reprinted with permission from Ref. 46. Copyright 1992 American Chemical Society.)

different from those expected in the absence of the following reaction (solid lines in Fig. 31). These differences increase with an increase in K′, suggesting the possibility of the extraction of the K′ value from steady-state data. Two families of working curves representing steady-state tip current and collection efficiency as functions of K′ and L are shown in Figure 32. The K′ value can be obtained by fitting experimental I_T-L and I_S-L curves to the theoretical data in Table 5. The upper limit for a second-order rate constant accessible from such measurements is at least 4×10^8 M^{-1} s^{-1}, that is, over one order of magnitude greater than the analogous value for RRDE.

D. Adsorption/Desorption Processes

The use of the SECM in studies of adsorption/desorption kinetics and surface diffusion at the solid/liquid interface is based on its abilities to perturb the local adsorption equilibrium and measure the resulting flux of adsorbate

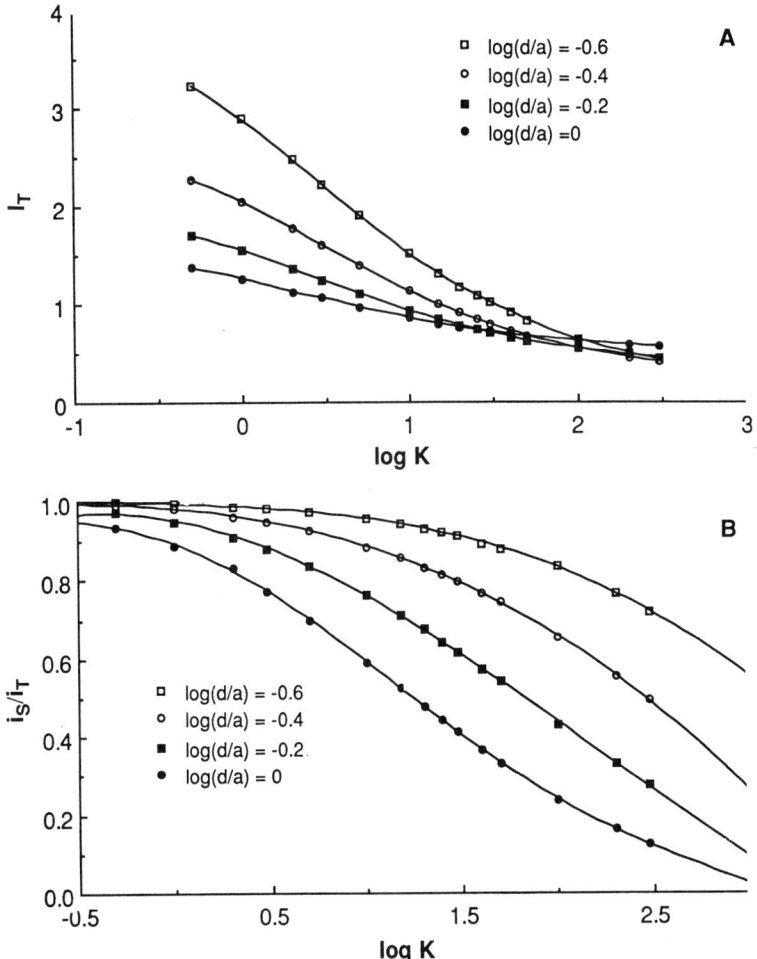

FIG. 32. Normalized tip current (A) and collection efficiency (B) for the E_rC_{2i} mechanism at various tip-substrate separations. (Reprinted with permission from Ref. 46. Copyright 1992 American Chemical Society.)

leaving the surface. The principles of the relevant technique, named scanning electrochemical microscope-induced desorption (SECMID) [48], are outlined in Figure 33. The UME tip is positioned close to the surface of the single crystal substrate covered with an adsorbate (H^+ in Fig. 33). Following the application of the potential step to the tip electrode, proton reduction begins

TABLE 5

Theoretical Normalized Tip and Substrate Steady-State Currents for the E_rC_{2i} Mechanism

log(d/a)	K = 1 $i_T/i_{T,\infty}$	K = 1 $i_S/i_{T,\infty}$	K = 5 $i_T/i_{T,\infty}$	K = 5 $i_S/i_{T,\infty}$	K = 10 $i_T/i_{T,\infty}$	K = 10 $i_S/i_{T,\infty}$	K = 20 $i_T/i_{T,\infty}$	K = 20 $i_S/i_{T,\infty}$	K = 50 $i_T/i_{T,\infty}$	K = 50 $i_S/i_{T,\infty}$	K = 100 $i_T/i_{T,\infty}$	K = 100 $i_S/i_{T,\infty}$	K = 200 $i_T/i_{T,\infty}$	K = 200 $i_S/i_{T,\infty}$	K = 500 $i_T/i_{T,\infty}$	K = 500 $i_S/i_{T,\infty}$	K = 1000 $i_T/i_{T,\infty}$	K = 1000 $i_S/i_{T,\infty}$	K → ∞ $i_T/i_{T,\infty}$
−0.8	4.255	4.236	2.778	2.756	2.186	2.144	1.672	1.630	1.166	1.099	0.864	0.797	0.629	0.577	0.433	0.365	0.335	0.254	0.102
−0.7	3.490	3.479	2.289	2.256	1.813	1.756	1.396	1.336	0.977	0.898	0.728	0.655	0.552	0.470	0.392	0.296	0.311	0.205	0.127
−0.6	2.889	2.873	1.909	1.860	1.517	1.448	1.180	1.098	0.834	0.734	0.639	0.536	0.498	0.382	0.369	0.239	0.303	0.164	0.159
−0.5	2.418	2.392	1.616	1.544	1.296	1.200	1.021	0.908	0.726	0.601	0.581	0.439	0.464	0.311	0.361	0.192	0.307	0.131	0.198
−0.4	2.050	2.011	1.392	1.291	1.131	1.001	0.906	0.754	0.673	0.501	0.546	0.359	0.451	0.252	0.367	0.153	0.324	0.102	0.246
−0.3	1.766	1.708	1.225	1.089	1.012	0.840	0.828	0.629	0.638	0.413	0.535	0.292	0.458	0.202	0.389	0.119	0.356	0.078	0.303
−0.2	1.549	1.468	1.105	0.925	0.931	0.708	0.779	0.526	0.628	0.339	0.545	0.236	0.484	0.160	0.430	0.091	0.403	0.058	0.371
−0.1	1.386	1.276	1.023	0.791	0.882	0.600	0.760	0.439	0.639	0.277	0.574	0.189	0.527	0.125	0.485	0.069	0.466	0.042	0.448
0.0	1.265	1.123	0.972	0.680	0.860	0.508	0.764	0.365	0.670	0.224	0.620	0.148	0.585	0.095	0.556	0.050	0.542	0.030	0.534
0.1	1.178	0.997	0.946	0.586	0.859	0.429	0.786	0.302	0.715	0.179	0.680	0.115	0.655	0.072	0.635	0.036	0.626	0.021	0.624
0.2	1.123	0.892	0.939	0.503	0.874	0.361	0.821	0.247	0.770	0.141	0.746	0.088	0.729	0.053	0.701	0.026	0.711	0.015	0.710

Source: Ref. 46.

Scanning Electrochemical Microscopy

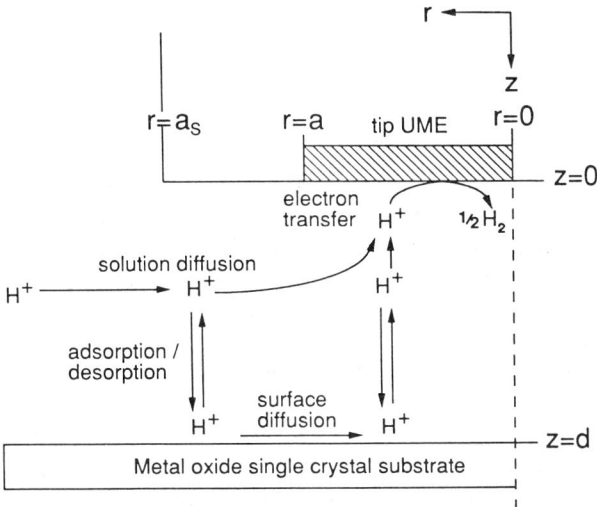

FIG. 33. Schematic representation of the transport processes coupled with a reversible adsorption/desorption process at the substrate. Following the application of a potential step at the tip UME, the transport of protons to the tip/substrate domain occurs via both solution diffusion and surface diffusion of adsorbate. (Reprinted with permission from Ref. 48. Copyright 1992 American Chemical Society.)

leading to the depletion of H^+ in the tip-substrate gap. The depletion in turn results in two competing processes: the desorption of protons from the substrate and surface diffusion driven by the surface concentration gradient. The desorbed species can diffuse to the tip and react at its surface, thus contributing to the measured tip current. The rates of both desorption and surface diffusion processes can be deduced from the chronoamperometric SECM response [48].

The related theory is more cumbersome than that in any case previously considered. The time-dependent diffusion equation, initial conditions, and boundary conditions for the tip surface are the same as previously for a diffusion-controlled process. The substrate boundary condition (52) accounts for both cylindrical surface diffusion and reversible adsorption/desorption (assuming Langmuirian characteristics):

$Z = L; 0 < R < \rho;$

$$\gamma \frac{\partial \theta'}{\partial \tau} = D_a \gamma \left[\frac{\partial^2 \theta'}{\partial R^2} + \frac{1}{R} \frac{\partial \theta'}{\partial R} \right] - K_d^{app} \theta' + K_a^{app} C(L) (1 - \theta') \qquad (52a)$$

$$0 < R < \rho; \tau = 0; \theta' = \left(1 + \frac{K_a^{app}}{K_d^{app}}\right)^{-1} \tag{52b}$$

$$\rho < R; \theta' = \left(1 + \frac{K_a^{app}}{K_d^{app}}\right)^{-1} \tag{52c}$$

where $\tau = tD_{sol}/a^2$, $K_a^{app} = k_a^{app} a/D_{sol}$, $K_d^{app} = k_d^{app} a/D_{sol}c°$, $\gamma = N°/c°a$, $D_a = D_{sur}/D_{sol}$, D_{sol} and D_{sur} are the diffusion coefficient of the species of interest in solution and the surface diffusion coefficient of the adsorbate, respectively, θ' is the fractional surface coverage, k_a^{app} (cm s^{-1}) and k_d^{app} (mole/cm^2/sec) are the apparent adsorption and desorption rate constants, and $N°$ is the maximum density of adsorption sites.

The results of the numerical solution of this problem by the ADI method [48] were summarized with a few families of working curves (Figs. 34 and 35). Figure 34 illustrates the effects of K_a^{app}, L, K_d^{app}, and γ on the shape of tip chonoamperograms in absence of surface diffusion. For small K_d^{app} and K_a^{app} ($K_a^{app} = K_d^{app} = 0.1$ in Figs. 34A and 34B), the adsorption/desorption kinetics are slow compared to the rate of diffusional mass transport. The response of the adsorption/desorption process to the perturbation (depletion) in the concentration of the solution component of the adsorbate adjacent to the substrate-solution interface is sluggish, and thus, the current-time behavior is close to that predicted for an inert substrate. For faster adsorption/desorption processes, the response to the perturbation in the interfacial equilibrium is more rapid; the resulting desorption process contributes to a tip current that is larger than that for an inert substrate at short times. For a given value of γ, the faster the adsorption/desorption kinetics, the larger the current at short times, up to the limit where the kinetics are sufficiently fast that the adsorption/desorption process is essentially always at equilibrium on the time scale of the SECM measurements. The UME current response becomes increasingly sensitive to the surface processes as the tip-substrate separation is decreased (Fig. 34B), since the effect of this is both to hinder solution diffusion into the gap and to increase the ratio of effective substrate surface area to solution volume probed by the technique. In general, the closer the tip is located to the substrate, the greater is the depletion of the solution component of the adsorbate, and thus, the larger is the overall perturbation to the adsorption-desorption equilibrium. Processes involving much lower surface coverages can also be studied with SECMID, if the kinetics are sufficiently fast (see Fig. 34C for $K_a^{app}/K_d^{app} = 0.1$, i.e., $\theta'|_{\tau=0} = 0.091$). In this case, the lower initial surface coverage leads to currents for the various kinetic cases approaching the value for an inert substrate more rapidly. However, the various kinetic cases can be distinguished in the short-time region. Finally, increasing the surface concentration relative to that in solution (i.e., increasing γ, Fig. 34D) serves to increase the overall charge passed during the transient.

At times sufficiently long for a true steady state to prevail (typically $\tau >$ 200), I_T for all kinetic cases attains the value for an inert insulating substrate, the adsorption-desorption process reaches a new equilibrium, and the tip current depends only on the rate of diffusion of the species in solution.

Surface diffusion provides an additional path for the transport of the adsorbed mediator into the tip-substrate domain. Its main effect is to increase the magnitude of the current flowing in the longer time region of the transient and, in particular, to enhance the final steady-state current as compared to that at an inert substrate (Fig. 35). Clearly, the larger the surface diffusion coefficient, as compared to that in solution, the larger the steady-state current at the tip UME. Moreover, the larger the effective surface concentration (i.e., the larger the value of γ or θ'), the more pronounced the effect of surface diffusion on the UME current. The steady-state current becomes increasingly sensitive to the surface diffusion process as L is minimized (see Figs. 35A and 35B).

One can deduce from the above that the most suitable route to obtaining adsorption/desorption kinetic information is from the short-time transient behavior. Under these conditions the effect of surface diffusion is negligible. Alternatively, under steady-state conditions, the adsorption/desorption process is at equilibrium, and thus, the UME current depends on the solution and surface diffusion rates. Such measurements may provide a means for evaluating the surface diffusion rate, if the tip-substrate separation can be determined independently.

E. SECM with a Nondisk Tip and Tip Shape Characterization

Two aspects of the use of non-disk-shaped microtips in SECM experiments are the evaluation of the UME shape and the utilization of advantages associated with a specific tip geometry. Approximate expressions were obtained for the steady-state current in a thin-layer cell formed by two electrodes, for example, one a plane, and the second a cone or hemisphere [49]. It was shown that the normalized steady-state, diffusion-limited current, as a function of the normalized separation for thin-layer electrochemical cells, is fairly sensitive to the geometry of the electrodes. However, the thin-layer theory does not describe accurately the steady-state current between a small disk tip and a planar substrate because the tip steady-state current, $i_{T,\infty}$ was not included in the approximate model [49]. A more realistic approximate theory for SECM with a tip shaped as a cone or spherical segment was presented in

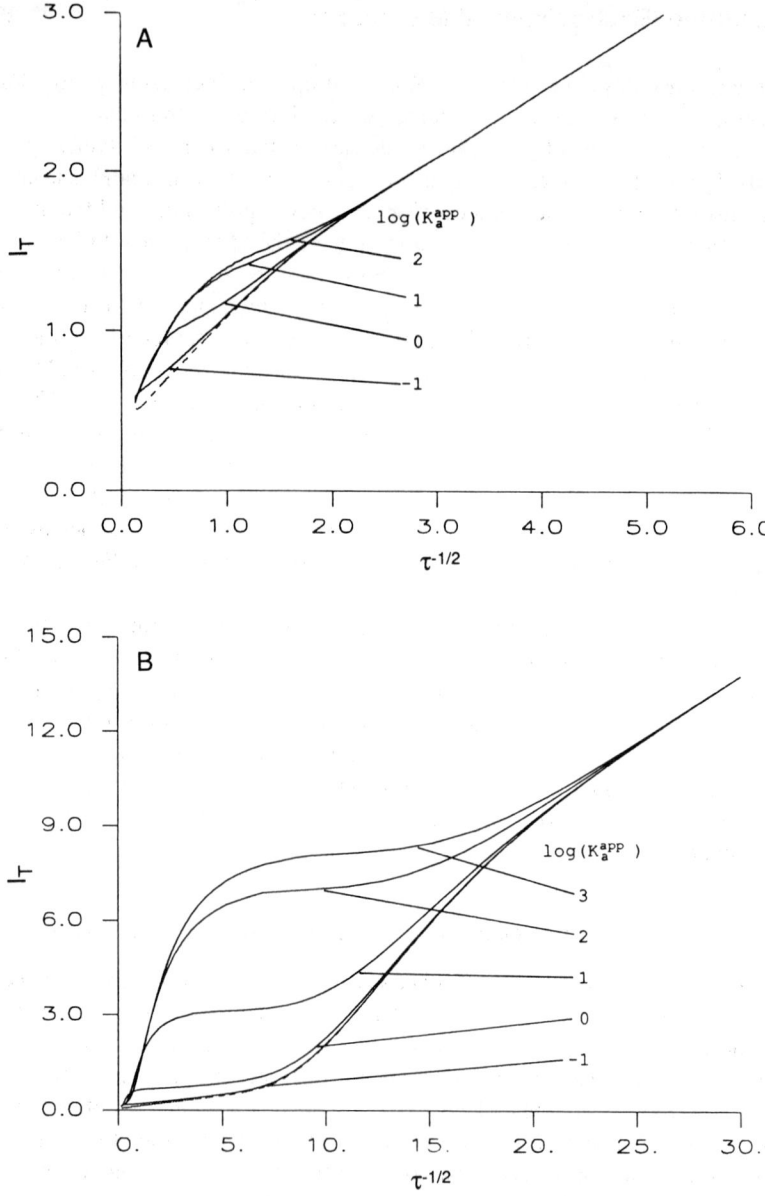

FIG. 34. Theoretical tip chronoamperograms for a substrate absorption/desorption process as a function of various parameters. (A) $\gamma = 8$, $K_d^{app} = K_a^{app}$, $\rho = 10$, and L = 1. (B) Same as (A), but L = 0.1. (C) $\gamma = 8$, $K_d^{app}/K_a^{app} = 10$, $\rho = 10$, and log L = -0.5. (D) $K_d^{app} = K_a^{app} = 10$, $\rho = 10$, log L = -0.5, and $\gamma = 2, 8$, and 25 (as indicated). (Reprinted with permission from Ref. 48. Copyright 1992 American Chemical Society.)

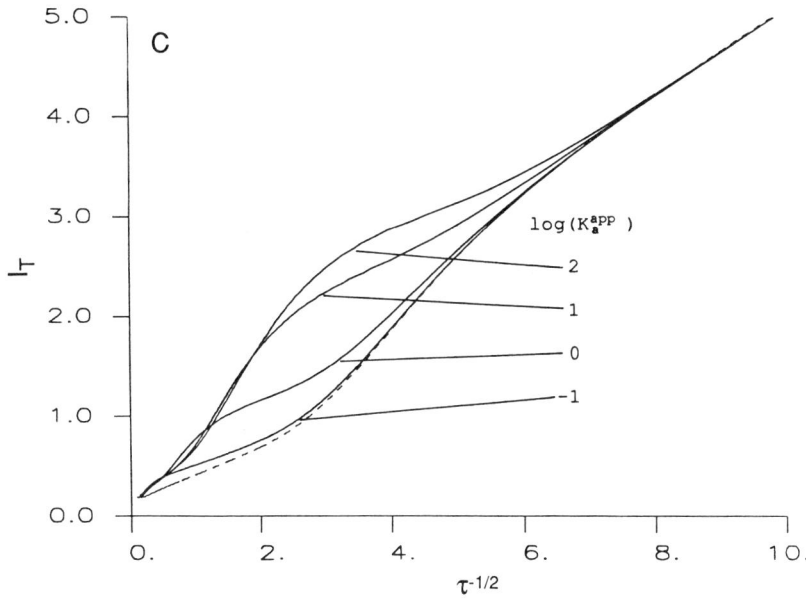
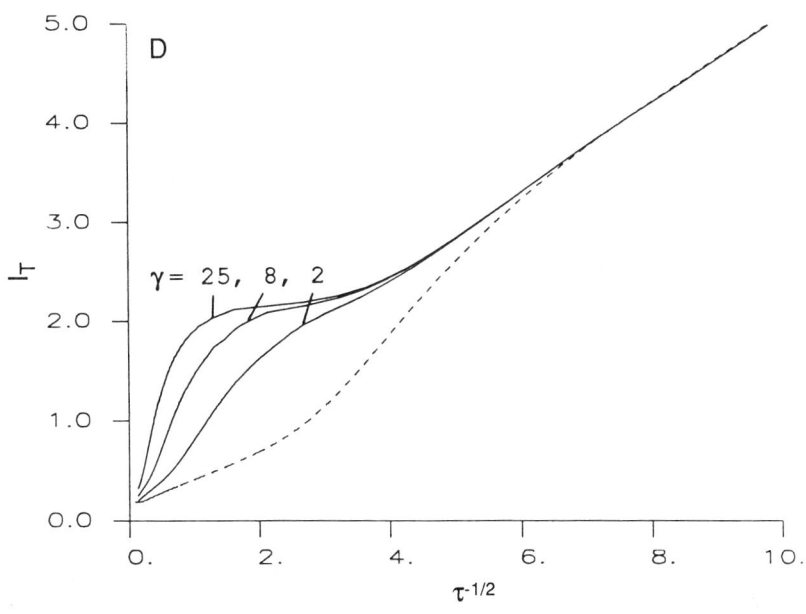

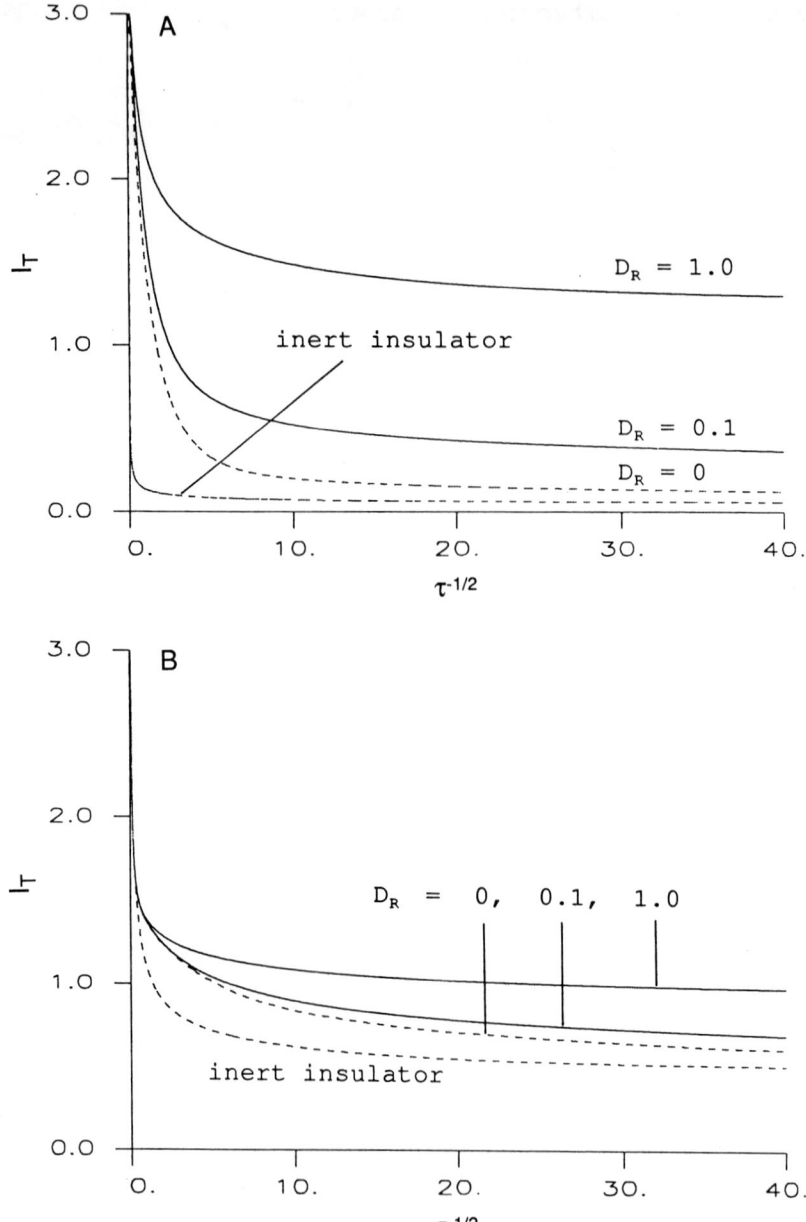

FIG. 35. The effect of the surface diffusion coefficient on the shape of tip chronoamperograms for a substrate adsorption/desorption process. $\gamma = 8$, $K_d^{app} = K_a^{app} = 10$, $\rho = 10$, and $L =$ (A) 0.1 and (B) 1. (Reprinted with permission from Ref. 48. Copyright 1992 American Chemical Society.)

Ref. 25. The surface of the nonplanar electrode was considered to be a series of thin circular strips, each of which is parallel to the planar electrode (Fig. 36). The diffusional flux to each strip was calculated from Eq. (25) for a conductive substrate or Eq. (27) for an insulating substrate. The normalized current to the whole surface of the segment can be expressed as

$$I_{sp}(L) = \frac{2}{\sin^2(\alpha_o)} \int_{\cos\alpha_0}^{1} I_{disk}(z) y \, dy \qquad (53)$$

for a spherically shaped tip and

$$I_{cone}(L) = \frac{2}{k^2} \int_0^k I_{disk}(x) y \, dy \qquad (54)$$

for a conical one, where L is the distance between the substrate and the point of tip closest to it normalized by the tip radius (i.e., $L = d_o/a_o$), $z = L + (1 - \cos \alpha)/\sin \alpha_o$, $x = 2 + y$, $I_{disk}(z)$ is the SECM current function given by Eq. (25) or Eq. (27), and k is the ratio of the cone height to its radius; other parameters are explained in Figure 36. The families of working curves for conductive and insulating substrates were computed from Eqs. (53) and (54) for spherical and conical (Fig. 37) tip shapes, respectively. These working curves represent I vs. L dependencies for various sizes and shapes of electrodes. One can see that the different working curves possess substantially different curvatures, thus a unique curve can be found to obtain the best fit with the experimental data. The upper curve in Figure 37A and the lower one in Figure 37B represent the theory for a microdisk tip [27]. One can also see from Figure 37C that the current in a cone/plane cell (unlike that in a disk/plane cell) tends to some limiting value as $L \to 0$. Obviously, as $k \to 0$ the conical working curves approach that computed for a disk-shaped tip.

The recent discussion of the applicability of nm-size microelectrodes ("nanodes" [26]) to electrochemical measurements [50,51] called attention to the importance of a thorough evaluation of the shape of such electrodes. The conclusion was that any electrode in which the conductor tip is withdrawn inside a "lagoon" formed by insulating glass is hardly suitable for kinetic experiments. The alternative is an inlaid microelectrode, such as a microdisk, or a tip with a conductor protruding from the insulating sheath, such as a cone or a spherical cap. While it is difficult to construct very small (nm) insulated electrodes with a true disk or hemispherical shape, the preparation of spherically or conically shaped ultramicroelectrodes is easier [25]. The shape of such a tip can be characterized by fitting the experimental current-distance curves to the theoretical ones; the shape and size of a conical tip with a radius

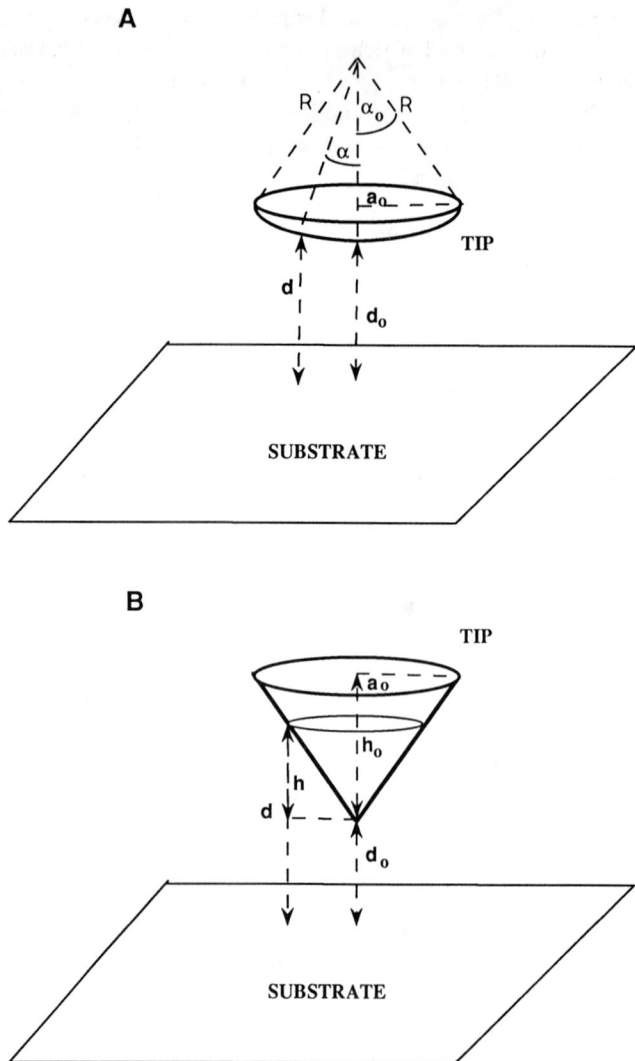

FIG. 36. Schematic diagrams of thin layer cells formed by a spherical segment tip and a planar substrate (A) and by a conical tip and a planar substrate (B). (Reprinted with permission from Ref. 25. Copyright 1992 Elsevier.)

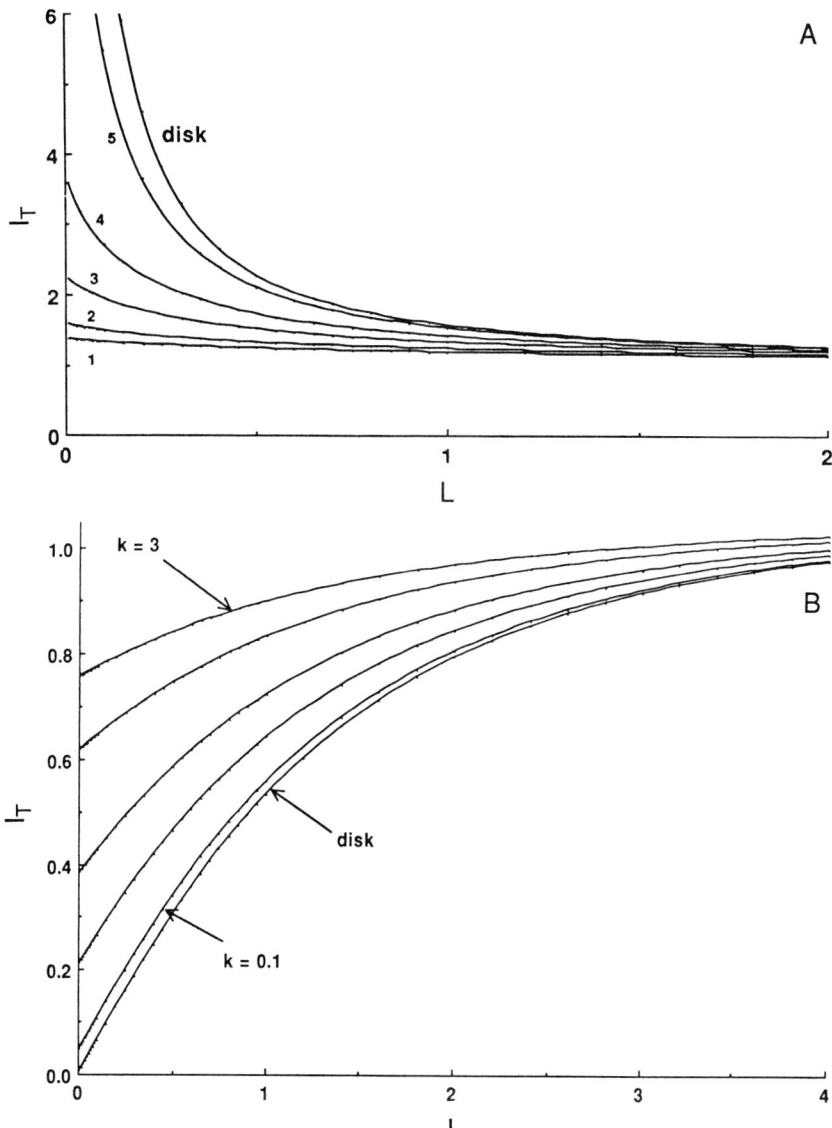

FIG. 37. Steady-state current-distance curves for a conical tip over conductive (A and C) and insulating (B) substrates corresponding to different values of the parameter $k = h_0/a_0$. (A) $k = 3$ (curve 1), 2 (curve 2), 1 (curve 3), 0.5 (curve 4), and 0.1 (curve 5). The upper curve was computed for a disk-shaped tip from Eq. (25). (B) From top to bottom, $k = 3, 2, 0.5$, and 0.1. The lower curve was computed for a disk-shaped tip from Eq. (27). (C) The same data as in (A) but in a semi-logarithmic scale. Triangles are computed from Eq. (9) in Ref. 49 with the same parameter values as in curve 4. (Reprinted with permission from Ref. 25. Copyright 1992 Elsevier.)

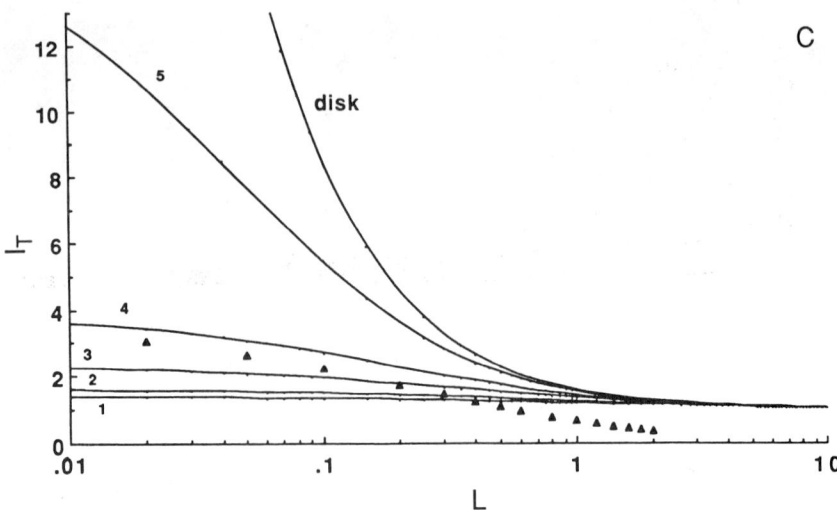

FIG. 37 *(Continued)*

as small as 80 nm can be evaluated [25] (Fig. 38) (See also Sec. II.C). A similar procedure allowed the characterization of a 30-nm-radius conical tip inside the Nafion film loaded with Os(bpy)$_3^{2+}$ [52] (see Sec. IV.F).

IV. APPLICATIONS

A. SECM Images

A three-dimensional SECM image is obtained by scanning the tip in the x-y plane and monitoring the tip current, i_T, as a function of tip location. The image can be converted to a plot of z-distance (distance between tip and substrate), d, vs. x-y position via an i_T vs. d calibration plot or can be plotted as a gray-scale image, where high values of i_T are shown as light colors and small values as dark colors. For example, different presentations of the SECM image of an interdigitated electrode array (IDA), which consists of Pt bands (3-μm wide and 0.1-μm thick) deposited on an SiO_2 substrate and spaced 5 μm apart, are shown in Figure 39 [22]. For this image, a Pt disk (0.2-μm diameter) coated with glass served as the tip, and this tip and the IDA were immersed in a solution of 0.2 M $MVCl_2$ and 2 M KCl (MV^{2+} is the methyl viologen dication). The tip was held at −0.76 V vs. SCE, where the reaction $MV^{2+} + e^- \rightarrow MV^+$ occurred. A single scan across the bands is shown in Figure 39A. When the tip is over the insulating SiO_2, the current is much smaller than when it is over Pt, where positive feedback occurs, because the

Scanning Electrochemical Microscopy

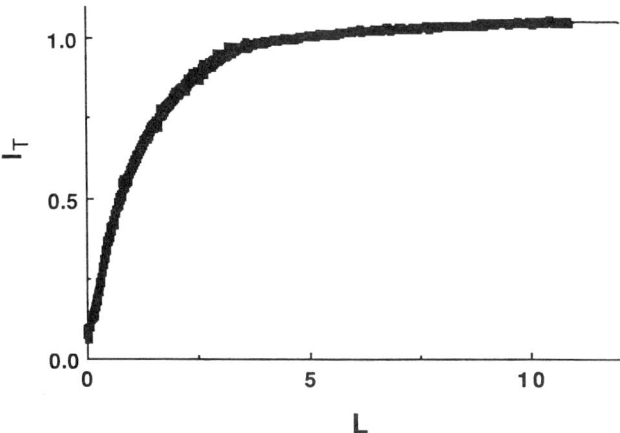

FIG. 38. Normalized tip current vs. normalized distance for a tip electrode over a nonconductive substrate, n-WSe$_2$ (2 mm diameter), at open circuit in a solution containing 0.5 M K$_4$Fe(CN)$_6$ and 0.5 M Na$_2$SO$_4$; $i_{T,\infty}$ = 2.24 nA. Thin line shows the theoretical curve computed for a conical 80-nm-radius tip, k = 0.2, offset = −3 nm. (Reprinted with permission from Ref. 25. Copyright 1992 Elsevier.)

MV$^+$ produced at the tip is reoxidized to MV^{2+}. In this example, the changes in current do not represent topographical changes, but rather differences in the chemical nature (as reflected by the conductivity) of the surface. A three-dimensional image of the IDA electrode and its gray-scale presentation are shown in Figures 39B and 39C.

SECM has also been used to obtain topographic information about biological samples immersed in an electrolyte solution, either by using the feedback mode or by detecting a substrate-generated electroactive species (e.g., oxygen) at the tip [53]. For example, the upper surface of a blade of grass that was immersed in an aqueous solution of Fe(CN)$_6^{4-}$ could be imaged by SECM (Fig. 40A). The image, taken over an area of 188 μm × 142 μm, shows the parallel venation pattern characteristic of monocot leaves. An SECM image over the bottom surface of a *Ligustrum siense* leaf is shown in Figure 40B. Several open stomata structures are shown in the gray-scale image, since the guard cells (pairs of specialized epidermal cells surrounding each stomata) are swollen and protrude above the surrounding epidermal cells. In Figure 40C, the image of Figure 40B after image processing by the use of the LOG filter [20] is shown. Better resolution is clearly obtained in Figure 40C, where a large, craterlike feature at middle-left shows finer struc-

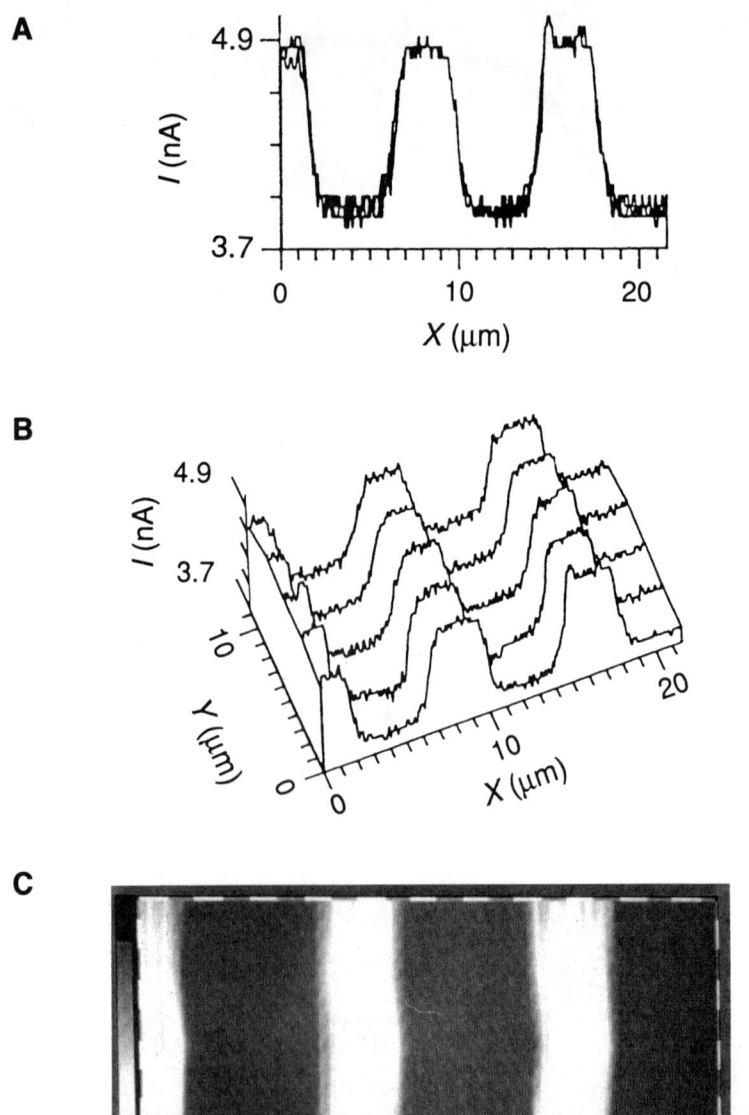

FIG. 39. Three presentations of the SECM image of an IDA (see text). (A) Single scan. (B) Three-dimensional view. (C) Gray-scale image ($i_{T,\infty}$ = 3.7 nA). (Reprinted with permission from Ref. 60. Copyright 1991 American Association for the Advancement of Science.)

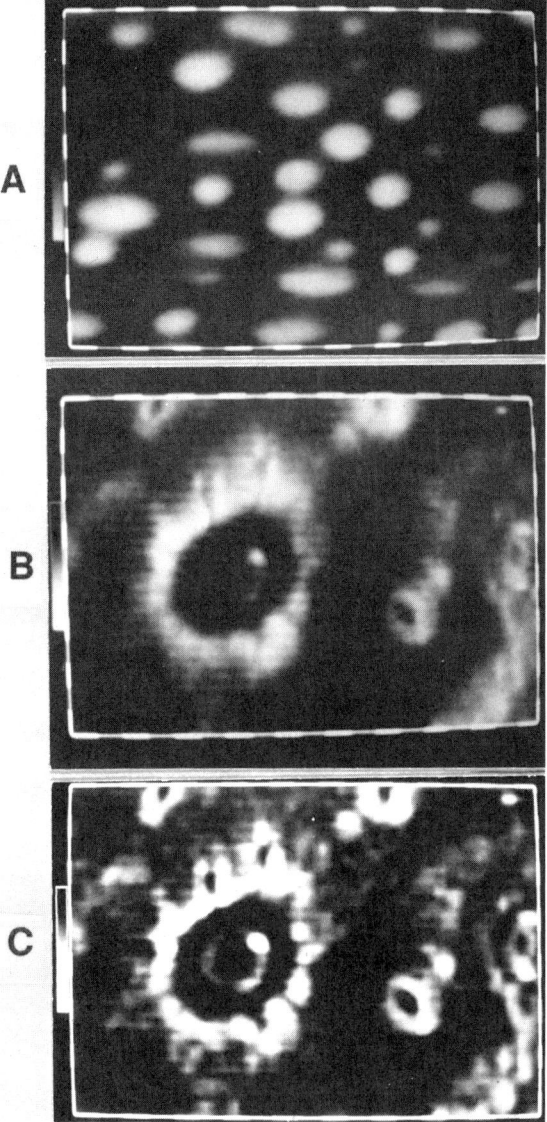

FIG. 40. Gray-scaled images of biological substrates in a 20 mM $K_4Fe(CN)_6$ and 0.1 M KCl solution scanned with a 1-μm-radius Pt tip: scan area 188 × 142 μm (white markers around edges = 10 μm). (A) Top surface of grass. (B) Bottom surface of L. sinense leaf. (C) The image in (B) after image processing by use of Eq. (6) in Ref. 20. (Reprinted with permission from (A,B) Ref. 9, Copyright 1990 American Chemical Society, and (C) Ref. 20, Copyright 1991 American Chemical Society.)

ture and some stomata at the upper-right and middle-right edges show their structures more clearly. The scanning tip can also be used to detect electrochemical products generated at a biological substrate. For example, the detection of oxygen during illumination of an *Elodea* leaf has been carried out in a 10 mM KCl solution saturated with CO_2. Changes in local concentration of substrate-generated oxygen under illumination were detected with a carbon microdisk electrode [53].

SECM has also been used to image membranes [54], polymers [47,52,55–59], and enzyme locations in a membrane [60,61]. A technique called reaction-rate imaging is particularly useful in imaging the areas on a surface where reactions occur, e.g., catalytic sites on an electrode [36]. An example is shown in Figure 41. Here, the substrate is a composite electrode of Au regions embedded in a glassy carbon (GC) matrix. For the mediator solution used, Fe^{3+} ion in 1 M H_2SO_4, the Au regions are significantly more active than the GC regions for the oxidation of tip-generated Fe^{2+} at intermediate potentials. The SECM image in Figure 41A was obtained when a Pt tip (10 μm diameter) was scanned over a 200 μm × 130 μm region of the composite, with the composite biased at 0.57 V vs. the $Fe^{2+/3+}$ formal potential. The bright regions represent the larger feedback currents over the Au regions due to the larger electron-transfer rate at these locations compared to the GC regions. Since the reaction rate is a function of electrode potential, one can increase the oxidation rate of Fe^{2+} at GC by applying a sufficiently positive potential to the composite. The image in Figure 41B was scanned over the same regions as that in Figure 41A but with the composite biased at a potential of 0.82 V. Here, the electron-transfer rate at both phases is at the diffusion limit, and a more uniform image results. However, topographic features are retained in both images. For example, the round depression in the upper right of both Figures 41A and 41B is caused by a pit in the GC surface.

Reaction rate imaging is unique to SECM and clearly illustrates its "chemical imaging" nature. By proper choice of solution components to control the tip reaction and the chemistry at the substrate/solution interface, differential reaction rates at different sorts of surfaces can be probed. For example, the location of enzyme sites in a membrane or organelle, where a particular reaction is catalyzed, can be seen. A typical study involved rat liver mitochondria, which contain in their outer membrane the enzyme cytochrome b_5 NAD reductase, which catalyzes the oxidation of NADH by a suitable oxidant (D. T. Pierce and A. J. Bard, unpublished). The solution chosen to probe this system contained NADH and N,N,N',N'-tetramethyl-*p*-phenylene diamine (TMPD) with the tip scanned over mitochondria on a glass slide with the tip held at a potential where the reaction $TMPD - e^- \rightarrow TMPD^+$ occurs.

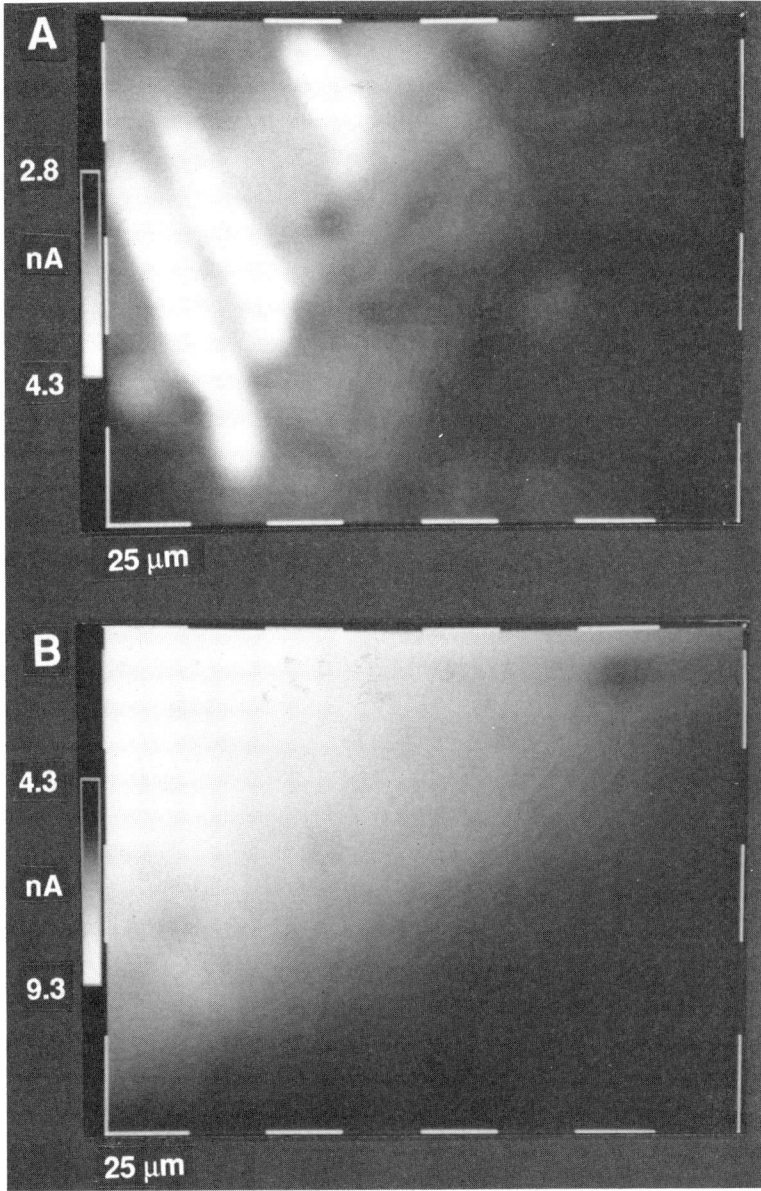

FIG. 41. Gray scale images of identical regions of a composite GC/Au surface. The tip was a 10-μm-diameter Pt disk held at −0.45 V, tip scan speed was 10 μm s^{-1}. Imaged region is 200 μm × 130 μm. Note the different current scales. (A) Substrate potential is +0.55 V. (B) Substrate potential is +0.8 V. (Reprinted with permission from Ref. 36. Copyright 1991 The Electrochemical Society.)

When the tip is over the glass or regions in the mitochondrion where there is no enzyme, only negative feedback is observed. However, when the tip is over enzyme regions, the interaction between TMPD$^+$ and NADH is possible, producing a positive feedback of TMPD to the tip and enhanced tip current (Fig. 42).

Note that in addition to qualitative imaging of relative reaction rates, the SECM is capable of determining the magnitude of the heterogeneous reaction rate at the surface, e.g., by a study of the i_T vs. d (approach) curve, as described in Sec. IV.B. This kind of application, i.e., a study of the rate of glucose oxidation at glucose oxidase, has been reported [61]. Note that enzyme activity, especially of enzymes that are not oxidoreductases, can also be probed with ion selective potentiometric tips (see Sec. IV.E).

As mentioned previously, the use of the tip-position modulation (TPM) technique can substantially improve the sensitivity and image resolution of the

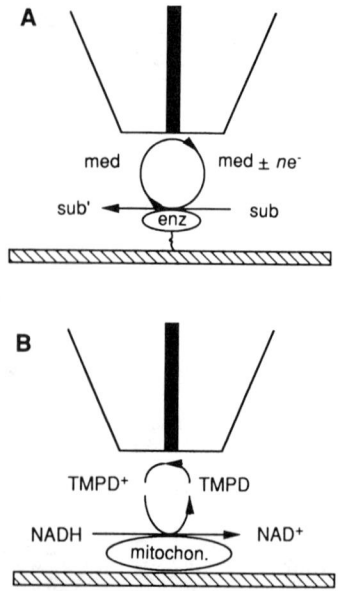

FIG. 42. Schemes depicting the principles of positive-feedback detection of enzymes: (A) SECM tip near a surface-constrained enzyme (enz) with mediator (med) turnover generated in the presence of substrate (sub); (B) SECM tip imaging of a single mitochondrion (mitochon.) through the use of the outer membrane enzyme, NADH-cytochrome b_5 reductase. (Reprinted with permission from Ref. 60. Copyright 1991 American Association for the Advancement of Science.)

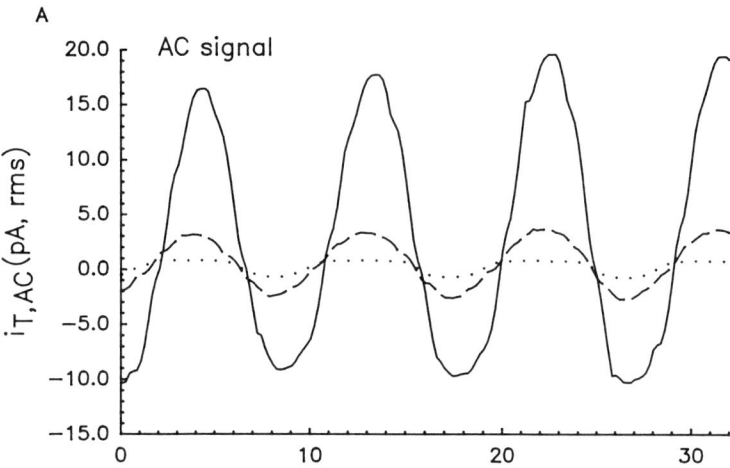

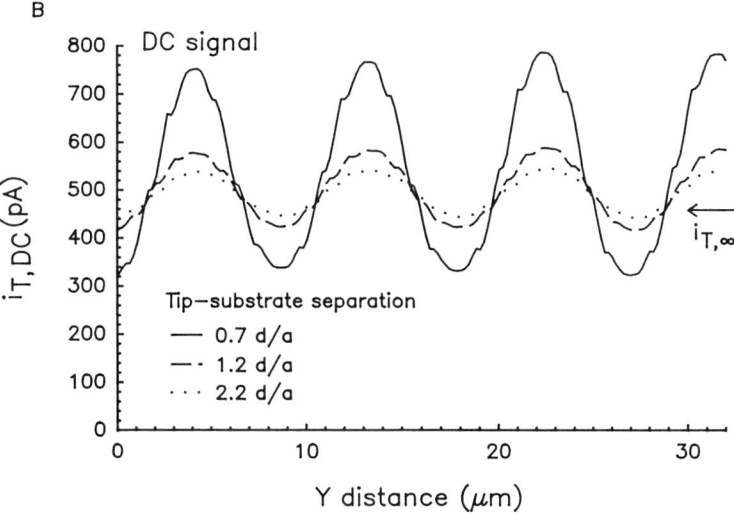

FIG. 43. Line scans at various distances with TPM SECM in-phase current and SECM at an IDA electrode (see text). Conditions: mediator 1.5 mM $Ru(NH_3)_6^{3+}$ in pH 4.0 buffer, tip electrode, 1-μm-radius Pt; f_m = 160 Hz, δ/a = 0.1; scan speed, 2 μm/sec. (A) TPM SECM response; (B) dc SECM response. (Reprinted with permission from Ref. 21. Copyright 1992 American Chemical Society.)

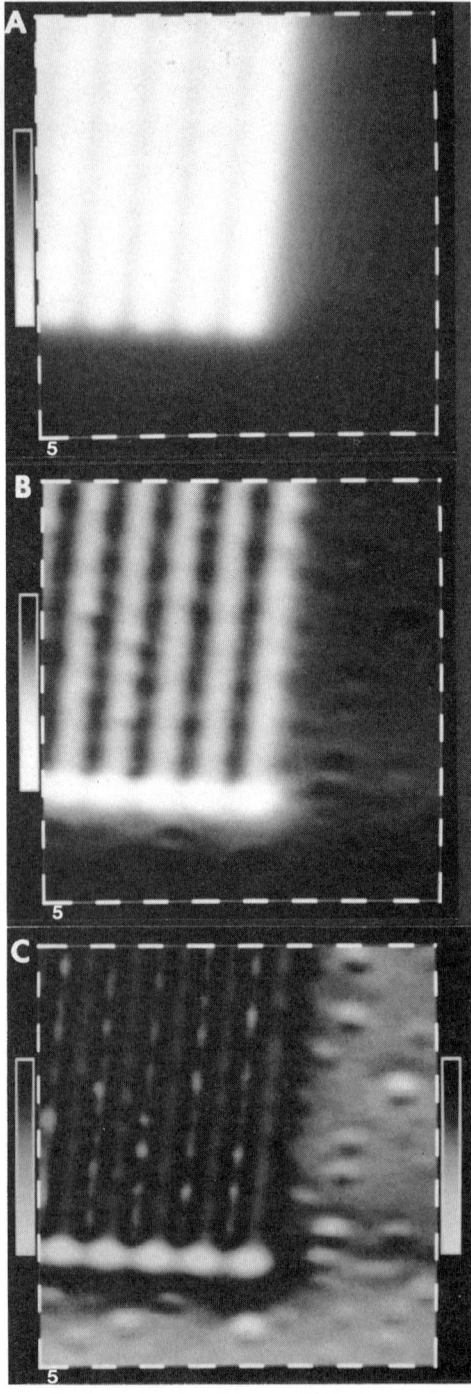

SECM and provide a method of distinguishing between conductive and insulating areas on the substrate surface being examined. Figure 43 shows several line scans over the surface of an IDA electrode using TPM SECM [21]. This IDA consists of 3-μm-wide bands of Pt separated by 5 μm of SiO_2. Note that, as expected, the TPM response of the line scan is bipolar and rapidly falls off with increasing distance. This contrasts with the dc signal, which is present on top of the $i_{T,\infty}$ level and shows a more gradual decrease.

Another important distinction is the increased sensitivity in the insulator region. This is especially noticeable at large tip-substrate separations, when the dc insulator signal is nearly swamped by the larger conductor signal. A common problem in both in-phase TPM and dc SECM is the difficulty in imaging insulating regions adjacent to conducting regions. Even when the tip is completely over the insulator, the mediator species can provide some positive feedback by diffusion from the tip to the conductor edge making the observed insulator response more positive and obscuring insulator features. Images of a portion of the IDA electrode acquired with dc and TPM SECM are shown in Figure 44 [21]. The band structures observed in the upper left of the image are Pt bands, and the remainder of the image is the insulating SiO_2 substrate. Figure 44A is the image acquired with the dc signal, and Figures 44B and 44C are TPM images. As shown, the images acquired with the TPM signal are significantly more detailed than the dc image. In particular, note the improved detail in the insulating region of the image due to the better sensitivity of the TPM signal at insulators.

B. Heterogeneous Electron Transfer and Reaction Rate Imaging

The similarity between the SECM and TLC measurements [12] suggests that the feedback current is quite sensitive to the rates of heterogeneous reactions at both the tip and substrate electrodes. Thus, SECM can be employed in kinetic studies and for imaging surface features with different chemical/electrochemical reactivities.

FIG. 44. TPM and dc SECM images of a 80-μm × 80-μm portion of an IDA substrate. Conditions: mediator 1.5 mM $Ru(NH_3)_6^{3+}$ in pH 4.0 buffer, tip electrode, 1-μm-radius Pt; d ≈ 0.9 μm; f_m = 160 Hz, δ/a = 0.1; scan speed, 10 μm/sec. (A) SECM image, scale is 200–850 pA, $i_{T,\infty}$ = 470 pA. (B) TPM SECM image, scale −7 to 15 pA_{rms}. (C) Absolute value TPM SECM image, left scale = 0–15; right scale = 0–7 pA_{rms}. (Reprinted with permission from Ref. 21. Copyright 1992 American Chemical Society.)

1. Studies of Heterogeneous Kinetics

a. Kinetic Measurements Employing SECM with a Solid Substrate

The dependence of the feedback SECM current on the rate of the heterogeneous reaction at the substrate was studied experimentally [62] with the $Fe^{3+/2+}$ redox mediator in 1 M H_2SO_4. The oxidized form of the mediator, Fe^{3+}, was reduced at a tip electrode (diffusion-controlled), and Fe^{2+} was oxidized at a biased glassy carbon (GC) substrate adjusted to different potentials. The essentially irreversible voltammetric behavior of this system allowed a variation of the backward (oxidation) reaction rate at the substrate by changing the potential over a wide range, while the forward reaction rate at these potentials remained negligibly small. The current-distance curves obtained at different substrate potentials (Fig. 45A) reflect the gradual change in substrate behavior between "perfect conductor," when all Fe^{2+} species coming from the tip are reoxidized at the substrate (upper curve), and insulator, when the substrate reaction rate is small (lower curve). In a later paper [30], the whole family of i_T-L curves given in Figure 45A was compared to the theoretical working curves (Fig. 45B) and yielded the kinetic parameter values, $k° = 2 \times 10^{-5}$ cm/sec and $\alpha = 0.69$, in good agreement with the literature data.

The irreversible chemical regeneration of the SECM mediator is formally similar to the above. Experiments with glucose oxidase (GO) (the enzyme from the mold *Aspergillus niger*, frequently used in glucose biosensors [63]) immobilized on several insulating substrates allowed a study of the kinetics of catalytic oxidation of β-D-glucose to D-glucono-δ-lactone [61]:

$$\beta\text{-D-glucose} + GO^{ox} \rightarrow \delta\text{-gluconolactone} + GO^{Red} \quad (55)$$

$$GO^{Red} + 2\,Med^{ox} \xrightarrow{k_f} GO^{ox} + 2\,Med^{Red} \quad \text{(substrate)} \quad (56)$$

$$Med^{Red} - e^- \rightarrow Med^{ox} \quad \text{(tip)} \quad (57)$$

where the mediators (Med) employed were FcCOOH (Fc = ferrocene), $Fe(CN)_6^{4-}$, or H_2Q (hydroquinone).

Although the fit between theory and experiment in this case was not as good as in experiments with flat metallic substrates (mostly because of the uncertainties associated with transport in a μm-thick porous enzyme layer), it was possible to establish zero-order enzyme-mediator kinetics [Eq. (56)] and to determine apparent heterogeneous rate constants for several mediators.

Two approaches to the study of quasi-reversible heterogeneous kinetics by SECM have been reported. First, the analysis of the tip current as a function of the substrate potential (Fig. 46), while the tip potential was held

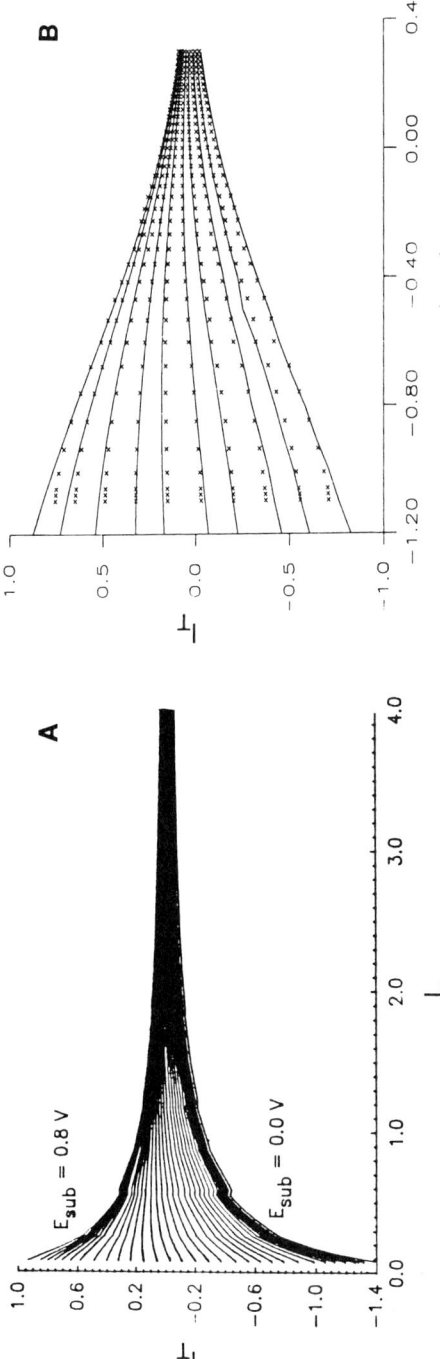

FIG. 45. Steady-state current-distance curves for the $Fe^{3+/2+}$ system (solid lines in A and symbols in B) and the corresponding best theoretical fits to the data (solid lines in B). (A) The tip electrode (5.5-μm-radius carbon fiber) was held at a potential of −0.6 V, and the GC substrate electrode was held at potentials of 0–800 mV in 25 mV increments. (B) The tip electrode (5.5-μm-radius carbon fiber) was held at a potential of −0.6 V, and the GC substrate electrode was held at potentials of 300–750 mV in 50 mV increments. ((A) Reprinted with permission from Ref. 62. Copyright 1991 The Electrochemical Society. (B) Reprinted with permission from Ref. 30. Copyright 1992 American Chemical Society.)

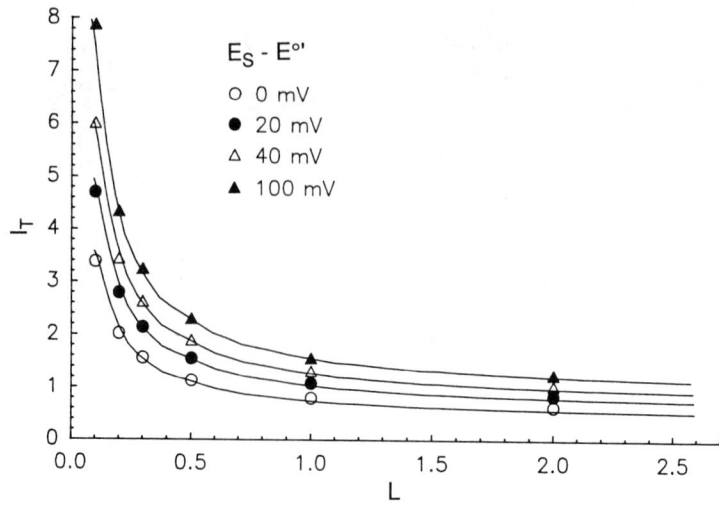

FIG. 46. Experimental and simulated current-distance data for a quasi-reversible substrate process. The solid lines are experimental data for a 1.7 mM solution of $Ru(NH_3)_6^{3+}$ in a pH 4.0 buffer. The tip is a 12.5-μm-radius Au disk. The substrate is a glassy carbon electrode biased at potentials of 0, 20, 40, and 100 mV positive of the formal potential. The symbols are computed from Eqs. (17) and (18) with L = 15 at the potentials indicated. (Reprinted with permission from Ref. 30. Copyright 1992 American Chemical Society.)

constant and the tip process was diffusion-controlled, yielded the rate constant for the $Ru(NH_3)_6^{3+/2+}$ couple at a glassy carbon substrate (0.076 cm/sec) [30]. In the feedback mode, only the process at a small portion of the substrate confronting the tip (approximately of the same size as a tip, as stated above) contributes to the tip current. Thus, one can make a part of the large substrate work as a pseudo-microelectrode. This may be helpful for studying electrochemical kinetics at various materials (such as single crystals of semiconductors) which are unsuitable for manufacturing microelectrodes. At the same time, this arrangement allows one to get rid of parasitic processes (e.g., corrosion of a semiconductor, oxygen reduction). Most of such irreversible processes would not contribute to the tip current, and thus, a separation of the redox reaction from parallel processes can be achieved.

The tip kinetics can also be studied from a steady-state CV obtained by scanning the tip potential at a constant substrate potential. The measurement of the very fast kinetics of the oxidation of ferrocene at a Pt UME tip by this approach has been described [64]. Five steady-state voltammograms obtained at different d-values are shown in Figure 47 along with the theoretical curves calculated with the values of the kinetic parameters extracted from the quartile potentials (see Sec. IV.B.1.c and Table 6). The k° value obtained (3.7 ± 0.6 cm/sec) is about two to four times higher than that determined by fast scan voltammetry [65–67], indicating that even very careful compensation of IR-drop cannot guarantee the desired accuracy of measurements when the heterogeneous kinetics are rapid. On the other hand, the value obtained is very close to that found from impedance analysis (2.6 cm/sec at 285 K [68]). The transfer coefficient found ($\alpha = 0.37 \pm 0.02$) was somewhat lower than the theoretically expected value ($\alpha = 0.5$). The much larger rate constant (k° $\cong$ 10 cm/sec) predicted in Ref. 67 from steady-state voltammetry at 0.3- to 0.5-μm-radius microdisks may result from the imperfect geometry of such small UME.

Figure 47 illustrates a simple way to check the validity of the experimental results and the reliability of the kinetic analysis. While the mass-transfer rate increases with a decrease in the tip-substrate separation (i.e., from 3 to 5 in Fig. 47 and in Table 6), the calculated heterogeneous rate constant and transfer coefficient should remain constant within the range of experimental error, as shown in Table 6. Such a check is not possible for heterogeneous

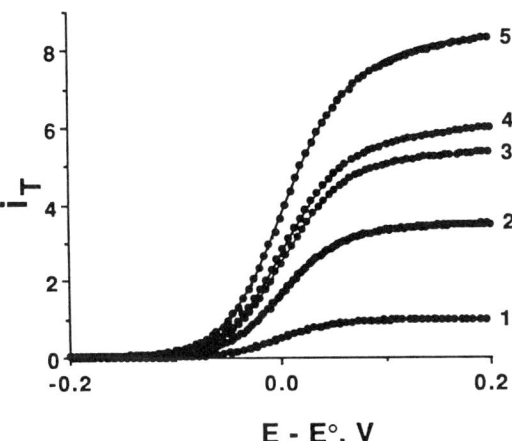

FIG. 47. Tip steady-state voltammograms for the oxidation of 5.8 mM ferrocene in 0.52 M TBABF$_4$ in MeCN at a 1.08-μm-radius Pt tip. Solid lines calculated from Eq. (38). Tip-substrate separation decreases from 1 to 5; see Table 6 for parameter values. The numbers on the curves correspond to those in Table 6. (Reprinted with permission from Ref. 64. Copyright 1993 American Chemical Society.)

TABLE 6
Kinetic Parameters for Oxidation of Ferrocene in Acetonitrile at a Pt Tip[a] Electrode from SECM Steady-State Voltammograms

No.	$\Delta E_{1/4}$[b]	$\Delta E_{3/4}$[b]	L	i_T	$k°$ (cm/sec) Eq. (32)	$k°$ (cm/sec) Eq. (38)	α Eq. (32)	α Eq. (38)	$\Delta E^{o'}$[b] Eq. (32)	$\Delta E^{o'}$[b] Eq. (38)
1	28.6	28.6	∞	1.0	(process is essentially reversible)					
2	30.5	32.3	0.27	3.55	3.4	2.4	0.48	0.49	5.4	5.5
3	31.4	34.7	0.17	5.47	4.5	3.5	0.38	0.39	6.5	6.5
4	32.3	36.9	0.14	6.10	4.1	3.3	0.36	0.36	7.9	8.0
5	32.9	38.5	0.10	8.53	5.1	4.3	0.35	0.35	9.1	9.1

[a] $a = 1.08$ μm.
[b] $\Delta E_{1/4} = E_{1/4} - E_{1/2}$, $\Delta E_{3/4} = E_{1/2} - E_{3/4}$, $\Delta E^{o'} = E^{o'} - E_{1/2}$.
Source: Ref. 64.

kinetics measurements at a single microelectrode, since the steady-state mass-transport rate can only be changed by using different sizes of UME.

b. Experiments with a Hg Substrate

A Hg substrate is interesting for several reasons:

1. An extremely close tip/substrate separation can be obtained, and consequently, the upper limit for the determinable rate constant can be extended, because the liquid substrate is essentially atomically smooth and horizontal and the insulator surrounding the tip electrode can penetrate slightly into its volume without disturbing the measurements or destroying the tip.
2. Mercury is known to be a classic example of a smooth uniform electrode. It is also a very good substrate for studying adsorption and new phase formation phenomena.

Two types of the SECM experiments with a liquid Hg substrate were described [69]: (1) conventional steady-state feedback measurements with Hg behaving as a conductive substrate and (2) measurements in a thin layer of electrolyte brought inside the mercury pool by the UME tip. The transition from the first regime to the second is shown in Figure 48. The initial part of the i_T-d curve in Figure 48A displays an exponential type of growth of the tip current as the tip approaches the mercury pool. This part of the curve corresponds to the known theory for SECM with a conductive substrate [27], thus demonstrating the suitability of a Hg substrate for conventional SECM experiments. However, after the point of inflection of d ≅ 0 (Fig. 48B),

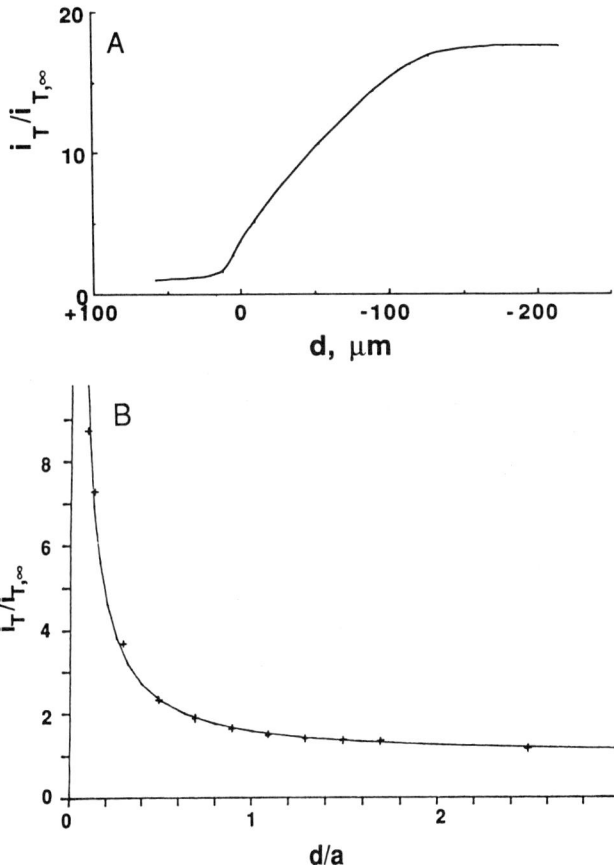

FIG. 48. Current-distance curves for a 25-μm-diameter Pt tip approaching a mercury substrate. The solution contained 5 mM $Ru(NH_3)_6Cl_3$ and 0.2 M KNO_3. Zero points were found from a fit of the experimental results to the SECM theory [Eq. (25)]. (A) Positive distances correspond to the tip approaching the Hg surface; negative distances correspond to tip penetration into the mercury pool. (B) Points on the initial part of the I_T-d curve (before penetration and thin layer formation). The solid line was computed from Eq. (25). Scan rate was 5 μm/s. (Reprinted with permission from Ref. 69. Copyright 1992 The Electrochemical Society.)

where the tip penetrates the Hg, the slope of the i_T-d curve decreases, and the tip current tends to a limiting value independent of distance. This suggests that rather than directly contacting the Hg, which would cause an immediate large increase in i_T, the tip traps a thin layer of the electrolyte between the tip and the Hg substrate electrodes, whose limiting thickness is independent of the tip position. This tip/electrolyte/substrate configuration behaves as a twin electrode TLC whose thickness, l, can be evaluated from the diffusion-limiting current [12].

$$i_{TLC} = nFAc° \frac{2D_O D_R}{(D_O + D_R)l} \tag{58}$$

Comparing Eq. (58) with Eq. (1) for the microdisk steady-state current in solution far from substrate, and assuming $D_O = D_R$, one obtains

$$l = \frac{\pi a i_{T,\infty}}{4 i_{TLC}} \tag{59}$$

Values of l from 200 nm to several μm were reported [69,70]. This first value is about one order of magnitude lower than the thickness of any conventional TLC reported to date, thus allowing quite fast steady-state measurements of electrode kinetics. The lowest value of l obtained so far with a 1-μm-radius Pt tip and C_{60}-mediator dissolved in benzonitrile was only 10 nm (M. V. Mirkin and A. J. Bard, unpublished). Additional efforts are needed to overcome the problem of current oscillations, probably resulting from vibration, which prevents quantitative measurements with layer thicknesses < 200 nm.

The SECM/TLC technique was used to determine kinetic parameters of moderately fast electroreduction of $Ru(NH_3)_6^{3+}$ at a carbon fiber electrode. The steady-state voltammograms (Fig. 49) were analyzed as described in the next section. In order to validate the results, the values of kinetic parameters obtained ($k° = 0.15$ cm/sec, $\alpha = 0.44$) were compared to those obtained from steady-state CV at the same 11-μm-diameter carbon microdisk ($k° =$

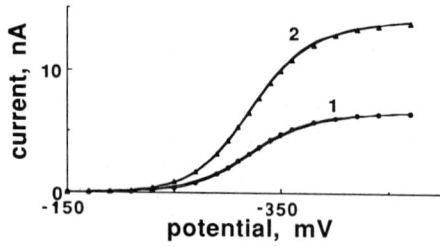

FIG. 49. Steady-state cyclic voltammogram of a solution containing 5 mM $Ru(NH_3)_6Cl_3$ in 0.2 M KNO_3 at a carbon UME: (1) with the tip far from a mercury substrate; (2) with the tip within the mercury, i.e., in thin-layer cell configuration. $v = 5$ mV/sec. Circles were calculated according to Eq. (35) and triangles according to Eqs. (32) and (34). See text for values of kinetic parameters. (Reprinted with permission from Ref. 69. Copyright 1992 The Electrochemical Society.)

0.11 cm/sec and $\alpha = 0.40$). The agreement between the two sets of parameters is very good, even with the high value of $k°$, which is in fact close to the upper limit for measurements with this size microdisk.

The application of this technique to study the fast first stage of the reduction of C_{60} in two highly resistive solvents, o-dichlorobenzene and benzonitrile [70], demonstrated that the SECM can be helpful in a situation when reliable kinetic data cannot be obtained by either fast scan voltammetry or steady-state CV at a microdisk electrode. The use of fast scan voltammetry required very accurate compensation of the ohmic potential drop, which was significant because of the high resistivities of both solvents and low solubility of the supporting electrolyte (TBABF$_4$). With the Fc$^+$/Fc couple as an internal standard, IR-compensation was achieved for scan rate values up to 50 V/sec. This allowed bracketing, rather than quantitative determination, of the kinetic parameters. The analysis of steady-state voltammograms of C_{60} obtained with the smallest available microdisk electrode (1-μm radius) also yielded only a lower limit for the rate constant. By using the SECM/TLC, the mass-transport rate was increased sufficiently for quantitative characterization of the electron-transfer kinetics, preserving the advantages of steady-state methods, i.e., the absence of problems associated with the ohmic drop, adsorption, and charging current. The steady-state CV obtained in 0.2- to 0.4-μm-thick TLC (Fig. 50) were analyzed as described in Ref. 35 (see also following section), and two sets of kinetic parameters were calculated: $k° = 0.46 \pm 0.08$ cm/sec, $\alpha = 0.43 \pm 0.05$ (o-dichlorobenzene) and $k° = 0.12 \pm 0.02$ cm/sec, $\alpha = 0.52 \pm 0.05$ (benzonitrile). Similar values of kinetic parameters were obtained in benzonitrile [71] by impedance measurements; however, in this latter case, the authors had to perform additional measurements of known rate constants to prove the adequacy of the IR compensation.

c. Extraction of the Heterogeneous Kinetic Parameters

There are two distinct approaches to studying fast electrode kinetics. One utilizes various relaxation techniques, such as a fast scan voltammetry [5], ac, or potential/current pulses [17]. The second one achieves a fast mass-transport rate by decreasing the thickness of the diffusion layer at steady state. This can be done by using a small working electrode [5,26], TLC [12], or rotating electrodes [17]. The SECM can work in both of these regimes. The tip current caused by either potential step, or fast sweep of the potential, or an ac signal applied to the substrate can be monitored [8,11] and analyzed theoretically [29]. Similarly, fast scan tip CV and potentiostatic transients can be obtained at a constant substrate potential. However, the reported short-time (or high-frequency) distortions of the measured current [8,11], probably owing to the capacitive coupling between the tip and the substrate, discourage such appli-

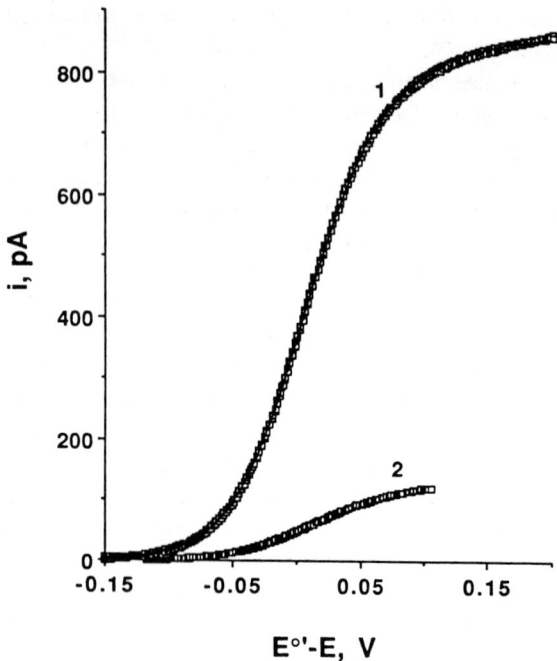

FIG. 50. Steady-state voltammograms of the first reduction of C_{60} in (1) *o*-dichlorobenzene and (2) benzonitrile in a thin-layer cell formed by a 1-μm-radius Pt tip and a Hg substrate. v = 20 mV/sec. Squares are experimental data, and solid lines are theoretical curves calculated according to Eqs. (32) and (34) with the kinetic parameters given in the text. (Reprinted with permission from Ref. 70. Copyright 1993 American Chemical Society.)

cations. Steady-state measurements are probably more appropriate for SECM, especially with the very small tip-substrate separations described above. Here we consider only the analysis of steady-state data.

In principle, kinetic parameters can be extracted by fitting experimental current-distance curves to the theory as was done in Ref. 30 (see Figs. 45B and 46). The fitting process is more straightforward in the case of irreversible kinetics where the feedback current depends on one, forward rate constant. For a quasi-reversible reaction, the presence of an additional fitting parameter, the backward rate constant, makes the fitting very difficult.

A much simpler approach [35] is based on the analysis of the steady-state tip voltammogram recorded at a constant substrate potential (where the substrate process is diffusion-controlled). The steady-state current at any uni-

Scanning Electrochemical Microscopy

formly accessible electrode is a function of two parameters, κ and θ [35] (the generalization for non–uniformly accessible electrodes is straightforward; see, for example, Ref. 72). Consequently, three parameters, $k°$, α, and $E°'$, can be determined from a steady-state voltammogram using three values of quartile potentials: $E_{1/2}$, $E_{1/4}$, and $E_{3/4}$. A single table containing the above parameters for all possible pairs of $\Delta E_{1/4} = E_{1/4} - E_{1/2}$ and $\Delta E_{3/4} = E_{1/2} - E_{3/4}$ (Table 1 in Ref. 35) is suitable for any kind of uniformly accessible electrode. To analyze voltammograms obtained with a non–uniformly accessible working electrode, one needs to calculate an analogous table for that particular electrode geometry (such a table for a microdisk is also given in Ref. 35). For SECM, this is not simple because I_T is a function of L, so that a special table has to be established for any particular L value. Although the available computer program* allows one to generate such a table for any value of L, the use of the uniform approximation is certainly advisable.

For the case of a liquid substrate (Hg), the exact TLC geometry is unknown; however, the symmetry of the cell restricted by the limits of solvent droplet trapped inside the Hg suggests uniform accessibility of tip surface. The mass-transfer coefficient can be calculated without knowledge of cell geometry:

$$m_O = \frac{i_{dif}}{nFAc°} \tag{60}$$

After determination of $\Delta E_{1/4}$ and $\Delta E_{3/4}$, the values of Λ, α, and $E°' - E_{1/2}$ are available immediately from Table 1 in Ref. 35, and $k° = \Lambda m_O$. For example, $\Delta E_{1/4} = 32.6 \pm 0.2$ mV and $\Delta E_{3/4} = 36.6 \pm 0.4$ mV were found [70] for electroreduction of C_{60} in o-dichlorobenzene (Fig. 50, curve 1). The set of parameters found for this reaction from Table 1 in Ref. 35 was given in the previous section.

Voltammograms of the oxidation of ferrocene at a Pt UME over a Pt substrate (Fig. 47) were analyzed with and without the uniform approximation (Table 6) [64]. For a large L (the first voltammogram in Fig. 47 and the first set of data in Table 6), the voltammogram at a 1-μm-radius microdisk far from the substrate was essentially nernstian. This suggests that the dimensionless parameter, Λ_T, is greater than 10 [35], and the lower limit for the rate constant is $k° > 1.6$ cm/sec. Other voltammograms obtained with the tip in the proximity of the substrate were analyzed with and without uniform approximation [Eqs. (32) and (38), respectively]. For curve 2, the tip-substrate distance was fairly large (L = 0.274). The mass transfer in this

---

*The program is available from the authors.

case, although significantly faster than that for the microdisk above, was still not sufficiently fast to satisfy well the criterion $\Lambda' < 5$, and the kinetic parameters found were not very reliable. However, even at this relatively large L, an analysis based on either Eq. (32) or Eq. (38) led to essentially the same values of α and $E^{o\prime}$, and the variation in k^o was small. The last three sets of data, obtained at smaller tip-substrate separations, yielded very similar values for the kinetic parameters. The k^o values obtained from Eq. (32) are about 20% higher, but this is probably within the range of experimental error. Thus both Eq. (32) (with Table 1 in Ref. 35) and Eq. (38) (with a corresponding table computed for a given L) are suitable for the evaluation of electrode kinetics, which are too fast for simple UME voltammetry (i.e., where essentially nernstian voltammetric response is found at a tip UME far from the substrate). Obviously, the uniform approximation is most suitable for analysis of the SECM data obtained with a solid substrate, when $L << 1$. It is just these conditions that make the SECM experiment advantageous compared to simpler measurements with a microdisk electrode alone.

2. Imaging Surface Reactivity

Because the SECM response is a function of the rate of the heterogeneous reaction at the substrate, it can be used to image the reactivity of surface features. The first example of mapping of surface activity by SECM was reported in Ref. 10 where the tip response to the substrate potential steps in a G/C mode was presented as a function of the tip position over an array of 100-μm-diameter inlaid Pt microelectrodes (Fig. 51). The changes in the tip current allow one to distinguish between the conductive Pt surface and the surrounding insulator.

The steady-state feedback mode is more suitable for imaging, and it is possible not only to distinguish between conductors and insulators but also to detect the differences in reactivities of two conductors such as gold and glassy carbon (Fig. 41A) [36], or even different portions of a polyelectrolyte coating loaded with a redox mediator (Fig. 52) [59]. The data presented in Figure 19 suggest the possibility of distinguishing between substrate components with a ratio of rate constants as low as 2:1, when L is sufficiently small [29].

The possibility of mapping surface reactivity, i.e., the local electron-transfer kinetics, in a SG/TC mode was explored [73]. The spatial resolution, however, was only of the order of tens of micrometers because of the thick diffusion layer at the substrate electrode and non–steady-state behavior of the system (see Sec. III.B.2). The differences in reactivities of Pt and C portions of the model platinum-carbon heterogeneous substrate were detected, but no quantitative information was obtained, and the tip-substrate distance scale in

Scanning Electrochemical Microscopy

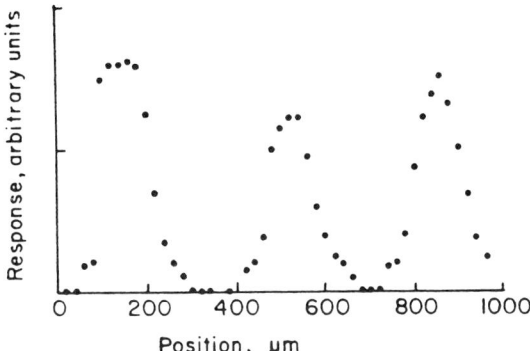

FIG. 51. Response to substrate potential steps from -0.2 to 0.6 V vs. SCE. The tip UME was moved in one dimension over an array of three 100-μm-diameter Pt microelectrodes. Data points were taken at approximately 20-μm intervals. (Reprinted with permission from Ref. 10. Copyright 1986 American Chemical Society.)

the z-direction was not established. The authors suggest that better resolution of SG/TC imaging could be achieved by using unstable species as a mediator to minimize the thickness of the diffusion layer.

C. Studies of Homogeneous Chemical Reactions and Adsorption/Desorption Processes

1. First-Order Homogeneous Reactions

An attempt was made [11] to use the SECM in a SG/TC mode to establish the mechanism for the 4-electron oxidation of epinephrine to adrenochrome (with adrenalinequinone as an intermediate). Although the authors achieved qualitative agreement between experimental and simulated data (Fig. 53), the fit was not good enough to distinguish unambiguously among first-order, ECE, or second-order mechanisms. The lack of fit was probably caused by the complexity of the process and the approximate character of G/C theory used in that work (see Sec. III.B).

The process of the oxidation of N,N-dimethyl-p-phenylenediamine (DMPPD) in aqueous basic solution, which follows an E_rC_i mechanism (a reversible two-electron oxidation followed by a first-order deamination reaction [74], was studied by feedback steady-state measurements (Fig. 54) with a conductive substrate as well as by chronoamperometry (Fig. 55) [32]. The i_T vs. d curves (Fig. 54) obtained at different pH values were converted to the

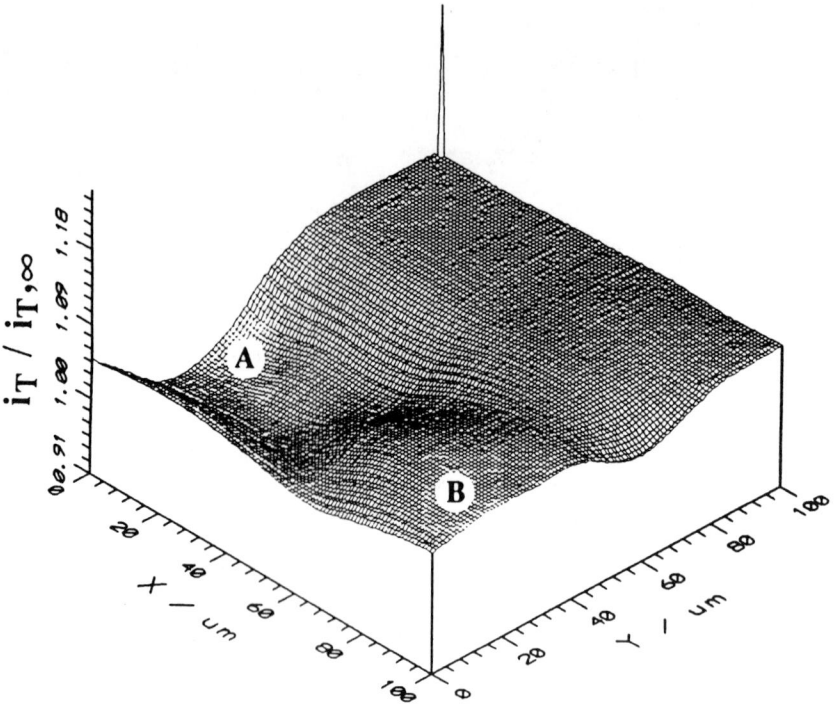

FIG. 52. SECM image of an age protonated poly(4-vinylpyridine) coating in a 10.8 mM solution of $IrCl_6^{2-}$. "Topographic features" of the image represent spatial variations in film conductivity. (Reprinted with permission from Ref. 59. Copyright 1992 American Chemical Society.)

values of the parameter K via working curves (Fig. 29) to calculate the following k_c values: 19 s^{-1} (pH = 11.2), 5.8 s^{-1} (pH = 10.8), and 1.4 s^{-1} (pH = 10.2). The good coincidence of theoretical tip current transients calculated using these values of k_c with the experimental ones (Fig. 55) supports the validity of the results.

The SG/TC mode of the SECM was also employed in the analysis of the mechanism of sodium borohydride oxidation [75] to find out whether the intermediates generated in the first two-electron wave react on the anode surface or diffuse into the solution and participate in a homogeneous reaction. The G/C experiment was expected to show a reduction current at the tip electrode if the intermediates left the substrate. The tip voltammogram (Fig. 56), obtained by sweeping the substrate potential in the region of BH_4^-

Scanning Electrochemical Microscopy

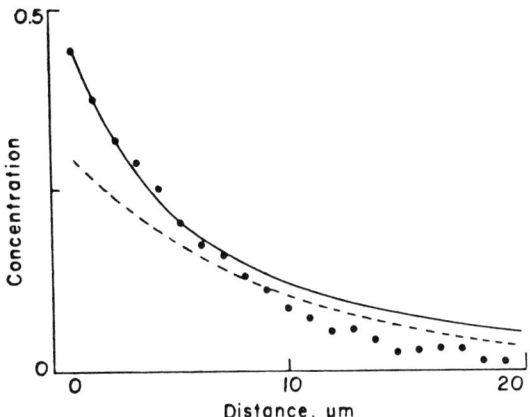

FIG. 53. Dependence of adrenalinequinone concentration on the distance from the conductive substrate in a SG/TC experiment. Filled circles, experimental; dashed line, first-order simulation; solid line, second-order simulation. (Reprinted with permission from Ref. 11. Copyright 1992 American Chemical Society.)

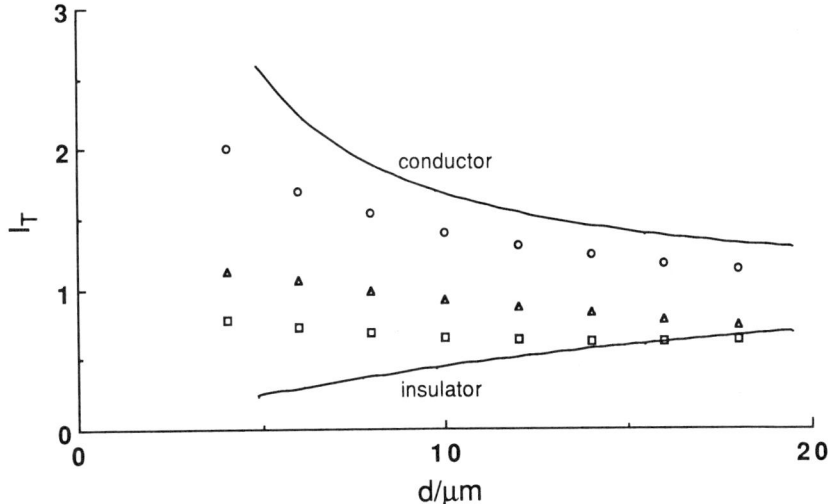

FIG. 54. Steady-state I_T-d curves for the oxidation of DMPPD (0.5 mM) at a 25-μm Pt tip over an unbiased substrate at pH 10.20 (open circle), pH 10.78 (open triangle), and pH 11.24 (open square). The solid lines represent the theoretical behavior for insulating and conductive substrates. (Reprinted with permission from Ref. 32. Copyright 1991 American Chemical Society.)

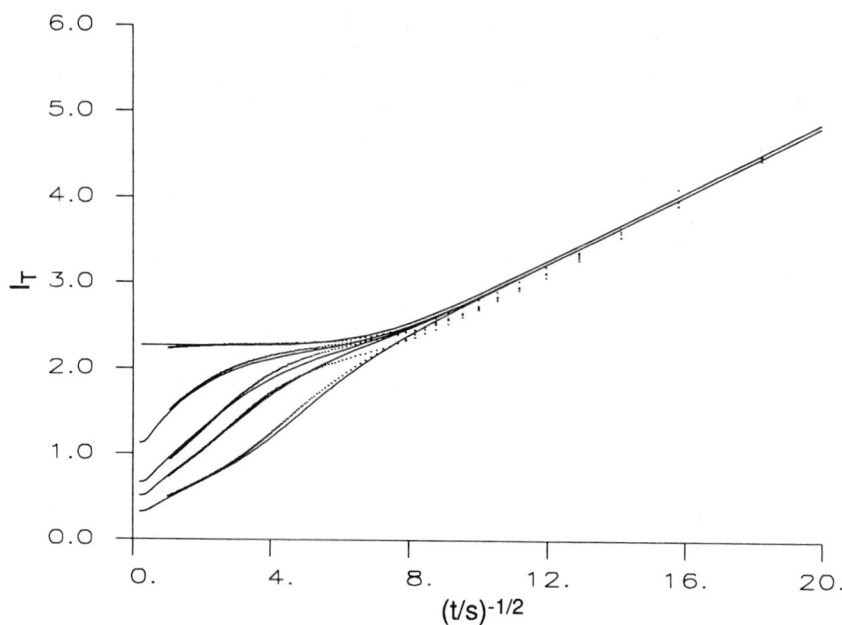

FIG. 55. Experimental chronoamperometric characteristics for the oxidation of DMPPD (dotted line) with the corresponding best theoretical fits (solid lines). pH values (from top to bottom): 7.80, 10.90, 11.38, 11.63, and 12.42. Theoretical curves were calculated with k_c values derived from steady-state feedback results (see Fig. 54). (Reprinted with permission from Ref. 32. Copyright 1991 American Chemical Society.)

oxidation while the constant tip potential was held at the value corresponding to the reduction of the intermediate proved the homogeneous character of the following chemical reaction. The rate constant of this reaction was estimated to be of the order of 500 s^{-1}, in agreement with the result obtained from fast scan voltammetry.

2. Second-Order Following Reactions

The quantitative TG/SC theory developed for a process with a product dimerization [46] was utilized in studies of the reductive coupling of both dimethyl fumarate (DF) and fumaronitrile (FN) in N,N-dimethylformamide. While the first dimerization reaction is relatively slow (k_c was estimated to be between 220 and 320 M^{-1} s^{-1} [76,77], the second one is quite fast (k_c was previously bracketed between 6 × 10^5 and 1.4 × 10^6 M^{-1} s^{-1} [76,77]) and

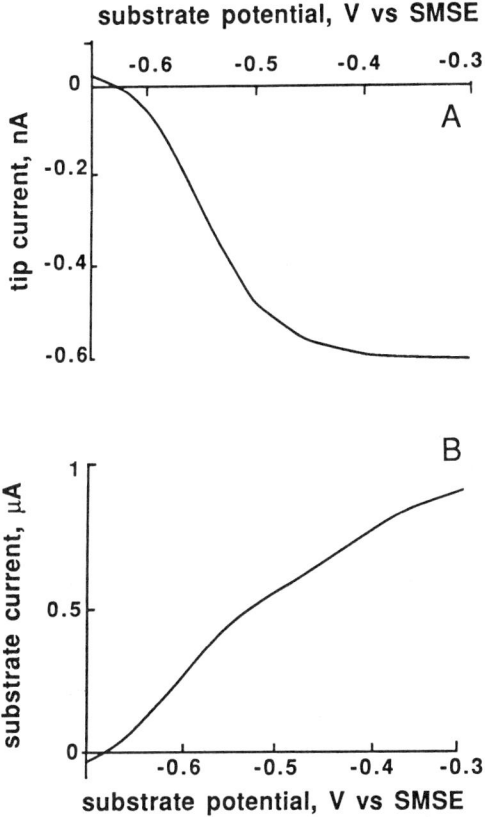

FIG. 56. Dependencies of the steady-state tip (A) and substrate (B) currents vs. substrate potential for the oxidation of 10 mM NaBH$_4$ in 1 M NaOH in the SG/TC mode of the SECM. 5-μm-radius Au tip biased at −1.3 V vs. a Hg/Hg$_2$SO$_4$/sat. K$_2$SO$_4$ reference electrode (SMSE) and 50-μm-radius Au substrate were separated by ~1 μm. (Reprinted with permission from Ref. 75. Copyright 1992 The Electrochemical Society.)

is difficult to study with conventional electrochemical methods. Therefore, the dimerization of DF could be studied with a relatively large tip electrode (a = 12.5 μm) and tip-substrate separations in a range from 3 to 20 μm. Both steady-state tip and substrate current-distance experimental curves (Fig. 57) fit the theory quite well, and the value of k$_c$, 180 M^{-1} s^{-1}, obtained agreed with the rate constants measured by other methods [76,77].

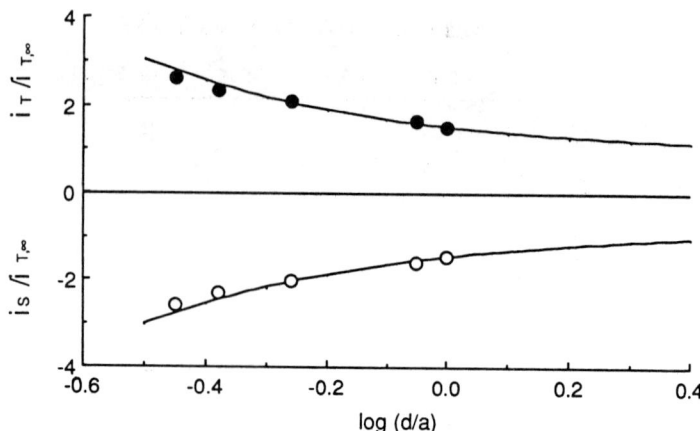

FIG. 57. Steady-state tip (filled circles) and substrate (open circles) current-distance curves for DF (5.15 mM) reduction in TG/SC mode. Solid lines calculated for K = 14 are close to those predicted for a simple diffusion-controlled process, demonstrating that the following dimerization reaction is slow on the time scale of the SECM experiment. (Reprinted with permission from Ref. 46. Copyright 1992 American Chemical Society.)

The faster rate of FN dimerization required a significantly smaller tip radius ($a = 5$ μm) and smaller d-values (1–10 μm). Tip and substrate steady-state voltammograms typical for the TG/SC regime were recorded (Fig. 58), and comparable values of both of the plateau currents indicated that the mass-transport rate was sufficiently fast to study this rapid homogeneous reaction. From the I_T-L and I_S-L curves (Fig. 59) obtained at various FN concentrations (from 1.5 to 121 mM), a rate constant $k_c = 2.0 \,(\pm\, 0.4) \times 10^5 \, M^{-1} \, s^{-1}$ was determined.

Clearly, independently determined tip-substrate separation values were required to construct Figures 57 and 59. Such values could be obtained in two ways: either by bringing the tip in contact with the substrate to establish the zero-separation point with the subsequent use of the piezo calibration to establish z-scale or by using a calibrating redox mediator. The shortcomings of the first approach include the possibility of errors caused by insulating glass touching the substrate as well as the danger of damaging the tip. On the other hand, the presence of a redox mediator may influence the kinetics of the process being studied. The half-wave potential of the redox couple, TMPD/TMPD$^+$, used as a mediator in Ref. 46 was about 1 V more positive than those of both studied processes, and no interference of the mediator with either process was detected.

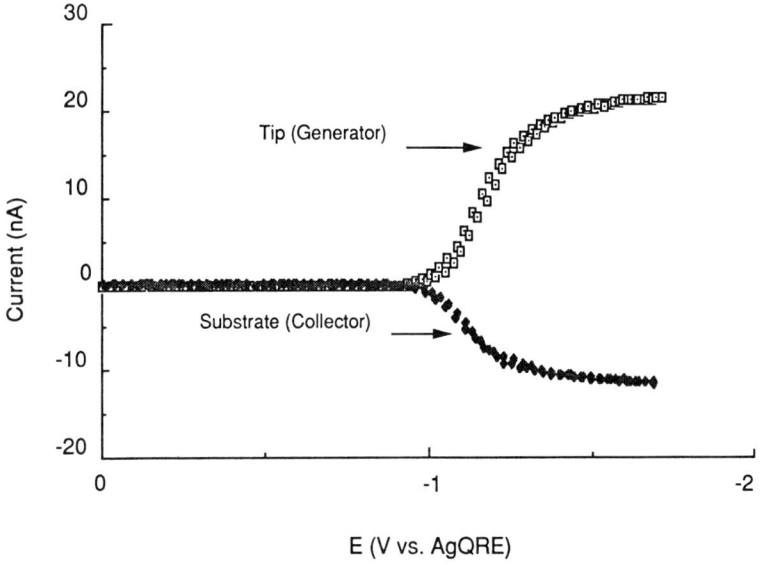

FIG. 58. SECM voltammograms for FN (28.2 mM) reduction in TG/SC mode. d = 1.8 μm. E_T was scanned at 100 mV/sec with E_S = 0.0 V vs. Ag quasi-reference electrode. (Reprinted with permission from Ref. 46. Copyright 1992 American Chemical Society.)

3. Adsorption/Desorption Processes

The adsorption of protons on (001) TiO_2 (rutile) and (010) $NaAlSi_3O_8$ (albite) surfaces was studied by SECMID in dilute acid solutions (e.g., 2×10^{-4} HCl) [48]. As described in Sec. III.D, the reduction of protons at a 25-μm-diameter Pt UME positioned closely to a single crystal (either rutile or albite) substrate induced H^+ desorption from the substrate surface. Steady-state current-distance curves (Fig. 60) were compared with ferrocyanide oxidation tip-substrate calibration data, and the theoretical behavior was predicted for an inert substrate. The good agreement found between the theoretical data and the results for H^+ reduction indicates that surface diffusion of adsorbed H^+ is essentially negligible under the conditions of these experiments. The effect of desorption as an additional source of protons is seen from the short-time parts of chronoamperograms obtained at close tip-substrate separations (Figs. 61A and 61B). The short-time currents are far greater for both albite and rutile substrates than those predicted theoretically for an inert insulating substrate. The deduced values of the intrinsic adsorption and desorption rate constants for rutile, 0.3–0.5 cm/sec and 4.1×10^{-8} to 6.9×10^{-8} mole/cm²/sec,

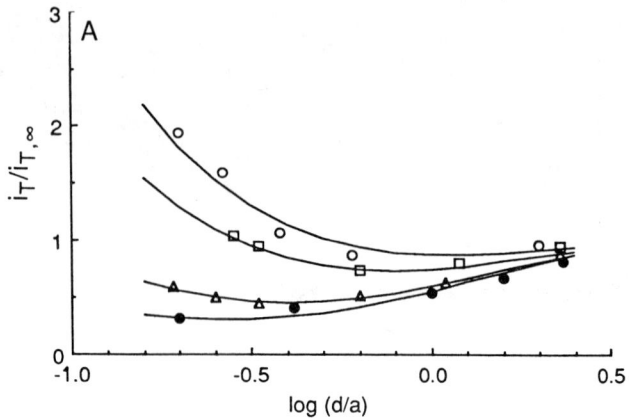

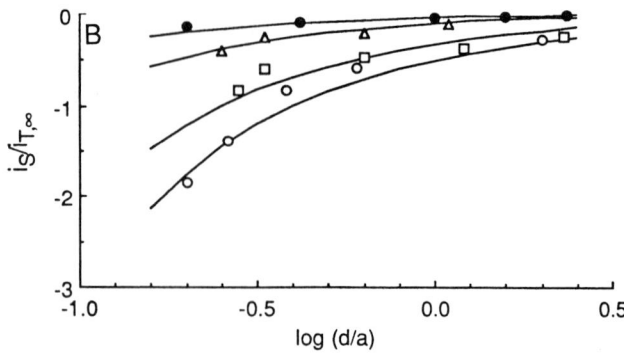

FIG. 59. Normalized tip (generation, A) and substrate (collection, B) current-distance behavior for FN reduction. FN concentration: (open circle) 1.50 mM, (open square) 4.12 mM, (open triangle) 28.2 mM, and (filled circle) 121 mM. a = 5 μm, a_S = 50 μm. The solid lines represent the best theoretical fit for each set of data. (Reprinted with permission from Ref. 46. Copyright 1992 American Chemical Society.)

respectively, are very close to the values reported in the literature for the anatase form of TiO_2 [78]. Although the values of k_a^{app} = 0.002–0.005 cm/sec and k_d^{app} = 4.7 × 10^{-10} to 1.2 × 10^{-9} mole/cm²/sec provide a reasonable fit to the experimental data for albite (Fig. 61B), there was a systematic decrease in the rate of the adsorption/desorption process at longer times in the SECMID transient, as indicated by the trend in the current from the theoretical behav-

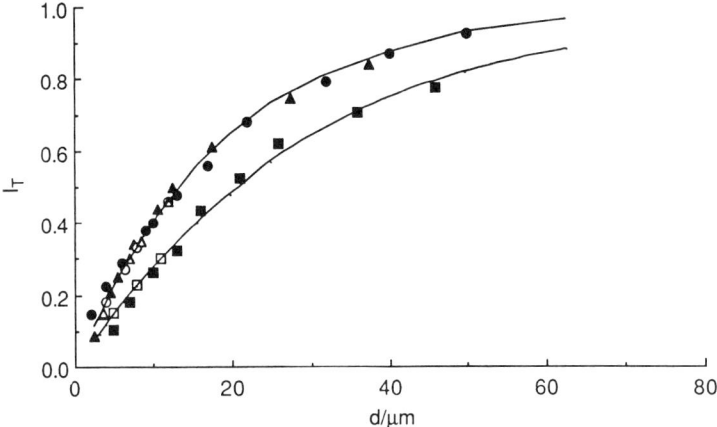

FIG. 60. Current-distance curves with substrates of Plexiglas (inert substrate, filled and open circles), rutile (001) (filled and open triangles), and albite (010) (filled and open squares). In each case, the filled symbols are data for H^+ reduction, and open symbols are tip-substrate distance calibration data for ferrocyanide oxidation. The solid lines show the behavior for an inert substrate with $\rho = 10$ (Plexiglas and rutile experiments) and $\rho = 20$ (albite experiments). (Reprinted with permission from Ref. 48. Copyright 1992 American Chemical Society.)

ior. Possible reasons for such behavior were discussed [48]. One must be cautious in the interpretation of the experimental results, because the theory for this process includes at least six adjustable parameters.

The SECM was also employed in measurements of equilibrium adsorption in small volumes of solution deposited on macroscopic adsorbents [15]. In this experiment, however, the SECM functioned only as a micromanipulator aimed to bring the microtip electrode inside the small drop of liquid (3.5–20 μl volume). The adsorption was determined through the change in concentration of adsorbing species in a liquid phase. In this way the isotherm of H^+ adsorption on the (010) surface of the mineral albite was measured.

D. Fabrication

The SECM can be used to fabricate microstructures on surfaces by deposition of metals or other solids or by etching of the substrate. Two different approaches have been used, the direct mode [79] and the feedback mode [80]. Typically, in the direct mode the tip, held in close proximity to the substrate, acts as a working electrode (in deposition reactions) or as the counterelectrode (in etching processes). The direct deposition of a metal is illustrated in Figure

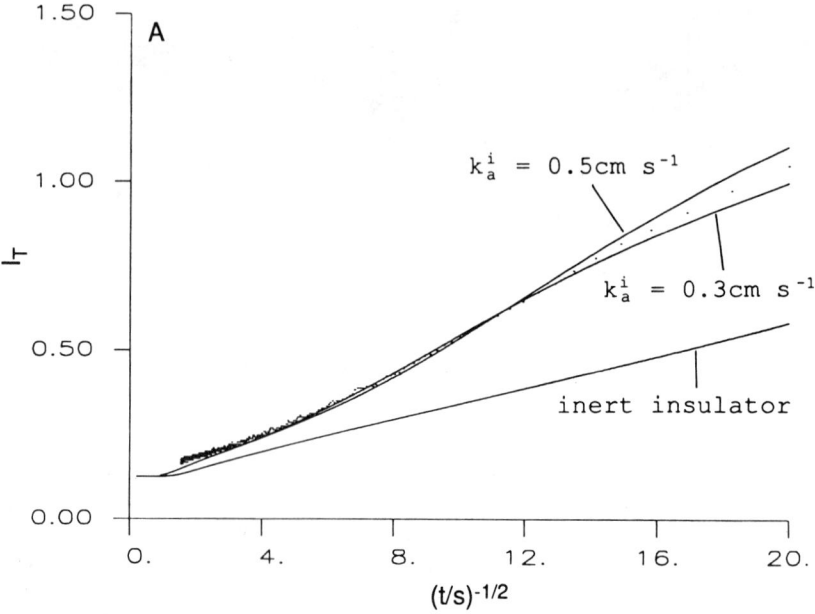

FIG. 61. SECM chronoamperograms for H^+ reduction (dotted lines) (A and C) with rutile substrate and (B) with albite substrate. d = (A) 2.6 μm, (B) 2.8 μm, and (C) 22.0 μm. The solid lines in each figure show the behavior for an inert substrate and a substrate absorption/desorption process characterized by k_a^{app} = 0.3 cm/sec and 0.5 cm/sec (rutile) and k_a^{app} = 0.002 cm/sec and 0.005 cm/sec (albite). At d = 22 μm (C), the current responses for the defined adsorption/desorption process and for an inert insulator are identical. (Reprinted with permission from Ref. 48. Copyright 1992 American Chemical Society.)

62 [79,81,82]. A thin film of an ion exchange polymer (polyelectrolyte), e.g., Nafion or poly(vinylpyridine) (PVP), is coated on the surface of the substrate, which is then immersed in a solution of the metal ion to be reduced. The metal ion is in the cationic form, e.g., Ag^+ or Cu^+, with Nafion (with an anionic backbone) or in the anionic form, e.g., $AuCl_4^-$ or $PdCl_4^-$, with PVP (with a cationic backbone). The substrate film is then removed from the metal ion solution, placed in the SECM, and scanned in air, as shown in Figure 62A. In this application, the conical tip need not be insulated, and the resolution is determined by the extent of tip penetration into the film. This can be controlled by monitoring the current and holding it constant. The smaller the tip current, typically on the order of nA, the smaller the depth of penetration and

Scanning Electrochemical Microscopy

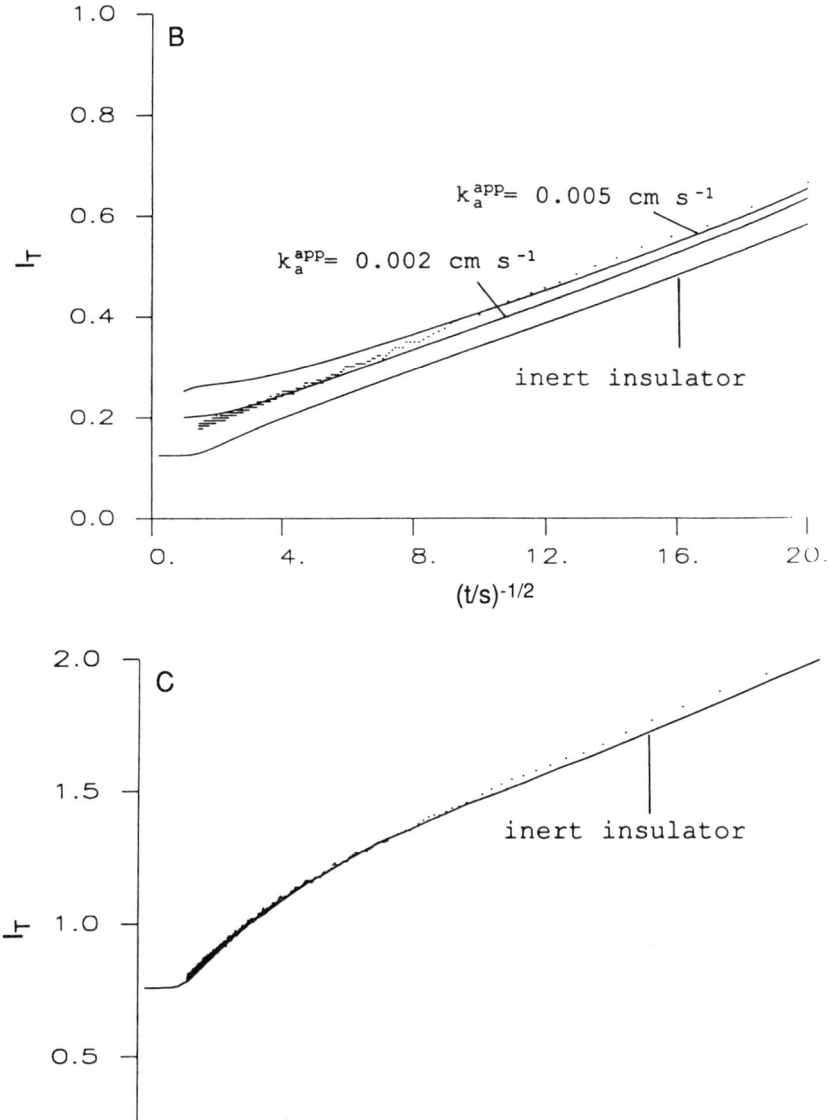

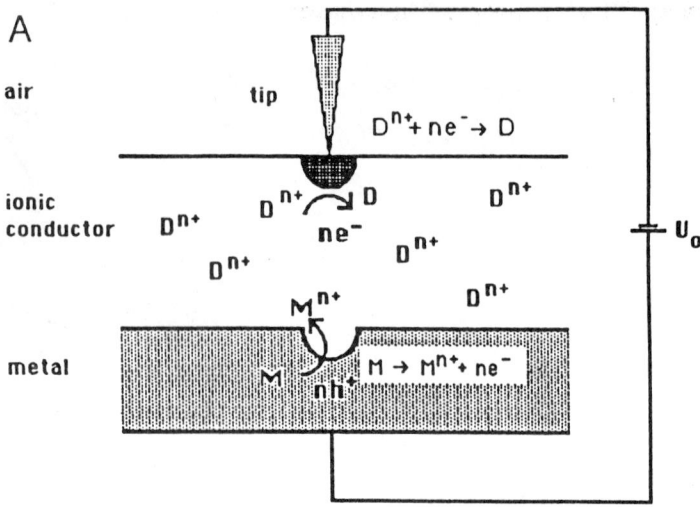

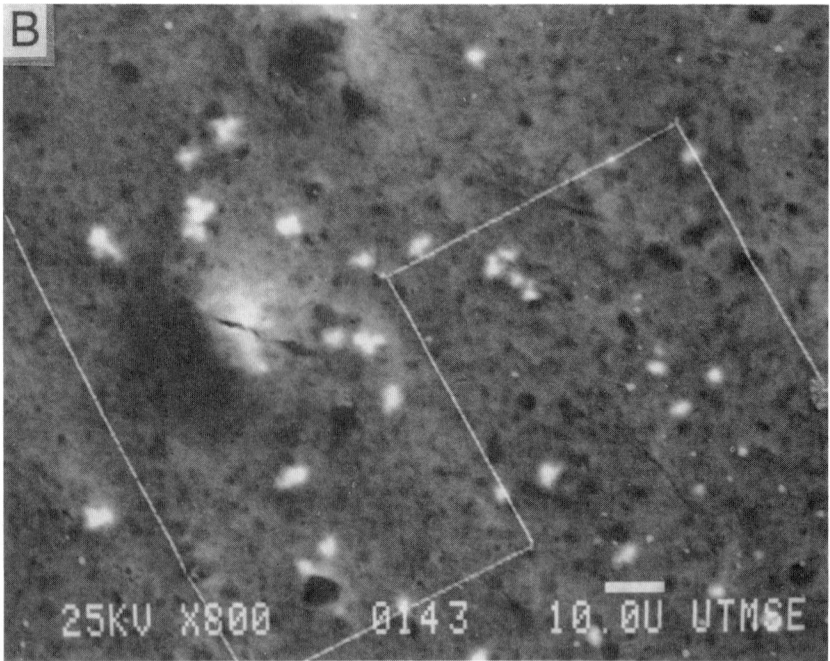

FIG. 62. (A) Schematic representation of the SECM operating in the direct mode. (B) Scanning electron micrograph of silver lines deposited in a Nafion film by using the SECM in the direct mode. Tip material, tungsten; bias, 5 V; tip current 0.5 nA; and scan rate, 900 Å/sec. (Reprinted with permission from Ref. 82. Copyright 1989 The Electrochemical Society.)

the smaller the area of deposited metal. The deposition of Ag deposited in a Nafion film is shown in Figure 62B. The W tip was moved in the desired pattern at 90 nm/sec across the Nafion/Ag$^+$ surface while maintaining a constant current of 0.5 nA. Lines as narrow as 0.3 μm were produced by this procedure. Similar lines could be produced by deposition of Au and Pd in polymer films at the tip. Interestingly, the thickness of the deposited line does not continue to grow during the scan, showing that the main electrodeposition current density is directly beneath the tip. By changing the polarity of the tip, direct metal deposition on the substrate, presumably with the oxidation of water in the polymer occurring at the tip, is possible. Patterns of metals like Au deposited in polymer films by this technique have been suggested as possible masks for X-ray lithography [81,82]. A similar approach has also been used to deposit polyaniline by oxidation of aniline on a Pt substrate covered with a Nafion film, with the tip acting as the cathode [83].

Direct etching of a metal substrate (e.g., Cu) covered with a Nafion film was accomplished by biasing the substrate at a positive potential vs. the tip [81]. In this case, a mediator, such as MV^{2+}, was introduced into the film and was reduced at the tip. The etching resolution attainable by this technique is lower than that for deposition at the tip because the electric field distribution, even in a thin film, yields a lower current density at the substrate than at the tip. The same holds true for deposition on the substrate. Note that some etching reactions that have been described with the STM in air may well occur by the direct SECM mode in the thin film of water that forms on the substrate surface. For example, pits can be etched in highly oriented pyrolytic graphite (HOPG) by biasing it positive (by at least 2.3 V) with respect to the STM tip [84]. Etching is not seen, however, when the HOPG is biased negative to the tip, nor is it seen when the tip is moved closer to the positively biased HOPG (by increasing the current at the same bias), where tunneling can occur, nor is it seen for HOPG in vacuum with any bias. Moreover, the minimum bias required is about that expected for the occurrence of electrode reactions at both tip and substrate. Once a small pit is formed, it can be expanded and converted to a desired shape (e.g., line, square) by continued scanning at a smaller positive bias.

The resolution attainable in direct deposition and etching depends on a number of factors: tip size, depth of penetration, tip currents, tip scan speed. Faster scan speeds tend to yield smaller features. However, fast scanning across the surface produces a higher contribution from charging current. Typical maximum scan speeds applied are about 500 nm/sec. Instrumental factors, e.g., vibration-damping, feedback response, and temperature control, are also important to high resolution, as are the reactions at the tip and substrate.

The feedback mode of fabrication utilizes the same arrangement as in SECM imaging. In this case, however, the tip reaction is selected to generate a species that reacts at the substrate to promote the desired reaction, i.e., deposition or etching (Fig. 63) [79,85]. For example, a strong oxidant, like Br_2, generated at the tip can etch the area of the substrate, e.g., GaAs directly beneath the tip [86]. The mediator reactant is chosen to be one that reacts completely and rapidly at the substrate, thus confining the reaction to a small area on the substrate and producing features of area near that of the tip. Clearly, small tip size and close tip-substrate spacing are required for high resolution. Note that in many cases (as shown in Fig. 63), the reaction at the substrate regenerates the species that reacts at the tip and yields the typical SECM positive feedback response. Thus, the usual approach curves can be employed to estimate the tip-substrate distance.

In metal deposition reactions, the tip-generated reactant reacts with a thin film of metal precursor on the substrate surface [85]. For example, (Fig. 63A) Au or Pd were electrodeposited in PVP films containing $AuCl_4^-$ or $PdCl_4^{2-}$ by generating a suitable reductant at the tip. A similar approach should be applicable with other types of films, such as films of insoluble metal salts. The feedback mode of deposition has the advantage that the substrate need not be a conductor.

Etching is carried out in the feedback mode by generating an appropriate oxidant (etchant) [79]. For example, Cu is etched by generating $Os(bpy)_3^{3+}$ at the tip (Fig. 63B), where the etching reaction is

$$2Os(bpy)_3^{3+} + Cu \rightarrow 2Os(bpy)_3^{2+} + Cu^{2+} \tag{61}$$

Note that in the feedback mode the tip is held at a positive potential, so there is no tendency to plate Cu (from the Cu^{2+} produced in the etching reaction) on the tip. This can be a problem when etching is carried out by the direct mode. Moreover, since the substrate is maintained at a potential characteristic of the reduced mediator, except directly under the tip, there is less tendency for the copper to be oxidized at areas distant from the tip. It has been suggested that the resolution of SECM etching could be improved by generating an unstable oxidant at the tip [87]. This instability would limit the distance the oxidant can diffuse from the tip and minimize reactions at locations away from it. No actual reactants were suggested, and such a system has not yet been reduced to practice.

This feedback method is applicable with semiconductors and insulators, since the substrate is not an electrode in the electrochemical cell. Thus, GaAs can be etched by generation of Br_2 at the tip (Fig. 63C). Pits etched in GaAs by this method are shown in Figure 64. The size and depth of the pit depended

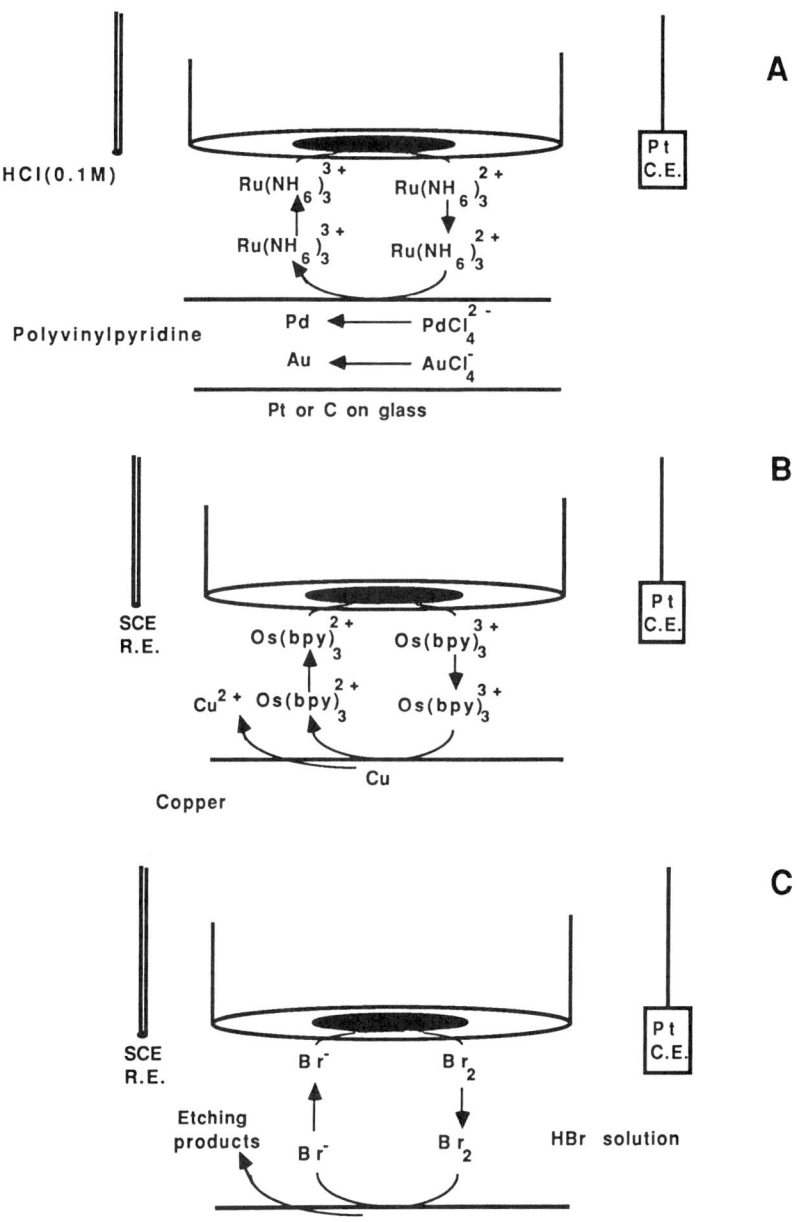

FIG. 63. Schematic representation of the SECM applied in the feedback mode for (A) metal deposition, (B) copper etching, and (C) semiconductor etching. (Reprinted with permission from Ref. 9. Copyright 1990 American Chemical Society.)

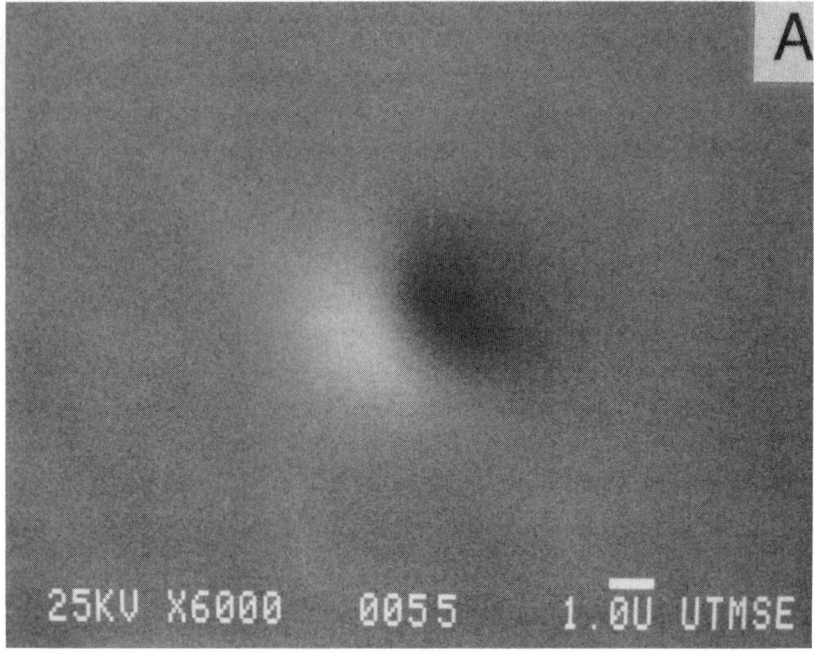

FIG. 64. Effect of UME diameter. SEM pictures of etching spots made with (A) 50-, (B) 25-, and (C) 2-μm Pt UME. (Reprinted with permission from Ref. 86. Copyright 1990 The Electrochemical Society.)

upon the length of time the biased tip was held above a particular point on the GaAs surface.

This feedback mode has also been to study the mechanism of the etching process of GaAs by using different electrogenerated oxidants, varying the pH, and contrasting the behavior of n-type, p-type, and intrinsic GaAs [88]. This study indicated that etching occurred only when the tip-generated oxidant was sufficiently energetic to inject a hole into the valence band of the GaAs. By using different mediators and noting the onset of positive feedback, the energetic location of the valence band edge could be estimated.

A very different use of the SECM was in forming a fluorescent micropattern in an ionically conductive polymer (Flemion) film (H. Sugimura et al., private communication). In this application, the fluorescer, Rhodamine-6G (R-6G), and a quencher, MV^{2+}, were introduced in a film of Flemion about 0.2 μm thick on a Pt substrate. The characteristic R-6G fluorescence was completely quenched under these conditions. However, when the MV^{2+} was

Scanning Electrochemical Microscopy 347

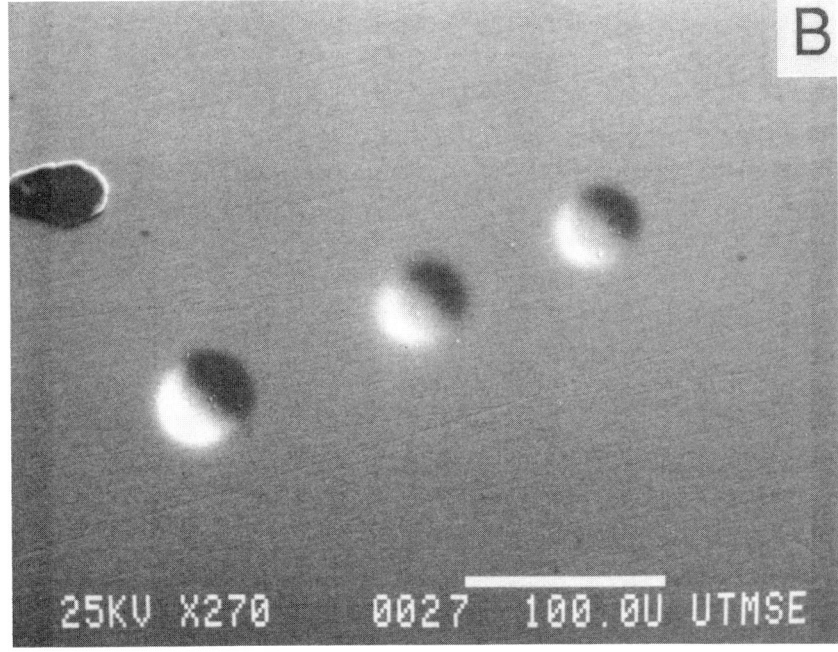

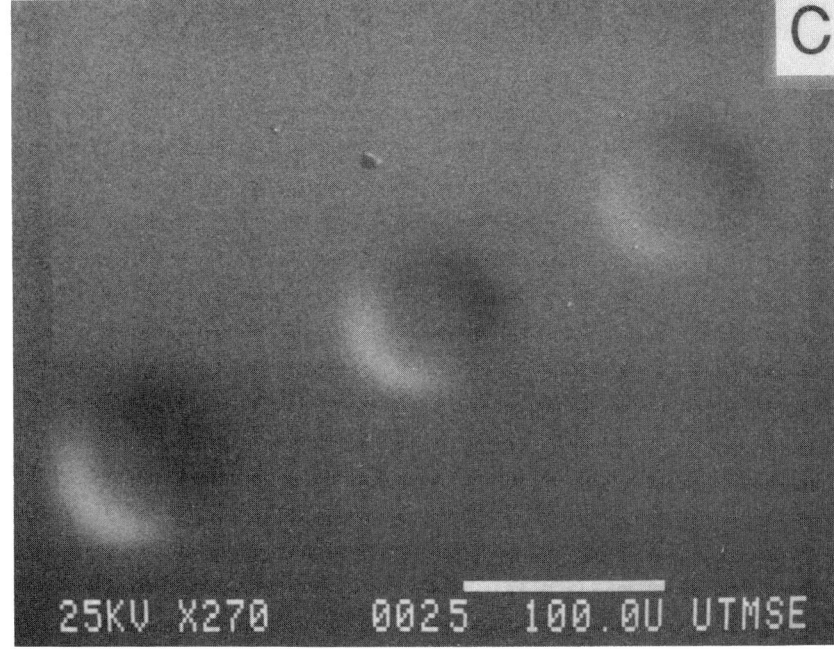

decomposed by a reduction process at the tip, the fluorescence of R-6G at this position was observed. Thus the tip could be used to write a fluorescent pattern in the Flemion layer with a resolution of at least 0.5 μm. The authors suggest that the actual features formed with a sharp tip may be smaller than this and not observable because of the limited resolution of the optical microscope used to observe the pattern.

E. Potentiometric and Other Tips

While most of the SECM work has been carried out with amperometric tips for measuring feedback currents or for use in the generation/collection mode, other types of tips are also possible. For example, reference electrode tips, such as Ag/AgCl, can be employed to map potential distributions, as in the corrosion studies described in Sec. I. For many interesting chemical systems, however, it may not be possible to devise an amperometric technique with an appropriate redox mediator to probe the behavior, e.g., when the major concentration changes in the system are those involving pH or a nonelectroactive ion (Na^+, Ca^{2+}). Profiles of such species are particularly important in the life sciences, and ion-selective microprobes have been used to probe such systems [89]. Such probes, if they are sufficiently small and have a rapid response, would be useful in SECM as well. The potentiometric tip, as opposed to the amperometric one, is a passive measuring probe which does not interact directly with the substrate. It can only be used to obtain information and to image a substrate when there are concentration gradients (ion activity profiles) in its vicinity. Moreover, while an amperometric tip can be used to determine d through the approach curve and is thus easily positioned with respect to the substrate, this is not true of a potentiometric tip. Thus positioning must be carried out by visual observation through a microscope [40], which is clearly inconvenient, especially for very small tips, or by using electrodes that can be alternately used in the potentiometric and amperometric mode [37]. Another approach is the use of multibarrel tips with separate potentiometric and amperometric functions; these are difficult to fabricate (D. T. Pierce and A. J. Bard, unpublished).

A Ag microelectrode was used as a potentiometric probe in an SECM to study the silver and chloride ion fluxes at electrodes [40]. In this study, the potential difference between a large reference electrode in the bulk solution and a Ag microdisk (10 and 50 μm diameter) positioned near the substrate was used to determine concentrations of silver ion diffusing away from a planar Ag electrode. In a separate experiment, a Ag/AgCl microdisk was used to monitor the Cl^- flux near a polyaniline layer deposited on a Pt substrate during oxidation and reduction. Measurements of current transients as a

Scanning Electrochemical Microscopy

function of d were reported, but no imaging results were given. One-dimensional pH and Cl⁻ profiles in corrosion pits on metals have also been obtained, but at a much lower resolution than that typical of the SECM [90].

An antimony tip can be used both as an amperometric probe (for positioning purposes) and as a potentiometric pH sensor [37]. The antimony tips were prepared by drawing down Sb-filled glass capillaries to yield Sb disks with diameters down to about 3 μm. Such tips showed a good pH response (40–50 mV per pH unit) over a pH range of 5–9.5. The tips were positioned by determining the current for the reduction of oxygen at -0.7 V vs. the Ag/AgCl reference using the approach curve for an insulating substrate (negative feedback). They were then brought back to 0 V to restore their pH sensitivity and were used to image pH profiles over different substrates. For example, an image of the pH profile near a 25-μm Pt disk where proton reduction is occurring in a pH 7.0 phosphate buffer is shown in Figure 65. The distribution of the pH gradient depends upon the buffer capacity as the base generated at the disks diffuses away and reacts with (titrates) the acidic component of the buffer. This same probe can be used to determine the activity of an enzyme whose reaction involves the production or consumption of protons. An example is shown in Figure 66 where urease immobilized in a glass capillary was imaged during the hydrolysis of urea:

$$CO(NH_2)_2 + H_2O \rightarrow 2\ NH_3 + CO_2 \tag{62}$$

which causes an increase in pH in the vicinity of the enzyme. Since pH changes occur frequently with enzyme systems, this approach should be very useful in imaging and studying enzymes by SECM that are not amenable to the amperometric approach discussed in Sec. IV.A.

Other types of tips, based on electrochemical sensors, should also be possible. For example, an enzyme electrode tip could be used to probe a particular species with high selectivity. Preliminary experiments with a peroxidase tip, for example, have been used to determine H_2O_2 produced during O_2 reduction on an electrode (B. R. Horrocks et al., unpublished). While the construction of small enzyme electrode tips is challenging, they could be especially useful in studies of biological systems.

F. Characterization of Thin Films and Membranes

SECM is also a useful technique for studying thin films on interfaces. For example, polyelectrolytes, such as Nafion and protonated poly(4-vinylpyridine), and electronically conductive polymers, such as polypyrrole, have been investigated by SECM [47,52,55–59].

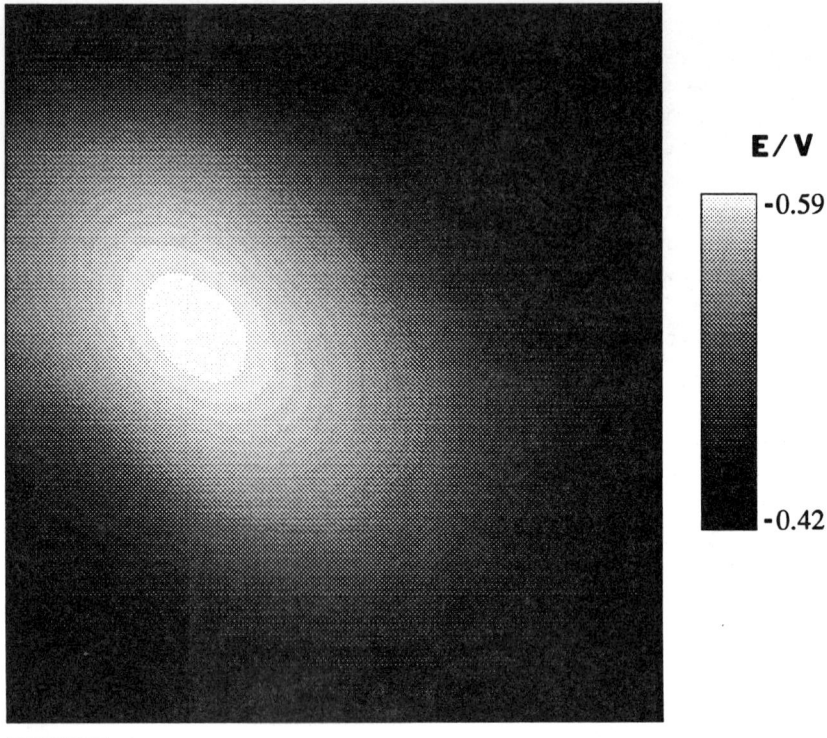

40.0 μm

FIG. 65. Image of the pH profile around a 25-μm platinum target reducing water. The gray scale shows the antimony tip potential; white corresponds to high pH. The potential of the target was −2 V vs. a platinum auxiliary and the current was 0.5 μA. The diameter of the antimony tip was 40 μm, the tip to surface distance was 33 μm, and the tip scan rate was 10 μm/sec. The solution contained 0.1 M KCl and 0.5 mM pH 7 phosphate buffer. (Reprinted with permission from Ref. 37. Copyright 1993 American Chemical Society.)

1. Cyclic Voltammetry

Lee et al. [56] investigated $Os(bpy)_3^{2+}$ incorporated into a Nafion film by SECM. A unique type of cyclic voltammetry, tip/substrate cyclic voltammetry (abbreviated T/S CV), was used to check the electrochemical response of the tip before SECM topographic scans were carried out. T/S CV involves monitoring the tip (T) current vs. the substrate (S) potential (E_S) while the tip

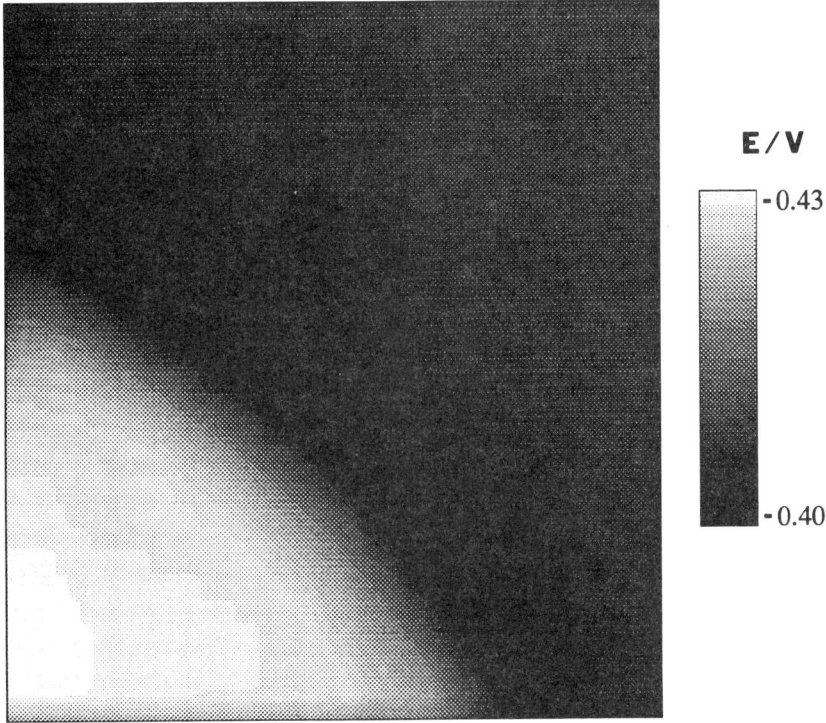

FIG. 66. Image of the pH profile around a 120-μm-diameter urease gel target. The gray scale shows the antimony tip potential, white corresponds to high pH. The tip diameter was 40 μm, the tip scan rate was 10 μm/sec, and the solution contained 10 mM urea, 0.1 M KCl, and 2 mM pH 7.0 phosphate buffer. (Reprinted with permission from Ref. 37. Copyright 1993 American Chemical Society.)

potential (E_T) is maintained at a given value and the tip is held near the substrate. As shown in Figure 67, the substrate cyclic voltammogram (i_S vs. E_S) of an Os(bpy)$_3^{2+}$-incorporated Nafion film covering a Pt disk electrode (5 mm diameter), at a scan rate v = 50 mV/sec in 10 mM K$_3$Fe(CN)$_6$, only shows a wave for the Os(bpy)$_3^{3+/2+}$ couple, indicating that the high negative charges prevent Fe(CN)$_6^{3-}$ from diffusion into the Nafion coating [91]. Thus, any reaction between Fe(CN)$_6^{4-/3-}$ and Os(bpy)$_3^{3+/2+}$ will be restricted to the solution/film interface. Typical T/S CV curves, where E_S is scanned at v =

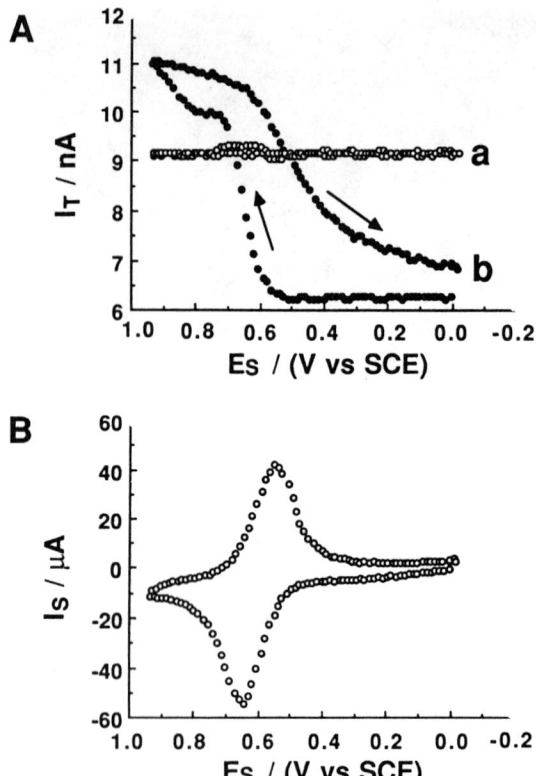

FIG. 67. T/S CVs (A) curve a, d = 500 μm; curve b, d = 10 μm, and S-CV (B) on Nafion/Os(bpy)$_3^{3+/2+}$ electrode in K$_3$Fe(CN)$_6$/Na$_2$SO$_4$, v = 50 mV/sec, E$_T$ = −0.4 V vs. SCE. (Reprinted with permission from Ref. 56. Copyright 1990 American Chemical Society.)

50 mV/sec, i$_T$ is measured at E$_T$ = −0.4 V vs. SCE, and Fe(CN)$_6^{3-}$ is reduced to Fe(CN)$_6^{4-}$, are shown in Figure 67A. When the tip electrode is far from the substrate (d = 500 μm), i$_T$ shows i$_{T,\infty}$ (9.2 nA) and is essentially independent of E$_S$. When the tip electrode is close to the substrate electrode (d = 10 μm), either negative or positive feedback effects arise, depending on the oxidation state of the Os(bpy)$_3^{2+/3+}$ couple in the Nafion. When E$_S$ is swept positive of the Os(bpy)$_3^{2+/3+}$ redox waves, i$_T$ > i$_{T,\infty}$. This positive feedback i$_T$ is the result of the reaction

$$Fe(CN)_6^{4-} + Os(bpy)_3^{3+} \rightarrow Fe(CN)_6^{3-} + Os(bpy)_3^{2+} \qquad (63)$$

at the solution/film interface leading to the regeneration of $Fe(CN)_6^{3-}$ in the solution gap region. When E_S is negative of the redox waves, the film behaves as an insulator and $i_T < i_{T,\infty}$, since the $Os(bpy)_3^{2+}$ formed is unable to oxidize tip-generated $Fe(CN)_6^{4-}$ back to $Fe(CN)_6^{3-}$ and the film blocks diffusion of $Fe(CN)_6^{3-}$ to the tip.

Figure 68 shows the CVs of an $Os(bpy)_3^{2+}$-incorporated Nafion film in an aqueous solution containing 10 mM $K_4Fe(CN)_6$ and 0.1 M Na_2SO_4 at v = 50 mV/sec and E_T = 0.6 V vs. SCE. As expected, the substrate CV of the $Os(bpy)_3^{2+/3+}$ couple in the presence of $Fe(CN)_6^{4-}$ in solution (Fig. 68B)

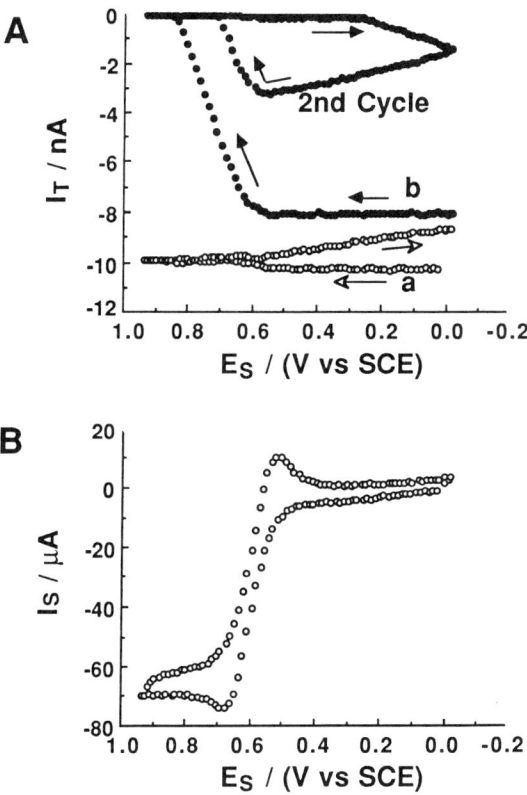

FIG. 68. T/S CVs (A) curve a, d = 220 μm; curve b, d = 10 μm, and S-CV (B) on Nafion/$Os(bpy)_3^{3+/2+}$ electrode in $K_3Fe(CN)_6/Na_2SO_4$, v = 50 mV/sec, E_T = 0.6 V vs. SCE. (Reprinted with permission from Ref. 56. Copyright 1990 American Chemical Society.)

shows the catalytic oxidation of $Fe(CN)_6^{4-}$ by electrochemically generated $Os(bpy)_3^{3+}$. For this solution, the substrate reaction causes the depletion of $Fe(CN)_6^{4-}$, rather than regenerating it from the tip-generated $Fe(CN)_6^{3-}$. The extent of this depletion is demonstrated in Figure 68A. Even at a relatively large distance (d = 220 μm), when E_S is scanned to the potentials where the oxidation of $Os(bpy)_3^{2+}$ takes place, i_T decreases because of depletion of $Fe(CN)_6^{4-}$ in the solution region between substrate and tip (curve a of Fig. 68A). The time, t_d, for i_T to decrease below $i_{T,\infty}$ in the cyclic scan is about one half of the time needed for the diffusion layer to grow out from the substrate to the tip, i.e., $t_d \approx d^2/2D$. This depletion effect can be seen for much larger tip distances as compared with the feedback mode, if sufficient time is allowed for substrate diffusion layer growth. As the tip is moved closer to the film surface (e.g., d = 10 μm), more rapid and greater depletion of the $Fe(CN)_6^{4-}$ species at the substrate is detected at the tip (curve b of Fig. 68A). i_T decreases rapidly to zero immediately after $Os(bpy)_3^{2+}$ is oxidized to $Os(bpy)_3^{3+}$ because the large modified substrate electrode (Pt disk, 5 mm diameter) depletes $Fe(CN)_6^{4-}$ near the surface of the film via reaction (63). The almost total depletion of $Fe(CN)_6^{4-}$ is maintained until the substrate potential scan is reversed and the $Os(bpy)_3^{3+}$ species in the Nafion film is reduced. i_T then increases because $Fe(CN)_6^{4-}$ from the bulk solution diffuses into the gap region. This increase continues until i_T approaches the value characteristic of that over an insulator. However, when E_S is scanned back into the region where the oxidation of $Fe(CN)_6^{4-}$ takes place, i_T decreases, again via the depletion effect.

2. Chronoamperometry and Chronopotentiometry

Potential-step and current-step experiments have been applied by Kwak and Anson [57] to the study of the ejection and incorporation of $Fe(CN)_6^{3-}$ and $Fe(CN)_6^{4-}$ counterions at a protonated poly(4-vinylpyridine) ($PVPH^+$) film. In the potential-step experiment, a microtip electrode was positioned near the surface of the substrate while the potential of the latter was stepped from $+0.6$ to -0.2 V vs. SCE in a solution containing $Fe(CN)_6^{3-}$, and the tip current was measured at a given potential. The chronoamperometric responses from the bare and coated substrate are shown in Figures 69A and 70A. With the tip electrode set at -0.2 V vs. SCE to detect (reduce) $Fe(CN)_6^{3-}$, a steady tip current flows before the potential of the substrate is stepped, and its magnitude depends upon the tip-substrate separation because of positive feedback. After the potential of the bare electrode is stepped to -0.2 V, where $Fe(CN)_6^{3-}$ is reduced to $Fe(CN)_6^{4-}$, the tip current decays toward zero. The decay is very rapid when the tip is close to the substrate

Scanning Electrochemical Microscopy

(Figs. 69B and 69C, dotted curves) and is small when the separation is increased. Note that when the potential of the coated substrate electrode is stepped to -0.2 V, increases, rather than decreases, in the concentration of $Fe(CN)_6^{3-}$ near the coated surface are observed initially, followed by a decay toward zero (Figs. 69B–69F, solid curves), indicating the ejection of $Fe(CN)_6^{3-}$ from the coating. The potential-step experiments with the tip potential set to monitor the concentration of $Fe(CN)_6^{4-}$ (Figs. 70B–70F) show that, after a delay, this anion is also ejected from the coating on reduction.

One set of the results for current-step experiments carried out on the same system as described in the potential-step experiments is shown in Figure 71 [57]. An interesting phenomenon is observed when the current direction is reversed so that the $Fe(CN)_6^{4-}$ anions generated at the substrate surface during the forward current step are oxidized back to $Fe(CN)_6^{3-}$ after current reversal. With a bare substrate, when the tip potential is maintained at -0.2 V vs. SCE, the cathodic tip current decays toward zero, as expected, during the forward current step and increases again immediately after current reversal (Figs. 71B and 71C, dotted curves). However, with the coated substrate, the cathodic tip current for small tip-substrate separations continues to decrease, and at an enhanced rate, instead of increasing immediately after the current reversal (Fig. 71B, solid curve). This unusual behavior has been attributed to the reincorporation of $Fe(CN)_6^{3-}$ anions by the coating as driven by the anodic current. Anion incorporation by the coating is required to maintain electroneutrality following current reversal. Apparently, the $Fe(CN)_6^{3-}$ anions present at the coating/solution interface reenter the coating despite the much higher concentration of chloride anions. Similar techniques have also been applied by Lee and Anson [58] to examine the ejection and incorporation of $Os(bpy)_3^{2+/3+}$ from Nafion coatings.

3. Direct Electrochemical Measurements

The SECM studies of polymer films described above rely on using the tip to probe the solution environment directly above the polymer to investigate electron transfer and ion ejection and incorporation into the film. In this section, we describe direct measurements of electrochemical parameters by recording the tip current as it is moved from the solution into the polymer phase and ultimately contacts the substrate [52].

In the tip current vs. tip displacement experiment, the tip current is monitored as a function of the relative tip displacement in the direction normal to a Nafion film incorporating $Os(bpy)_3^{2+}$. The experiment was carried out with an aqueous solution of 40 mM $NaClO_4$ as the supporting electrolyte with a small conical tip (30 nm radius, 30 nm height). During the scan of i_T vs. d,

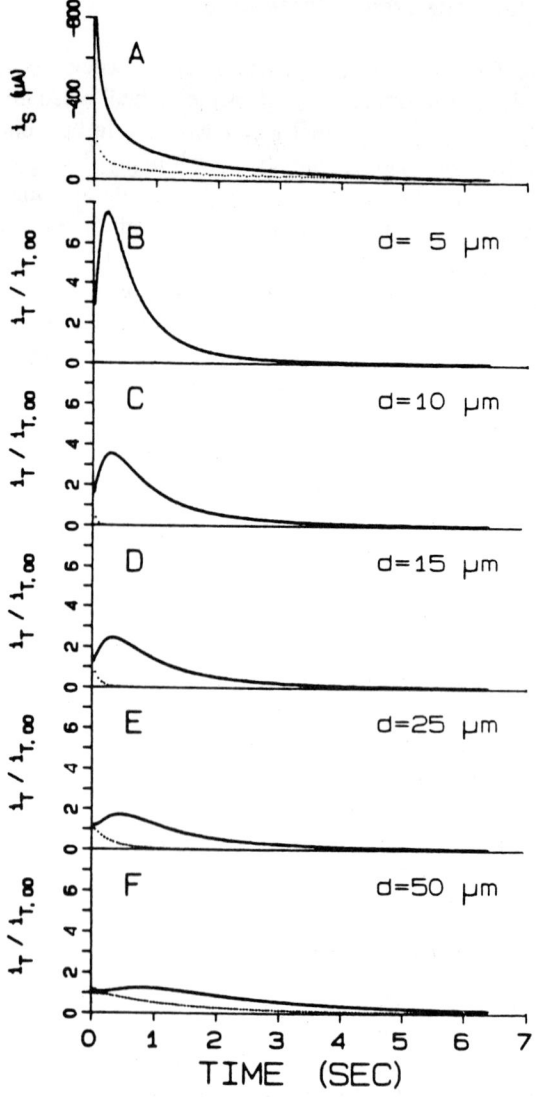

FIG. 69. Tip currents resulting from potential steps at the substrate electrode in solutions of $Fe(CN)_6^{3-}$. (A) Chronoamperometric substrate currents resulting from a potential step from $+0.6$ to -0.2 V with a bare substrate electrode (dotted curve, $[Fe(CN)_6^{3-}] = 1$ mM) or a $PVPH^+$-coated substrate electrode (solid curve, $[Fe(CN)_6^{3-}] = 0.2$ mM). Supporting electrolyte: 0.1 M KCl + 5 mM HCl. (B–F) Normalized cathodic tip currents obtained during the potential-step experiments in (A) with the tip maintained at -0.2 V to monitor the concentration of $Fe(CN)_6^{3-}$. The initial tip currents are distorted for 50–100 msec by coupling between the substrate and tip electronic control circuits, especially at the smallest values of d. The tip current corresponding to the bare substrate electrode decayed to zero too quickly to be seen in

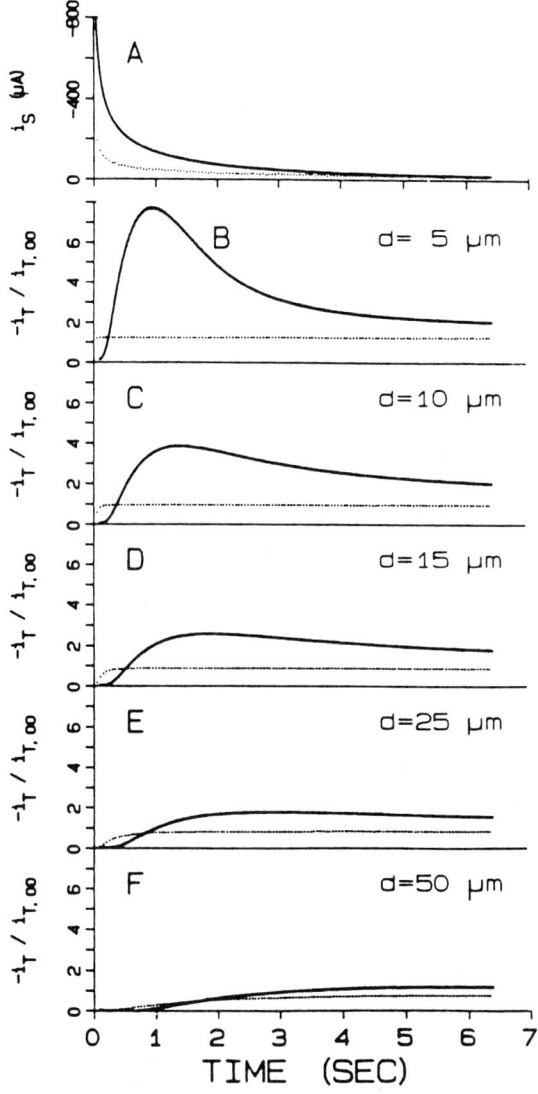

FIG. 70. Repeat of Figure 69 with the tip electrode maintained at 0.6 V to monitor the concentration of $Fe(CN)_6^{4-}$. (Reprinted with permission from Ref. 57. Copyright 1992 American Chemical Society.)

(B). The tip radius is 5.9 μm. The separation was calculated from the positive feedback currents measured with the substrate potential at 0.5 V. (Reprinted with permission from Ref. 57. Copyright 1992 American Chemical Society.)

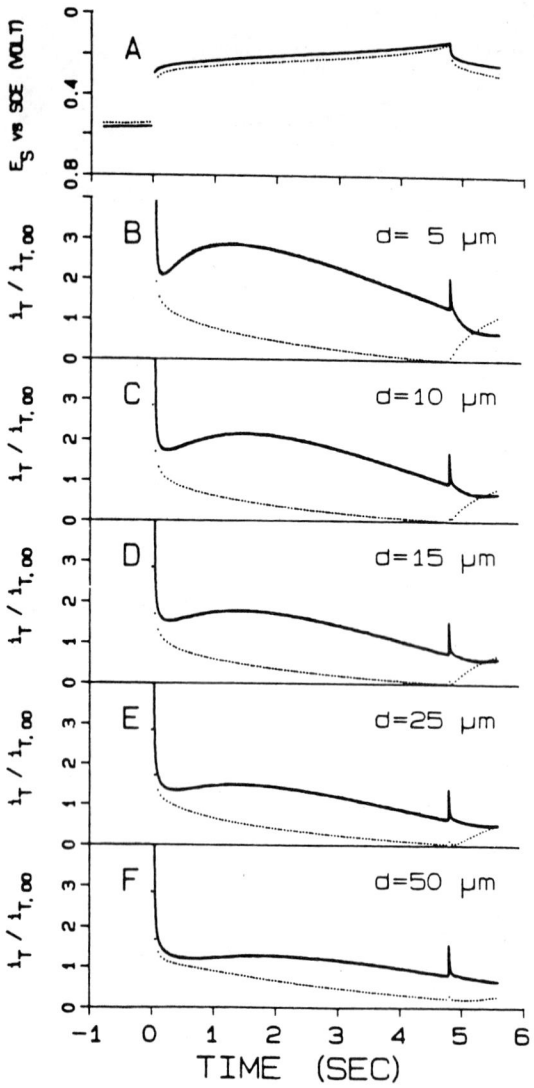

FIG. 71. Tip currents resulting from current steps at the substrate electrode in solutions of $Fe(CN)_6^{3-}$. (A) Chronopotentiograms recorded with a bare substrate electrode (dotted curve), $[Fe(CN)_6^{3-}] = 1$ mM) or a $PVPH^+$-coated substrate electrode (solid curve, $[Fe(CN)_6^{3-}] = 0.2$ mM). The cathodic current steps applied were -28 and -80 μA at the bare and coated electrodes, respectively. The current direction was reversed after 4.8 sec, which was before the transition time. The electrode was equilibrated at 0.6 V for 2 min and placed at open circuit for 0.8 sec

the tip was held at 0.80 V vs. SCE, where $Os(bpy)_3^{2+}$ oxidation is diffusion-controlled. The Nafion-coated indium tin oxide (ITO) substrate was biased at 0.20 V vs. SCE, where any $Os(bpy)_3^{3+}$ generated at the tip that reaches the substrate will be reduced back to $Os(bpy)_3^{2+}$, when the tip-ITO separation is small. A scheme of this SECM experiment, as shown in Figure 72, can be represented by five stages:

1. Initially (Fig. 72A), the Pt microtip is in the solution near the Nafion/electrolyte interface. Because the electrolyte contains no electroactive species, a negligibly small current is observed before the tip touches the film surfaces (see Fig. 73, the portion of the current-distance curve designated a).
2. When the tip starts to penetrate into the polymer (Fig. 72B), the anodic current increases gradually (Fig. 73, b) until it reaches some limiting value. This increase represents the increasing area of the tip exposed to Nafion as the conically shaped Pt penetrates the polymer film.
3. When the tip is completely immersed in the film, but still "far" (i.e., greater than a few tip diameters) from the ITO substrate, the tip current remains constant and independent of d (Fig. 73, c).
4. When the tip gets close to the substrate, the SECM positive feedback effect becomes important and the tip current increases (Fig. 73, d).
5. Finally, when the tip gets to within tunneling distance of the substrate, a large increase in current occurs (Fig. 73, e).

The thickness of the Nafion film, found as the difference in relative displacement between the film/solution interface coordinate and that for the onset of tunneling, has a value of 220 nm. The error in this determination, which results from neglecting the tunneling distance and uncertainty in the film boundary position, is within 10 nm, i.e., a relative error within 10%.

Voltammetry with the tip in the Nafion was performed to determine kinetic parameters ($k°$, α, and D). Because the steady-state voltammogram is fairly insensitive to the exact shape of the tip, it can be approximated by the equivalent size hemisphere. A voltammogram for $Os(bpy)_3^{2+}$ electro-oxidation in Nafion obtained at a scan rate, $v = 5$ mV/sec, with the tip

---

before the current step was applied. (B–F) Normalized cathodic tip currents obtained in response to the current steps in (A) with the tip electrode maintained at -0.2 V to monitor the concentration of $Fe(CN)_6^{3-}$. The tip currents at t = 0 are distorted for 50–100 msec by coupling between the substrate and tip electronic control circuits, especially by electronic coupling. Other conditions are as in Figure 69. (Reprinted with permission from Ref. 57. Copyright 1992 American Chemical Society.)

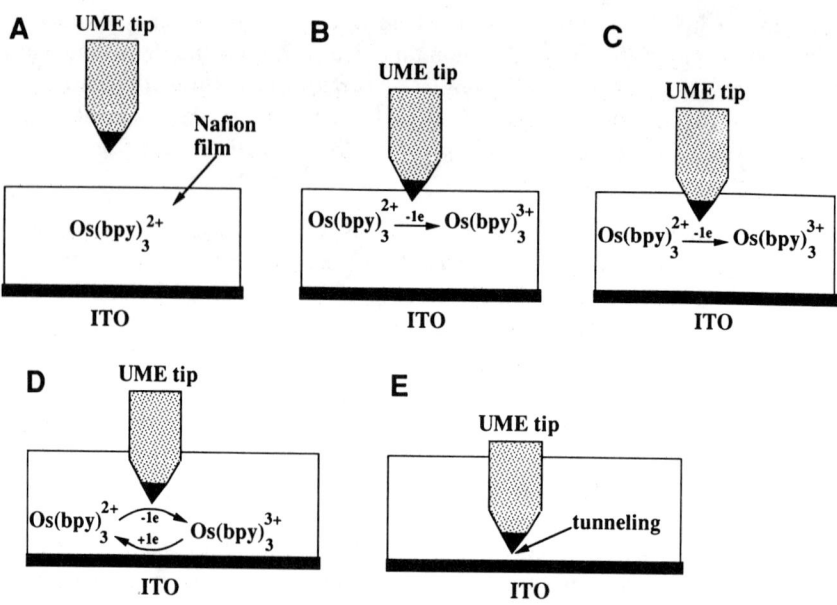

FIG. 72. A scheme representing five stages of the SECM current-distance experiment. (A) The tip is positioned in the solution close to the Nafion coating. (B) The tip has penetrated partially into Nafion and the oxidation of Os(bpy)$_3^{2+}$ occurs. The effective tip surface grows with penetration. (C) The entire tip electrode is in the film, but is not close to the ITO substrate. (D) The tip is sufficiently close to the substrate to observe positive SECM feedback. (E) The tunneling region. (Reprinted with permission from Ref. 52. Copyright 1991 American Association for the Advancement of Science.)

partially buried in the film as shown in Figure 74, shows a plateau current, $i(h) = 0.625$ pA, substantially lower than the steady-state current value observed in Figure 73, since only a part of the tip electrode is inside the Nafion film. From Figure 73, one can estimate the effective shape parameters of the conical tip and its equivalent size hemisphere with a radius, r_h, of 18.9 nm [52]. With the concentration of Os(bpy)$_3^{2+}$ in the Nafion film, $c = 5.7 \times 10^{-4}$ mol/cm^3 and the r_h value, the apparent diffusion coefficient of Os(bpy)$_3^{2+}$ is determined to be 1.2×10^{-9} cm^2/sec, which is in the range of previous studies [92,93] but is significantly larger than that for a Nafion film with a lower loading of Os(bpy)$_3^{2+}$ [94]. The kinetic parameters of the electrode reaction have also been evaluated by the three-point method, i.e., the half-wave potential, $E_{1/2}$, and two quartile potentials, $E_{1/4}$ and $E_{3/4}$ [35]. A

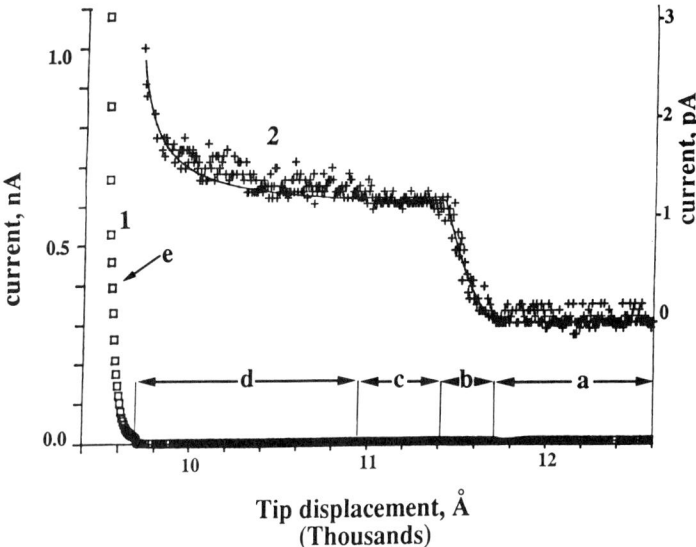

FIG. 73. Dependence of the tip current vs. distance. The letters a to e correspond to five stages in Figure 72. The displacement values are given with respect to an arbitrary zero point. The current observed during the stages a to d is much smaller than the tunneling current and therefore cannot be seen on the scale of curve 1 (the left-hand current scale). Curve 2 is at higher current sensitivity to show the current-distance curve corresponding to stages a to d (the right-hand current scale). The solid line is computed for a conically shaped electrode with a height h = 30 nm and a radius r_0 = 30 nm by Eq. (2) in Ref. 52 for zones a to c and SECM theory [25] for zone d. The tip was biased at 0.80 V vs. SCE, and the substrate was at 0.20 V vs. SCE. The tip moved at a rate of 30 Å/sec. (Reprinted with permission from Ref. 52. Copyright 1991 American Association for the Advancement of Science.)

transfer coefficient (α) of 0.52 and a rate constant ($k°$) of 1.6×10^{-4} cm/sec were obtained.

4. SECM Imaging of Films and Membranes

SECM has been used to image thin films and membranes, e.g., ionically or electronically conductive polymers, oxide films, and membranes [54–56,59,95]. Figure 75 shows a typical SECM image of an excised hairless mouse skin [95]. During imaging, an iontophoretic current density (i_{app}) of 2.0 mA/cm^2 is driven across the excised skin membrane separating two compartments of a diffusion cell, one of which contains a donor solution of

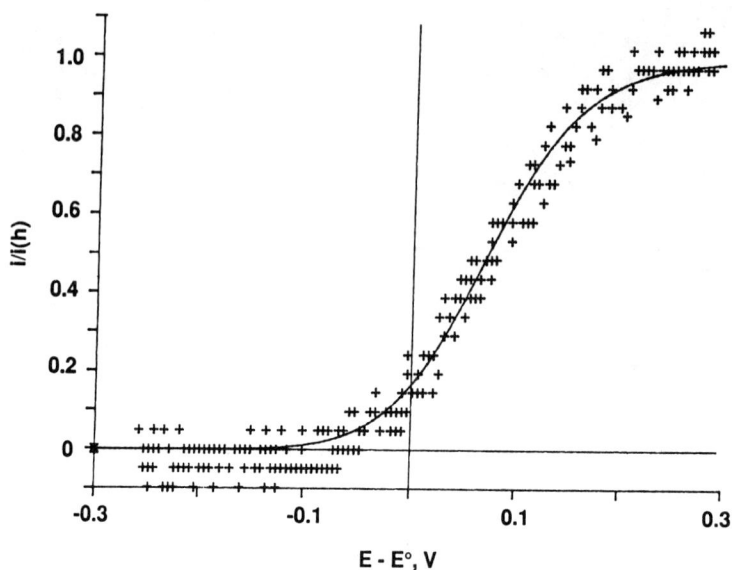

FIG. 74. Voltammogram at a microtip electrode partially penetrating a Nafion film containing 0.57 M $Os(bpy)_3^{2+}$. Scan rate, $v = 5$ mV/sec. The solid line is computed by substituting the kinetic parameters given in the text into Eq. (9) in Ref. 39. (Reprinted with permission from Ref. 52. Copyright 1991 American Association for the Advancement of Science.)

0.1 M $FeSO_4$ at pH 3.0, with a receptor solution (where the tip is located) of 0.1 M NaCl at pH 3.0 in the other. A current density of 2.0 mA/cm² had been applied for 12.5 hours previous to imaging. The tip, scanned over the skin surface in the receptor solutions, detects Fe^{2+} driven through pores in the skin by the applied current. Six current peaks of height greater than 50 pA over the average current level are clearly resolved in the image. Four smaller peaks are also seen, for a total of 10 peaks within the 1 mm² area scanned. With no applied iontophoretic current, no features in the SECM images above the background were resolved.

Quantitative measurements of the relative rates of Fe^{2+} and Fe^{3+} flux through a pore are shown in Figure 76. In this case, an iontophoretic current density of 0.4 mA/cm² was applied from a donor solution containing 0.05 M $FeCl_2$ and 0.05 M $FeCl_3$ at pH 1.6. Each of the 16 gaussian segments in Figure 76A corresponds to a 500-μm trace of the tip in the x-direction as it was scanned across the skin at fixed tip bias, E_T, and represents a flux profile

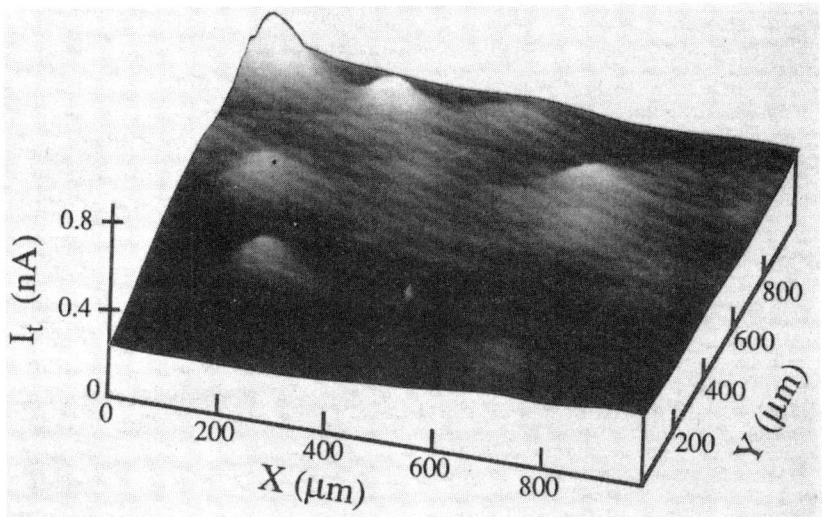

FIG. 75. Gray scale SECM image of Fe^{2+} flux emerging from a 1 mm^2 area of hairless mouse skin. Donor solution: 0.1 M FeSO$_4$, pH 3.0. Receptor solution: 0.1 M NaCl, pH 3.0. i_{app} 2.0 mA/cm^2, applied for 12.5 hr previous to and during imaging. Tip potential: 1.04 V vs. Ag/AgCl. (Reprinted with permission from Ref. 95. Copyright 1992 Elsevier.)

measured by the microelectrode. Each profile was measured over the spatially identical region of the skin. Traces were made at E_T of 0.20, 0.25, 0.30, . . . , 0.95 V, to show how varying E_T allows discrimination and quantitative measurement of the fluxes of different ions. At E_T negative of the $E_{1/2}$ of the Fe^{2+}/Fe^{3+} couple, the tip detects the flux of Fe^{3+}, indicated by the recording of cathodic currents (plotted downward). With E_T positive of $E_{1/2}$, Fe^{2+} is detected, as indicated by anodic currents. Both the cathodic and anodic current peaks initially increase with increasing departure from $E_{1/2}$, until they reach a limiting value, at which point the tip currents are limited by mass transport of the species to the tip. This observed voltammetric behavior closely mimics the behavior of an identically prepared tip during conventional cyclic voltammetry in a solution containing equal concentrations of Fe^{2+} and Fe^{3+} (Fig. 76B). These results suggest that the current peaks resolved in the SECM images are due to detection of the permeating Fe^{x+} species of interest and the flux of any electroactive species can be quantitatively measured by adjusting the tip potential to a value where the species is oxidized or reduced.

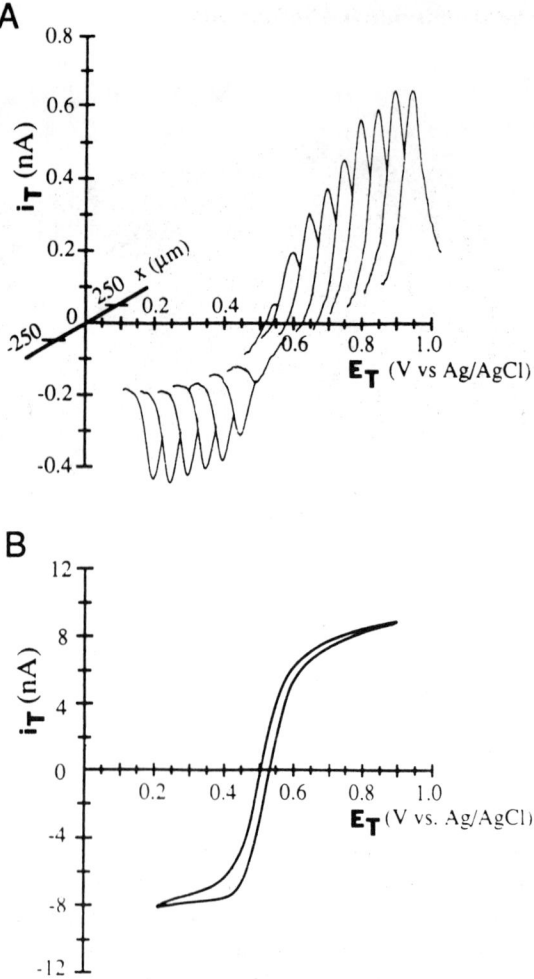

FIG. 76. (A) SECM flux profiles over identical regions of skin, measured at different tip biases. Each of the 16 gaussian segments is a plot of i_T as a function of tip position as the tip is scanned 500 μm in the x direction at a fixed potential E_T over an identical region of skin. Values of E_T range between 0.20 V and 0.95 V, in 0.05 V increments. Anodic current is positive; i.e., positive currents correspond to the oxidation of Fe^{2+}, while negative values correspond to the reduction of Fe^{3+}. Donor solution: 0.05 M $FeCl_2$, 0.05 M $FeCl_3$, pH 1.6. Receptor solution: 0.1 M NaCl, pH 1.6. i_{app}: 0.4 mA/cm² during imaging. A current density of 0.1 mA/cm² had been applied intermittently for an accumulated period of 20 hr previous to imaging, and the total time the skin was mounted in the cell was 48 hr. (B) Voltammetric response of a Pt-coated carbon fiber microelectrode immersed in a bulk solution of 0.01 M $FeCl_2$, 0.01 M $FeCl_3$, 0.08 M NaCl. Potential sweep rate: 200 mV/sec. (Reprinted with permission from Ref. 95. Copyright 1992 Elsevier.)

V. CONCLUSIONS

Although SECM is still a relatively young technique, it has already demonstrated a wide range of applications in electrochemical systems. It should also prove useful in other fields, for example, in studies of the surface reactions of minerals, membrane transport, and enzyme processes. High-resolution fabrication via etching and electrodeposition is also of interest. Future work will probably be directed toward improving resolution through the fabrication of tips in the 10-nm-diameter range. It would also be useful to devise a broader range of tip materials and new approaches to tip fabrication. The fabrication of ion-selective and enzyme electrode tips with high resolution and fast response would be especially valuable. Clearly, the field will advance more quickly and will be applied more broadly when commercial instruments are available.

ACKNOWLEDGMENTS

The support of our research in SECM by the Robert A. Welch Foundation, the National Science Foundation, and the Texas Advanced Research Program is gratefully acknowledged.

ABBREVIATIONS

ADI	alternating-direct implicit, as in ADI finite difference method
AFM	atomic force microscopy
CV	cyclic voltammetry
DF	dimethyl fumarate
DMPPD	N,N-dimethyl-p-phenylenediamine
ECE	electron transfer followed by a chemical reaction the product of which undergoes a second electron transfer
E_rC_i	reversible electron transfer followed by an irreversible chemical reaction
E_rC_{2i}	reversible electron transfer followed by an irreversible chemical (dimerization) reaction
E_qC_r	quasi-reversible electron transfer followed by a reversible chemical reaction
FN	fumaronitrile

GC	glassy carbon
GO	glucose oxidase
HOPG	highly oriented pyrolytic graphite
IDA	Interdigitated array, as in IDA electrode
ITO	indium tin oxide
MV^{2+}	methyl viologen dication
O	oxidized form of an electroactive couple
PMMA	poly (methyl methacrylate)
PVP	poly(vinylpyridine)
$PVPH^+$	protonated poly(4-vinylpyridine)
R	reduced form of an electroactive couple
RRDE	rotating ring-disk electrode
SECM	scanning electrochemical microscopy
	scanning electrochemical microscope
SECMID	SECM-induced desorption
SECM/TLC	scanning electrochemical microscopy/thin-layer cell
SPM	scanning probe microscopy
STM	scanning tunneling microscopy
$TBABF_4$	tetrabutylammonium tetrafluorobate
TLC	thin-layer cell
TMPD	N,N,N',N'-tetramethyl-p-phenylene diamine
TPM SECM	tip position modulation SECM
T/S CV	tip/substrate cyclic voltammetry
UME	ultramicroelectrode

LIST OF SYMBOLS

a	radius of electrode
a_S	radius of substrate

Scanning Electrochemical Microscopy 367

c	concentration of redox species
$c(t,r)$	surface concentration of electroactive species as a function of time and spatial variable
$c(t,r,z)$	concentration of electroactive species as a function of time and spatial variables
$c°$	bulk concentration of electroactive species
C	dimensionless variable equal to $1 - c/c_O°$
d	distance between tip and substrate
D	diffusion coefficient of redox species
D_a	D_{sur}/D_{sol}
D_{sol}	diffusion coefficient of the species of interest in solution
D_{sur}	surface diffusion coefficient of the adsorbate
E	electrode potential
$E°'$	formal potential
E_1	dimensionless substrate potential equal to $(E_S - E°')nf$
$E_{1/2}, E_{1/4}, E_{3/4}$	quartile potentials from a steady-state voltammogram used to determine $k°$, α, and $E°'$
F	the Faraday
f_m	frequency of sinusoidal tip movement in TPM SECM
i_{app}	iontophoretic current density driven across a membrane separating two compartments of a diffusion cell during SECM imaging of the membrane
i_S	current flowing through substrate
i_T	current flowing through tip
$i_{T,cond}, i_{T,ins}$	tip current at conductor and insulator (used in TPM SECM)
$i_{T,m}$	modulated tip current in TPM SECM
$i_{T,\infty}$	steady-state current at a tip electrode when the tip is far from the substrate

i_S/i_T	collection efficiency of substrate
i_T/i_S	collection efficiency of tip
i_{dif}	diffusion-limited current for any electrode geometry
I_T	normalized tip current equal to $i_T/i_{T,\infty}$
I_0	modified Bessel function of the first kind of order zero
$I_T(L)$	diffusion-limiting tip current at a normalized tip-substrate separation equal to L
$I_S(T)$	dimensionless substrate current, given by Eq. (16)
j	diffusion flux density
j^*	total flux at the substrate equal to $J_{AV}\pi c°Da_S$
J	dimensionless variable equal to $ja/(Dc°)$
J_S	dimensionless variable equal to $ja_S/(Dc°)$
J_0	the Bessel function of the first kind of order zero
k	the ratio of the cone height to its radius
$k°$	standard rate constant
k_b, k_f	heterogeneous rate constants for oxidation and reduction
$k_{b,S}$	heterogeneous rate constant at a substrate
$K_{b,S}$	dimensionless rate constant equal to $k_{b,S}a/D$
k_c	homogeneous rate constant in a first-order chemical reaction
k_c'	homogeneous rate constant in a dimerization reaction
K	kinetic parameter equal to a constant times k_c/D, e.g., a^2k_c/D
K'	dimensionless parameter equal to $k_c'a^2c°_O/D$
k_a^{app}, k_d^{app}	apparent adsorption and desorption rate constants, respectively
K_a^{app}, K_d^{app}	equal to $k_a^{app} a/D_{sol}$ and $k_d^{app} a/D_{sol}c°$, respectively
$K(p)$	complete elliptical integral of the first kind

Scanning Electrochemical Microscopy

L	normalized tip-substrate distance equal to d/a
m_O	effective mass transfer coefficient given by Eq. (34)
n	number of electrons transferred per redox event
$N°$	maximum density of adsorption sites
O, R, S, T	when used as subscripts, refer to oxidized form, reduced form, substrate, and tip, respectively
r,z	spatial variables
r*	hemisphere radius
r'	polar radius
R	dimensionless variable equal to r/a
Rg	radius of the insulating ring around a microtip
RG	dimensionless sheath radius equal to Rg/a
t	time
t_d	time for i_T to decrease below $i_{T,\infty}$, $\approx d^2/2D$
T	dimensionless variable equal to tD/a^2
v	scan rate
V_m	modulation voltage in TPM SECM
Z	dimensionless variable equal to z/a
α	transfer coefficient
$\delta/2$	amplitude of sinusoidal tip movement in TPM SECM
γ	equal to $N°/c°a$
κ	parameter equal to $k°\exp[-\alpha nf(E - E°')]/m_O$
κ'	parameter equal to $\pi a k°\exp[-\alpha nf(E - E°')]/(4D_O)$
Λ	variable equal to $ak°/D$
Λ'	variable equal to $dk°/D$
ρ	ratio of substrate radius to tip radius (a_S/a)
ρ^∞	the value of ρ (equal to 1 + 1.5L) at which the substrate behaves as an infinite one

τ	parameter equal to tD_{sol}/a^2
θ	parameter equal to $1 + \exp[nf(E - E^{o\prime})]D_O/D_R$
θ'	fractional surface coverage
θ_3, θ_4	theta functions
f	parameter equal to F/RT (F is the Faraday, R is the gas constant, and T is temperature)
l	the thickness of the thin-layer cell formed inside a liquid (Hg) substrate

REFERENCES

1. H. K. Wichramasinghe (ed.), Scanned Probe Microscopy, AIP Conference Proceedings, 241, American Institute of Physics, New York, 1992.
2. H. S. Isaacs and G. Kissel, J. Electrochem. Soc. *119*:1628 (1972).
3. H. S. Isaacs and M. W. Kendig, Corrosion *36*:269 (1980).
4. (a) L. F. Jaffe and N. Nuccitelli, J. Cell. Biol. *63*:614 (1974); (b) J. M. Hush and R. L. Overall, Biol. Bull. *176*(S):56 (1989).
5. R. M. Wightman and D. O. Wipf, in *Electrochemical Chemistry*, Vol. 15 (A. J. Bard. ed.), Marcel Dekker, New York, 1988, p. 267.
6. M. Fleischmann, S. Pons, D. R. Rolison, and P. P. Schmidt, *Ultramicroelectrodes*, Datatech Systems, Morgantown, NC, 1987.
7. M. I. Montenegro, M. A. Queirós, and J. L. Daschback (eds.), *Microelectrodes: Theory and Applications*, NATO ASI Ser. Appl. Sci. Vol. 197, Kluwer Acad. Publ., Dordrecht, 1991.
8. A. J. Bard, F.-R. F. Fan, J. Kwak, and O. Lev, Anal. Chem. *61*:132 (1989).
9. A. J. Bard, G. Denuault, C. Lee, D. Mandler, and D. O. Wipf, Acc. Chem. Res. *23*:357 (1990).
10. R. C. Engstrom, M. Weber, D. J. Wunder, R. Burgess, and S. Winquist, Anal. Chem. *58*:844 (1986).
11. R. C. Engstrom, T. Meaney, R. Tople, and R. M. Wightman, Anal. Chem. *59*:2005 (1987).
12. A. T. Hubbard and F. C. Anson, in *Electroanalytical Chemistry*, Vol. 4 (A. J. Bard. ed.), Marcel Dekker, New York, 1970, p. 129.
13. W. J. Albery and M. L. Hitchman, *Ring-Disc Electrodes*, Calrendon, Oxford, 1971.
14. A. J. Bard and F.-R. F. Fan, Faraday Discuss. (in press).
15. P. R. Unwin and A. J. Bard, Anal. Chem. *64*:113 (1992).
16. J. Kwak and A. J. Bard, Anal. Chem. *61*:1794 (1989).
17. A. J. Bard and L. R. Faulkner, *Electrochemical Methods, Fundamentals and Applications*, Wiley, New York, 1980.

18. J. Kwak, Ph.D. dissertation, The University of Texas at Austin, 1989.
19. K. Bartels, A. C. Bovik, C. Lee, and A. J. Bard, *Biomedical Image Processing II*, Vol. 1450 (A. C. Bovik and V. Howard, eds.), The International Society for Optical Engineering, Bellingham, Washington, 1991, pp. 30–39.
20. C. Lee, D. O. Wipf, A. J. Bard, K. Bartels, and A. C. Bovik, Anal. Chem. *63*:2442 (1991).
21. D. O. Wipf and A. J. Bard, Anal. Chem. *64*:1362 (1992).
22. C. Lee, C. J. Miller, and A. J. Bard, Anal. Chem. *63*:78 (1991).
23. A. A. Gewirth, D. H. Craston, and A. J. Bard, J. Electroanal. Chem. *261*:477 (1989).
24. L. A. Nagahara, T. Thundat, and S. M. Lindsay, Rev. Sci. Instrum. *60*:3128 (1989).
25. M. V. Mirkin, F.-R. F. Fan, and A. J. Bard, J. Electroanal. Chem. *328*:47 (1992).
26. R. M. Penner, M. J. Heben, T. L. Longin, and N. S. Lewis, Science *250*:1118 (1990).
27. J. Kwak and A. J. Bard, Anal. Chem. *61*:1221 (1989).
28. M. V. Mirkin and A. J. Bard, J. Electroanal. Chem. *323*:1 (1992).
29. M. V. Mirkin and A. J. Bard, J. Electroanal. Chem. *323*:29 (1992).
30. A. J. Bard, M. V. Mirkin, P. R. Unwin, and D. O. Wipf, J. Phys. Chem. *96*:1861 (1992).
31. A. J. Bard, G. Denuault, R. A. Friesner, B. C. Dornblaser and L. S. Tuckerman, Anal. Chem. *63*:1282 (1991).
32. P. R. Unwin and A. J. Bard, J. Phys. Chem. *95*:7814 (1991).
33. D. W. Peaceman and H. H. Rachford, J. Soc. Indust. Appl. Math. *3*:28 (1955).
34. M. Abramowitz and I. Stegun (eds.), *Handbook of Mathematical Functions*, Dover, New York, 1965.
35. M. V. Mirkin and A. J. Bard, Anal. Chem. *64*:2293 (1992).
36. D. O. Wipf and A. J. Bard, J. Electrochem. Soc. *138*:L4 (1991).
37. B. R. Horrocks, M. V. Mirkin, D. T. Pierce, A. J. Bard, G. Nagy, and K. Toth, Anal. Chem. *65*: 1213 (1993).
38. X. Liu, J. Lu and C. Cha, J. Electroanal. Chem. *295*:15 (1990).
39. K. B. Oldham and C. G. Zoski, J. Electroanal. Chem. *256*:11 (1988).
40. G. Denuault, M. H. Trise Frank, and L. M. Peter, Faraday Discuss. (in press).
41. M. Fleischmann and S. Pons, J. Electroanal. Chem. *250*:257 (1988).
42. L. Nanis and W. Kesselman, J. Electrochem. Soc. *118*:454 (1971).
43. J. Newman, in *Electroanalytical Chemistry*, Vol. 6 (A. J. Bard, ed.), Marcel Dekker, New York, 1973, p. 316.
44. R. C. Engstrom, R. M. Wightman, and E. W. Kristensen, Anal. Chem. *60*:652 (1988).
45. K. Aoki and M. Sakai, J. Electroanal. Chem. *267*:47 (1989).
46. F. Zhou, P. R. Unwin, and A. J. Bard, J. Phys. Chem. *96*:4917 (1992).
47. C. Lee, J. Kwak, and F. C. Anson, Anal. Chem. *63*:1501 (1991).
48. P. R. Unwin and A. J. Bard, J. Phys. Chem. *96*:5035 (1992).

49. J. M. Davis, F.-R. F. Fan, and A. J. Bard, J. Electroanal. Chem. *238*:9 (1987).
50. A. S. Baranski, J. Electroanal. Chem. *307*:287 (1991).
51. K. B. Oldham, Anal. Chem. *64*:646 (1992).
52. M. V. Mirkin, F.-R. F. Fan, and A. J. Bard, Science *257*:364 (1992).
53. C. Lee, J. Kwak, and A. J. Bard, Proc. Natl. Acad. Sci. USA *87*:1740 (1990).
54. E. R. Scott, H. S. White, and J. B. Phillips, J. Membr. Sci. *58*:71 (1991).
55. J. Kwak, C. Lee, and A. J. Bard, J. Electrochem. Soc. *137*:1481 (1990).
56. C. Lee and A. J. Bard, Anal. Chem. *62*:1906 (1990).
57. J. Kwak and F. C. Anson, Anal. Chem. *64*:250 (1992).
58. C. Lee and F. C. Anson, Anal. Chem. *64*:528 (1992).
59. I. C. Jeon and F. C. Anson, Anal. Chem. *64*:2021 (1992).
60. A. J. Bard, F.-R. F. Fan, D. T. Pierce, P. R. Unwin, D. O. Wipf, and F. Zhou, Science *254*:68 (1991).
61. D. T. Pierce, P. R. Unwin, and A. J. Bard, Anal. Chem. *64*, 1795 (1992).
62. D. O. Wipf and A. J. Bard, J. Electrochem. Soc. *138*:469 (1991).
63. A. Heller, Acc. Chem. Res. *23*:128 (1990).
64. M. V. Mirkin, T. C. Richards, and A. J. Bard, J. Phys. Chem., in press.
65. M. I. Montenegro and D. Pletcher, J. Electroanal. Chem. *200*:371 (1986).
66. D. O. Wipf, E. W. Kristensen, M. R. Deakin, and R. M. Wightman, Anal. Chem. *60*:306 (1988).
67. A. M. Bond, T. L. E. Henderson, D. R. Mann, T. F. Mann, W. Thormann, and C. G. Zoski, Anal. Chem. *60*:1878 (1988).
68. A. S. Baranski, K. Winkler, and W. R. Fawcett, J. Electroanal. Chem. *313*:367 (1991).
69. M. V. Mirkin and A. J. Bard, J. Electrochem. Soc. *139*:3535 (1992).
70. M. V. Mirkin, L. O. S. Bulhões, and A. J. Bard, J. Am. Chem. Soc., *115*: 201 (1993).
71. W. R. Fawcett, M. Opallo, M. Fedurco, and J. W. Lee, J. Am. Chem. Soc., *115*: 196 (1993).
72. K. B. Oldham, J. Electroanal. Chem. *323*:52 (1992).
73. R. C. Engstrom, B. Small, and L. Kattan, Anal. Chem. *64*:241 (1992).
74. L. K. J. Tong, J. Phys. Chem.*58*:1090 (1954).
75. M. V. Mirkin, H. Yang, and A. J. Bard, J. Electrochem. Soc. *139*:2212 (1992).
76. J. V. Puglisi and A. J. Bard, J. Electrochem. Soc. *119*:829 (1972).
77. I. B. Goldberg, D. Boyd, R. Hirasama, and A. J. Bard, J. Phys. Chem. *78*:295 (1974).
78. M. Ashida, M. Sasaki, H. Kan, T. Yasunaga, K. Hachiya, and T. Inoue, J. Coll. Interface Sci. *67*:219 (1978).
79. D. H. Craston, C. W. Lin, and A. J. Bard, J. Electrochem. Soc. *135*:785 (1988).
80. D. Mandler and A. J. Bard, J. Electrochem. Soc. *136*:3143 (1989).
81. O. E. Hüsser, D. H. Craston, and A. J. Bard, J. Vac. Sci. Technol. B *6*:1873 (1988).
82. O. E. Hüsser, D. H. Craston, and A. J. Bard, J. Electrochem. Soc. *136*:3222 (1989).

83. Y.-M. Wuu, F.-R. F. Fan, and A. J. Bard, J. Electrochem. Soc. *136*:885 (1989).
84. R. L. McCarley, S. A. Hendricks, and A. J. Bard, J. Phys. Chem. *96*:10089 (1992).
85. D. Mandler and A. J. Bard, J. Electrochem. Soc. *137*:1079 (1990).
86. D. Mandler and A. J. Bard, J. Electrochem. Soc. *137*:2468 (1990).
87. Z. Tian, Z. Fen, Z. Tian, X. Zhou, J. Mu, C. Li, H. Lin, B. Ren, Z. Xie, W. Hu, Disc. Faraday Soc., in press.
88. D. Mandler and A. J. Bard, Langmuir 6:1489 (1990).
89. D. Ammann, *Ion Selective Microelectrodes: Principles, Design and Application*, Springer, New York, 1986.
90. J. L. Luo, Y. C. Lu, and M. B. Ives, J. Electroanal. Chem. *326*:51 (1992).
91. M. Krishnan, X. Zhang, and A. J. Bard, J. Am. Chem. Soc. *106*:7371 (1984).
92. F. C. Anson, D. N. Blauch, J. M. Saveant, and C.-F. Shu, J. Am. Chem. Soc. *113*:1922 (1991).
93. M. Sharp, B. Lindhom, and E. L. Lind, J. Electroanal. Chem. *274*:35 (1989).
94. H. S. White, J. Leddy, and A. J. Bard, J. Am. Chem. Soc. *104*:4811 (1982).
95. E. R. Scott, H. S. White, and J. B. Phipps, Solid State Ionics *53*:176 (1992).

AUTHOR INDEX

Underlined numbers give the page on which the complete reference is listed.

Abbott, A. P., 43, 45, 56, 84
Abramowitz, M., 273, 282, 287, 371
Abrantes, L. M., 90, 121, 122, 203, 209, 234
Abrash, S., 101, 107, 236
Abruna, H. D., 90, 131, 231, 234
Adamson, A. W., 2, 5, 10, 14, 16, 81
Adeniyi, W. K., 57, 58, 86
Agarwal, R., 36, 38, 39, 65, 66, 84, 87
Ahmadi, M. F., 24, 83
Albery, W. J., 90, 121, 122, 185, 186, 187, 188, 189, 195, 201, 232, 237, 247, 370
Allara, D. L., 17, 42, 58, 82
Aloisi, G., 52, 85
Aminabhavi, T. M., 38, 84
Ammann, D., 348, 373
Anderson, S. E., 19, 83
Andreev, V. N., 90, 121, 209, 211, 234
Andrieux, C. P., 61, 87, 90, 95, 96, 105, 110, 111, 115, 121, 122, 159, 183, 188, 203, 204, 205, 209, 234, 235, 236, 239
Anson, F. C., 90, 99, 100, 102, 116, 124, 125, 126, 127, 130, 131, 132, 133, 136, 137, 138, 139, 234, 235,

238, 247, 285, 292, 314, 319, 326, 327, 330, 332, 349, 354, 355, 357, 359, 360, 361, 370, 371, 372, 373
Aoki, K., 62, 63, 87, 100, 123, 235, 238, 292, 371
Aoyagui, S., 29, 31, 41, 42, 43, 45, 56, 83, 84
Arena, J. V., 61, 87
Armstrong, R. D., 90, 132, 135, 136, 137, 138, 172, 190, 226, 232, 241
Arnaud, N., 66, 87
Ashida, M., 338, 372
Atwood, D., 2, 80
Audebert, P., 90, 121, 122, 203, 204, 205, 209, 234
Austin, R. G., 90, 231
Avery, S., 17, 82
Aveyard, B., 14, 15, 57, 82

Bácskai, J., 90, 148, 149, 150, 151, 152, 158, 159, 161, 162, 165, 166, 167, 168, 173, 174, 176, 177, 180, 181, 211, 218, 233, 239, 241
Baizer, M. M., 3, 52, 81
Baker, C. K., 90, 196, 234
Baldy, C. J., 115, 236
Ball, D., 54, 85

375

Author Index

Baranski, A. S., 307, 323, <u>372</u>
Barbero, C., 90, 212, 215, <u>234</u>, <u>241</u>
Bard, A. J., 14, 17, 25, 29, 31, 38, 40, 41, 42, 44, 53, 57, <u>82</u>, <u>84</u>, <u>86</u>, 90, 92, 95, 96, 121, 128, 129, 130, 137, 147, 148, 156, 172, 181, 184, 186, 200, 209, 211, 220, 223, <u>231</u>, <u>232</u>, <u>234</u>, <u>235</u>, <u>237</u>, <u>238</u>, <u>240</u>, <u>241</u>, 245, 246, 248, 250, 251, 252, 253, 254, 255, 257, 258, 259, 260, 262, 263, 264, 265, 268, 269, 270, 271, 272, 273, 274, 275, 276, 277, 278, 279, 280, 281, 282, 284, 285, 287, 289, 290, 291, 292, 293, 294, 296, 297, 298, 299, 300, 301, 302, 303, 304, 306, 307, 308, 309, 310, 311, 312, 313, 314, 315, 316, 317, 319, 320, 321, 322, 323, 324, 325, 326, 327, 328, 329, 330, 331, 332, 333, 334, 335, 336, 337, 338, 339, 340, 342, 343, 344, 345, 346, 348, 349, 350, 351, 352, 353, 354, 355, 360, 361, 362, <u>370</u>, <u>371</u>, <u>372</u>, <u>373</u>
Barendrecht, E., 121, <u>237</u>
Bargon, J., 197, <u>240</u>
Bartels, K., 253, 311, 313, <u>371</u>
Bartlett, P. N., 250
Bauer, H. H., 3, 16, <u>81</u>
Beck, F., 90, <u>232</u>
Beitz, J., 101, 107, <u>236</u>

Bellama, J. M., 79, <u>88</u>
Bencheikh-Sayarh, S., 48, 49, 50, <u>84</u>, <u>85</u>
Bennis, K., 52, <u>85</u>
Benzinger, J. B., 18, 27, <u>83</u>
Berthod, A., 53, 66, 70, <u>85</u>, <u>87</u>
Besio, G. L., 18, 27, <u>83</u>
Bidan, G., 191, 193, 219, <u>240</u>
Bijsterbosch, B. H., 15, <u>82</u>
Birke, R.L., 23, 24, 26, 28, <u>83</u>
Blackburn, A., 42, 58, <u>82</u>
Blauch, D. N., 99, 102, 116, 117, 137, <u>235</u>, 360, <u>373</u>
Blocman, C., 61, <u>87</u>
Bloor, D., 90, 121, 122, 185, 186, 187, 188, 189, 201, <u>232</u>
Blount, H. N., 29, 31, 45, 47, 48, 49, 53, 70, <u>83</u>, <u>84</u>, <u>85</u>
Bobalbhai, L., 67, 68, <u>87</u>
Bocarsly, A. B., 90, <u>231</u>
Bolikal, D., 51, <u>85</u>
Bolts, J. M., 90, <u>231</u>
Bond, A. M., 323, <u>372</u>
Bookbinder, D. C., 182, <u>240</u>
Botár, L., 98, 99, 130, <u>235</u>
Bovik, A. C., 253, 311, 313, <u>371</u>
Bowden, E. F., 225, 226, 227, <u>241</u>
Boyd, D., 334, 335, <u>372</u>
Brahimi, B., 59, <u>86</u>
Bratjer-Toth, A., 67, 68, <u>87</u>
Braun, H., 130, 132, 134, 137, 138, <u>238</u>
Bredas, J.-L., 121, 122, <u>237</u>
Briggs, J., 36, 38, <u>84</u>
Bright, T. B., 17, 42, 58, <u>82</u>
Brooks, M. Y., <u>83</u>
Bruckenstein, S., 90, 122, 148, 149, 151, 152, 173, 197, 220, 225, <u>233</u>, <u>239</u>

Author Index

Buck, R. P., 90, 99, 113, 115, 145, 183, 188, 220, 226, <u>234</u>, <u>235</u>, <u>236</u>
Bulhöes, L.O.S., 326, 327, 328, <u>372</u>
Bull, R. A., 90, 186, <u>232</u>
Bunding, K. A., 17, <u>82</u>
Bunton, C. A., 36, 38, <u>84</u>
Burgess, R., 247, 291, 331, <u>370</u>
Burgmayer, P., 193, 219, <u>240</u>
Buttry, D. A., 26, 27, 28, <u>83</u>, 90, 99, 100, 124, 125, 126, 127, 137, 148, 210, 211, <u>233</u>, <u>235</u>

Cairns, E. J., 139, 202, <u>238</u>
Candau, S. J., 38, <u>84</u>
Carlin, C. M., 121, 200, 220, <u>237</u>
Casselberry, R. L., 29, 31, 45, 47, 48, 49, <u>83</u>
Castaneda, F., 173, <u>240</u>
Castillo, J. I., 184, <u>240</u>
Cha, C., <u>371</u>
Chambers, J. Q., 90, 130, 148, 149, 151, 152, 157, 158, 159, 160, 161, 165, 173, 174, 175, 178, 179, 181, 182, 214, 218, 220, 225, 226, <u>234</u>, <u>238</u>, <u>239</u>
Chandar, P., 14, <u>82</u>
Chang, A.-C., 77, 78, <u>88</u>
Cheminat, B., 49, 50, 51, <u>85</u>
Chen, J.-W., 66, 68, <u>87</u>
Chen, X., 103, 104, 105, 129, 130, 131, 132, 137, <u>236</u>
Chen, Z., 90, 121, 122, 185, 186, 187, 188, 189, 201, <u>232</u>
Cheng, G., 24, 28, <u>83</u>
Chiang, J. C.-C., 198, <u>240</u>

Chiappardi, D. M., 29, 31, 45, 47, 48, 49, <u>83</u>
Chiba, K., 214, <u>237</u>
Chidsey, C. E. D., 17, <u>82</u>, 118, 119, 123, 136, 141, 156, <u>236</u>
Chien, J. C. W., 193, 219, <u>240</u>
Chokshi, K., 66, 68, <u>87</u>
Closs, G. L., 17, 46, <u>82</u>
Colaneri, N., 185, <u>240</u>
Colichman, E. L., 3, 16, 40, <u>81</u>
Collins, G. C. S., 173, <u>240</u>
Compton, R. G., 90, 121, 122, 185, 186, 187, 188, <u>232</u>
Connors, T. F., 61, 73, 74, <u>87</u>
Contamin, O., 90, 190, <u>232</u>
Cordoba-Torresi, S., 90, 210, <u>233</u>
Cotton, T. M., 76, <u>87</u>
Couture, E. C., 4, 19, 20, 22, 28, 29, 53, 56, 62, 73, 74, 76, <u>81</u>, <u>83</u>, <u>87</u>
Craston, D. H., 257, 339, 340, 342, 343, 344, <u>371</u>, <u>372</u>
Csillag, K., 94, 97, 99, <u>235</u>
Cuendet, P., 17, <u>82</u>
Curtiss, L. A., 101, <u>236</u>
Cushman, R. J., 199, 205, <u>240</u>
Cussler, E. L., 36, 38, <u>84</u>

Dahms, H., 94, 97, <u>235</u>
Daifuku, M., 90, 210, <u>233</u>
Dalton, E. F., 90, 99, 113, 145, 182, 220, <u>234</u>
Danly, D. E., 3, <u>81</u>
Darlington, J., 58, <u>86</u>
Daschback, J. L., 244, 327, <u>370</u>
Daum, P., 147, 148, 218, <u>238</u>
Dautartas, M. F., 225, 226, 227, <u>241</u>

Davidovic, A., 50, 85
Davidovic, D., 50, 85
Davis, J. M., 303, 371
Day, R. A., 4, 81
Day, R. W., 90, 130, 158, 159, 161, 173, 214, 225, 226, 234, 238, 239
Dayalan, E., 31, 33, 38, 40, 67, 68, 84, 87
Deakin, M. R., 90, 233, 323, 372
de Gennes, P. G., 93, 219, 220, 235, 241
Degrand, C., 158, 159, 173, 239
Delahey, P., 17, 83
DeLong, J. J., 26, 28, 83
Demeter, K., 94, 97, 99, 235
Denault, G., 245, 246, 251, 269, 270, 271, 272, 287, 345, 348, 370, 371
Derouin, C., 90, 211, 233
Deslouis, C., 90, 121, 123, 232
Desmettre, S., 28, 29, 30, 31, 36, 38, 40, 41, 42, 43, 46, 55, 56, 83, 84, 86
Deutsch, E., 56, 86
Devaux, B., 49, 85
Devreux, F., 122, 237
Diaz, A. F., 27, 28, 53, 83, 85, 90, 120, 121, 122, 123, 184, 185, 191, 193, 200, 202, 203, 219, 234, 237, 240
Dixit, N. S., 36, 38, 39, 65, 66, 84, 87
Doblhofer, K., 100, 130, 132, 134, 135, 137, 138, 226, 236, 238, 241
Dobos, L., 100, 236
Dong, S., 24, 28, 83
Donohue, J. J., 26, 28, 83
D'Orazio, P., 78, 79, 88

Dornblaser, B. C., 269, 270, 271, 272, 371
Dorshow, R. B., 36, 38, 84
Duic, L., 50, 85
Dumas-Bouchiat, J.-M., 61, 87
Durst, R. A., 79, 88, 100, 236, 250
Dusek, K., 169, 170, 240

Eales, R. M., 90, 122, 148, 197, 233
Eddowes, M. J., 31, 38, 40, 41, 43, 46, 55, 84
Elliot, C. M., 90, 115, 121, 122, 185, 186, 187, 188, 189, 201, 232, 236
Engler, E. M., 95, 157, 173, 235, 239
Engstrom, R. C., 247, 291, 292, 327, 330, 331, 333, 370, 371, 372
Epstein, A. J., 121, 122, 198, 199, 237
Erabi, T., 4, 81
Evans, D. F., 10, 11, 12, 13, 36, 38, 82, 84
Evans, D. H., 29, 84
Evans, G. P., 90, 120, 197, 198, 234
Evans, J. F., 225, 226, 227, 241
Exner, H. J., 3, 16, 46, 81

Fabre, H., 36, 84
Facci, J. S., 25, 27, 83, 90, 111, 112, 113, 138, 145, 220, 234, 236
Fan, F.-R. F., 90, 186, 232, 234, 245, 248, 259, 265, 268, 270, 278, 282, 303, 307, 308, 309, 310, 311, 312,

314, 316, 327, 343, 349, 355, 360, 361, 362, 370, 371, 372
Farkas, J., 100, 236
Faulkner, L. R., 14, 17, 25, 44, 57, 82, 92, 105, 106, 108, 123, 130, 131, 132, 134, 135, 137, 138, 140, 149, 218, 227, 235, 236, 238, 241, 252, 269, 291, 327, 370
Fawcett, W. R., 323, 327, 372
Fay, H. G., 139, 238
Fedurco, M., 327, 372
Fekete, E., 100, 236
Feldberg, S. W., 101, 107, 115, 121, 123, 124, 186, 187, 194, 201, 203, 236, 237, 240
Feldman, B. J., 193, 219, 240
Fen, Z., 344, 373
Fendler, E., 3, 6, 8, 9, 59, 60, 81
Fendler, J. H., 3, 6, 7, 8, 9, 10, 12, 45, 59, 60, 68, 77, 78, 81, 82, 87, 88
Finklea, H. O., 17, 42, 58, 82
Fiordiponti, P., 90, 121, 122, 123, 171, 201, 232
Fleischmann, M., 39, 83, 244, 256, 288, 370, 371
Florence, A. T., 2, 80
Foresti, M. L., 16, 82
Forster, R. J., 90, 141, 172, 218, 233, 238
Franceschetti, D. R., 90, 190, 232
Francis, C. V., 158, 175, 182, 239
Franklin, T. C., 2, 16, 18, 53, 54, 56, 57, 58, 80, 85, 86
Friedrich, V. J., 94, 97, 99, 235
Friesner, R. A., 269, 270, 271, 272, 371

Fritsch-Faules, I., 106, 108, 227, 236, 241
Fujihira, M., 28, 29, 30, 31, 36, 38, 40, 41, 42, 56, 83, 90, 121, 234
Fujihira, Y., 60, 86-87
Fukui, M., 158, 239
Fultz, M. L., 250
Funt, B. L., 158, 239
Furtsch, T., 17, 82

Gabrielli, C., 90, 100, 141, 190, 210, 232, 233
Garcia, E., 71, 87
Garnier, F., 197, 240
Gaudiello, J. G., 129, 130, 238
Gavach, C., 90, 130, 232
Geniés, E. M., 53, 70, 85, 121, 122, 137, 185, 191, 193, 199, 200, 205, 213, 219, 236, 237, 240
Genoud, F., 122, 185, 237
Georges, J., 28, 29, 30, 31, 36, 38, 40, 41, 42, 43, 46, 53, 55, 56, 66, 68, 70, 83, 84, 85, 86, 87
Gewirth, A. A., 257, 371
Ghosh, P. K., 129, 130, 238
Ginden, J. M., 121, 122, 237
Gipson, S., 54, 85, 86
Giraudeau, A., 38, 40, 84
Glarum, S. H., 90, 121, 122, 171, 200, 220, 232, 237
Goldberg, I. B., 334, 335, 372
Gonzalez Cortes, A., 52, 85
Gordon Vergara, A., 55, 86
Gosser, D. K., 19, 20, 28, 45, 53, 61, 62, 83
Goto, M., 56, 86

Gottesfeld, S., 90, 100, 123, 127, 137, 203, 211, <u>232</u>, <u>233</u>, <u>237</u>
Graham, D. C., 17, <u>83</u>
Gratzel, M., 3, 6, 9, 12, 17, 31, 33, 38, 40, 41, 43, 46, 53, 55, 59, 60, <u>81</u>, <u>82</u>, <u>84</u>
Grimes, P. G., 2, <u>81</u>
Guglielmi, M., 122, 185, <u>237</u>
Guidelli, R., 16, 52, <u>82</u>, <u>85</u>

Haas, O., 90, 100, 139, 141, 190, 202, 212, 215, <u>231</u>, <u>232</u>, <u>233</u>, <u>234</u>, <u>238</u>, <u>241</u>
Hachiya, K., 338, <u>372</u>
Haga, M., 79, <u>88</u>
Halpern, M., 198, <u>240</u>
Hamnett, A., 90, 122, 148, 186, 197, 219, <u>233</u>, <u>240</u>
Hapiot, P., 90, 121, 122, 203, 204, 205, 209, <u>234</u>
Harima, Y., 56, <u>86</u>
Hartman, K. W., 101, 107, <u>236</u>
Hartzell, C. R., 60, <u>86-87</u>
Harwell, J. H., 14, 15, 16, 57, <u>82</u>
Hayano, S., 4, 16, 18, 19, 27, 28, 29, 30, 31, 36, 38, 40, 41, 42, 43, 56, <u>81</u>, <u>82</u>, <u>83</u>
Hayter, J. B., 19, 27, <u>83</u>
He, P., 103, 104, 105, 129, 130, 131, 132, 137, <u>236</u>
Heben, M. J., 262, 307, 327, <u>371</u>
Heeger, A. J., 186, <u>240</u>
Heineman, W. R., 56, <u>86</u>
Heller, A., 320, <u>372</u>
Henderson, T. L. E., 323, <u>372</u>
Hendricks, S. A., 343, <u>373</u>
Hermansky, C., 65, <u>87</u>
Heyrovsky, J., 2, 3, 16, 17, 43, 56, <u>80</u>
Hiemenz, P. C., 2, 7, 14, <u>80</u>

Higgins, S. J., 90, 122, 148, 197, <u>233</u>
Hillman, A. R., 90, 122, 148, 149, 151, 152, 173, 197, 220, 225, <u>233</u>, <u>234</u>, <u>239</u>
Hirasama, R., 334, 335, <u>372</u>
Hirsch, E., 38, <u>84</u>
Hitchman, M. L., 247, <u>370</u>
Hiura, H., 4, <u>81</u>
Ho, C., 90, 190, <u>232</u>
Hoang, P. M., 158, <u>239</u>
Hoclet, M., 90, 121, 123, 186, <u>232</u>, <u>237</u>
Hodges, A. M., 129, 130, 182, <u>238</u>
Holdcroft, S., 158, <u>239</u>
Holleck, L., 3, 16, 46, <u>81</u>
Honda, T., 54, <u>85</u>
Hoppa, F., 186, <u>240</u>
Horányi, G., 90, 121, 123, 124, 149, 151, 152, 158, 159, 161, 173, 174, 178, 179, 194, 195, 196, 198, 199, 205, 206, 208, 209, 211, 212, 213, 214, 225, 226, <u>234</u>, <u>240</u>
Horch, F. R., 38, 40, <u>84</u>
Horrocks, B. R., 90, 121, 122, 185, 186, 187, 188, 189, 201, <u>232</u>, 280, 287, 289, 290, 291, 348, 349, 350, 351, <u>371</u>
Hoshino, K., 56, <u>86</u>
Hoskins, J. C., 14, 15, 16, 57, <u>82</u>
Hoyer, H. W., 28, <u>83</u>
Hu, N., 74, <u>87</u>
Hu, W., 344, <u>373</u>
Huang, W.-S., 121, 122, 198, 205, 207, <u>237</u>, <u>240</u>
Hubbard, A. T., 19, <u>83</u>, 247, 285, 319, 326, 327, <u>370</u>
Huddleston, R. K., 101, 107, <u>236</u>

Author Index

Huggins, R. A., 90, 190, 232
Hülser, P., 90, 232
Humphrey, B. D., 121, 122, 205, 207, 237
Humphries, M. W., 19, 27, 83
Hunt, D., 57, 86
Hunter, R. J., 19, 27, 83
Hunter, T. B., 90, 100, 156, 171, 232
Hurst, J. K., 76, 77, 87, 88
Hurwitz, H. D., 90, 130, 232
Hush, J. M., 244, 370
Hussam, A., 31, 33, 66, 67, 68, 84, 87
Hüsser, O. E., 340, 342, 343, 372

Inoue, T., 130, 133, 137, 238, 338, 372
Inzelt, G., 90, 99, 100, 121, 123, 124, 127, 130, 148, 149, 150, 151, 152, 153, 157, 158, 159, 160, 161, 162, 163, 164, 165, 166, 167, 168, 169, 170, 171, 172, 173, 174, 176, 177, 178, 179, 180, 181, 194, 195, 196, 198, 199, 205, 206, 208, 209, 211, 212, 213, 214, 216, 218, 224, 225, 226, 232, 233, 234, 235, 236, 237, 238, 239, 240, 241
Isaacs, H. S., 244, 370
Ishii, Y., 56, 86
Itagaki, H., 79, 88
Ives, M. B., 349, 373
Iwakura, C., 156, 239
Iwunze, M. O., 54, 69, 70, 71, 73, 74, 85, 87

Jaeger, D. A., 51, 85

Jaffe, L. F., 244, 370
Jakobs, R. C. M., 121, 237
Janssen, L. J. J., 121, 237
Javadi, H. H. S., 121, 122, 237
Jeon, I. C., 314, 330, 332, 349, 361, 372
Jernigan, J. C., 90, 113, 145, 220, 234
Jimbo, T., 54, 85
Johansen, O., 129, 130, 182, 238
Jones, E. T. T., 130, 131, 238
Joo, P., 158, 165, 173, 175, 182, 239
Jordan, J., 250
Jozefowicz, M. E., 121, 122, 237

Kahlweet, M., 10, 82
Kaifer, A. E., 27, 28, 29, 31, 38, 40, 41, 42, 53, 76, 83, 84, 85, 87
Kaishev, V., 18, 83
Kaisheva, M., 18, 83
Kalaji, M., 90, 121, 122, 123, 124, 203, 205, 208, 209, 233, 234
Kalyanasundarum, K., 8, 45, 81
Kamau, G. N., 9, 32, 38, 43, 44, 56, 57, 61, 62, 63, 74, 81, 87
Kamenka, N., 36, 38, 84
Kan, H., 338, 372
Kanako, M., 183, 184, 240
Kanazawa, K. K., 90, 121, 122, 123, 191, 202, 209, 233, 237
Kaneko, M., 90, 130, 136, 137, 138, 182, 183, 185, 219, 234, 238, 240
Kannuck, R. M., 79, 88
Karimi, H., 158, 175, 181, 220, 239

Kattan, L., 330, 372
Kaufman, F. B., 95, 157, 173, 235, 239
Kaufman, J. H., 90, 121, 122, 185, 191, 209, 233, 240
Kawagoe, T., 90, 210, 233
Kawai, T., 156, 239
Kaya, M., 121, 185, 199, 237
Kazarinov, V. E., 90, 121, 209, 211, 234
Keddam, M., 90, 210, 233
Keene, F. R., 99, 102, 235
Kelly, A. J., 90, 139, 233, 238
Kemula, W., 3, 14, 16, 46, 47, 81
Kendig, M. W., 244, 370
Kepley, L. J., 121, 200, 220, 237
Kertész, V., 211, 241
Kesselman, W., 371
Khan, A., 36, 84
Kim, Y.-T., 211, 241
Kinstle, J. F., 130, 158, 159, 173, 238, 239
Kirchoff, J. R., 56, 86
Kissel, G., 244, 370
Kita, H., 209, 210, 211, 240
Kitani, A., 121, 158, 185, 199, 237, 239
Kobayashi, T., 121, 198, 199, 236, 240
Kötz, R., 90, 212, 215, 234, 241
Kramer, S. R., 157, 173, 239
Kratohvil, J. P., 38, 84
Krishnan, M., 373
Kristensen, E. W., 291, 292, 323, 371, 372
Krounbi, M. T., 121, 185, 237
Krygowski, T. M., 3, 14, 16, 46, 47, 81
Kumosinski, T. F., 4, 7, 19, 20, 28, 29, 31, 32, 33, 35, 36, 37, 38, 39, 53, 56, 62, 68, 81
Kurihara, K., 10, 82
Kuta, J., 2, 3, 16, 17, 43, 56, 80
Kutner, W., 90, 144, 145, 146, 234
Kuwana, T., 60, 86-87, 250
Kwak, J., 90, 234, 245, 251, 252, 256, 268, 270, 292, 307, 311, 314, 324, 327, 349, 354, 355, 357, 359, 361, 370, 371, 372

Labbe, P., 59, 86
LaCroix, J.-C., 121, 122, 123, 200, 202, 203, 237
Laidman, D. L., 17, 82
Láng, G., 90, 100, 127, 158, 161, 170, 171, 172, 218, 224, 232
Lang, J., 8, 11, 81
Lange, M. A., 158, 239
Lange, R., 130, 132, 134, 135, 137, 138, 238
Lapkowski, M., 53, 85, 121, 122, 137, 185, 199, 200, 205, 237
Laurent, E., 55, 56, 86
Laversanne, R., 121, 122, 237
Laviron, E., 91, 95, 96, 100, 111, 159, 161, 235, 236, 239, 240
Leddy, J., 128, 129, 137, 147, 148, 156, 172, 238, 360, 373
Lee, C., 90, 130, 234, 245, 246, 251, 253, 257, 258, 259, 260, 262, 263, 264, 292, 310, 311, 313, 314, 345, 349, 350, 352, 353, 355, 361, 370, 371, 372

Author Index

Lee, E. M., 15, 82
Lee, J. W., 327, 372
Lee, R., 17, 82
Lee, W.-Y., 184, 240
Legg, K. D., 90, 231
Lei, Y., 77, 88
Leider, C. R., 90, 144, 145, 146, 234
Leipert, T., 9, 32, 38, 43, 44, 56, 57, 63, 81
Leiva, E., 108, 236
LeMest, Y., 220, 221, 222, 223, 241
Lenhard, J. R., 147, 148, 218, 238
Lev, O., 245, 270, 327, 370
Levart, E., 90, 190, 232
Levich, V. G., 94, 100, 235
Lewis, N. S., 262, 307, 327, 371
Lewis, T. J., 17, 82
Li, C., 344, 373
Li, F., 122, 186, 195, 237
Lien, M., 90, 196, 233
Lin, C. W., 339, 340, 344, 372
Lin, H., 344, 373
Lind, E.-L., 105, 106, 113, 114, 127, 128, 137, 236, 360, 373
Lindblom, G., 36, 84
Lindholm, B., 90, 105, 106, 113, 114, 127, 128, 132, 135, 136, 137, 138, 172, 232, 233, 236, 360, 373
Lindman, B., 10, 36, 38, 82, 84
Lindsay, S. M., 259, 261, 371
Liu, H.-Y., 184, 240
Liu, X., 371
Loder, J. W., 129, 130, 182, 238
Logan, J. A., 184, 240
Lombardi, J. R., 23, 24, 26, 28, 83
Longin, T. L., 262, 307, 327, 371

Longmire, M. L., 90, 99, 102, 113, 145, 220, 224, 234, 235
Loveday, D. C., 90, 122, 148, 149, 151, 152, 173, 197, 220, 225, 233, 239
Lu, J., 371
Lu, T., 76, 87
Lu, Y. C., 349, 373
Luisi, P. L., 9, 10, 11, 13, 82
Luo, J. L., 349, 373
Lynch, M., 17, 82
Lyons, M. E. G., 122, 139, 237, 238

McCarley, R. L., 343, 373
MacDiarmid, A. G., 121, 122, 198, 199, 205, 207, 237, 240
Macdonald, J. R., 90, 190, 232
McIntire, G. L., 4, 29, 31, 45, 47, 48, 49, 53, 56, 70, 81, 83, 84, 85, 86
Mackay, R. A., 28, 31, 36, 38, 39, 40, 65, 66, 67, 68, 83, 84, 87
McLarnon, F. R., 139, 202, 238
McLennan, G., 17, 82
McManis, G. E., 223, 241
McManus, P. M., 199, 205, 240
Magid, L. J., 9, 10, 11, 13, 82
Magner, G., 90, 190, 232
Majda, M., 27, 28, 83, 123, 130, 131, 238
Manassen, J., 78, 88
Mandal, A. B., 31, 33, 40, 84
Mandler, D., 245, 246, 251, 339, 344, 345, 346, 370, 372, 373
Mann, D. R., 323, 372

Mann, T. F., 323, 372
Mao, H., 122, 192, 193, 237
Marcus, R. A., 16, 82
Mariani, P., 52, 85
Mark, H. B., 90, 120, 185, 234
Marque, P., 197, 240
Marshall, J. H., 90, 121, 122, 171, 200, 220, 232, 237
Martin, C. R., 90, 100, 122, 128, 171, 186, 190, 232, 238
Martinet, P., 48, 49, 50, 52, 84, 85
Martre, A.-M., 48, 49, 50, 84, 85
Mathew, S., 2, 16, 18, 54, 56, 80
Matsuda, H., 100, 130, 132, 136, 137, 138, 172, 182, 183, 184, 235, 238
Matsumoto, M., 18, 83
Mau, A. W.-H., 129, 130, 182, 238
Meaney, T., 247, 291, 327, 331, 333, 370
Meites, L., 3, 16, 17, 40, 43, 81
Meites, T., 3, 16, 40, 81
Mermilliod, N., 90, 121, 123, 186, 232, 237
Merz, A., 90, 231
Mesquita, J. C., 90, 121, 122, 203, 209, 234
Meyer, L., 53, 85
Meyer, P., 108, 236
Meyer, T. J., 105, 236
Miaw, C. L., 43, 45, 56, 84
Milioto, S., 38, 40, 84
Miller, C. A., 10, 82
Miller, C. J., 257, 258, 259, 260, 262, 263, 264, 310, 371
Miller, F., 17, 82
Miller, J. R., 17, 46, 82, 101, 107, 236

Miller, L. L., 90, 158, 159, 173, 231, 239
Miras, M. C., 90, 212, 215, 234, 241
Mirkin, M. V., 90, 234, 259, 265, 268, 269, 270, 272, 273, 274, 275, 276, 277, 279, 280, 281, 282, 284, 285, 287, 289, 290, 291, 294, 307, 308, 309, 310, 311, 314, 320, 321, 322, 323, 324, 325, 326, 327, 328, 329, 330, 332, 335, 348, 349, 350, 351, 355, 360, 361, 362, 371, 372
Mitchell, D. J., 10, 11, 12, 13, 36, 38, 82, 84
Moaz, R., 42, 58, 82
Momma, T., 90, 122, 123, 201, 233
Moncelli, M. R., 52, 85
Monkman, A. T., 90, 121, 122, 185, 186, 187, 188, 189, 201, 232
Montenegro, M. I., 244, 323, 327, 370, 372
Moraes-Kraus, E., 55, 86
Mount, A. R., 90, 121, 122, 185, 186, 187, 188, 189, 201, 232
Mousset, G., 48, 49, 50, 51, 84, 85
Mousty, C., 49, 50, 51, 85
Mowry, G., 17, 82
Mu, J., 344, 373
Mu, S.-L., 198, 240
Mukherjee, S., 36, 38, 84
Murakoshi, K., 209, 210, 211, 240
Murray, R. W., 90, 91, 99, 100, 102, 111, 112, 113, 118,

Author Index

119, 123, 136, 138, 141, 144, 145, 146, 147, 148, 152, 156, 157, 182, 193, 218, 219, 220, 221, 222, 223, 224, 231, 234, 235, 236, 238, 240, 241
Musiani, M. M., 90, 121, 123, 232, 233
Myers, S. A., 67, 68, 87

Nadjo, L., 53, 85, 90, 159, 231, 239
Nagahara, L. A., 259, 261, 371
Nagasaka, H., 99, 102, 224, 235
Nagy, G., 280, 287, 289, 290, 291, 348, 349, 350, 351, 371
Nagy, M., 169, 170, 240
Nair, B. U., 31, 33, 40, 84
Nakahama, S., 100, 148, 152, 218, 236
Nakajima, T., 90, 122, 123, 201, 233
Nakamura, M., 183, 184, 240
Nakamura, S., 183, 184, 240
Nakamura, T., 58, 86
Naleway, C. A., 101, 236
Nanis, L., 371
Naoi, K., 90, 122, 186, 192, 196, 232, 233
Nath, B., 51, 85
Nechtschein, M., 90, 121, 122, 185, 203, 204, 205, 209, 234, 237
Netzer, L., 42, 58, 82
Newman, J., 288, 371
Nichols, K. H., 157, 239
Nicoli, D. F., 36, 38, 84
Nielson, R. M., 223, 241

Ninham, B.W., 10, 11, 12, 13, 36, 38, 82, 84
Nishiki, Y., 100, 132, 133, 137, 138, 235
Niwa, K., 130, 132, 134, 135, 137, 138, 238
Nnodimele, R., 57, 58, 86
Noftle, R. F., 42, 84
Nome, F., 68, 87
Nordyke, L. L., 27, 28, 83
Novodoff, J., 28, 83
Nowak, M. J., 186, 240
Nowak, R., 90, 231
Nucci, L., 52, 85
Nuccitelli, N., 244, 370
Nyholm, L., 90, 121, 122, 203, 205, 208, 209, 234

Obeng, Y. S., 17, 82
Ochmanska, J., 122, 192, 193, 237
O'Connell, K. M., 161, 239
O'Dea, J. J., 83
Odin, C., 90, 121, 122, 203, 204, 205, 209, 234
Ogata, N., 99, 102, 224, 235
Oh, S.-M., 108, 130, 132, 134, 135, 137, 138, 140, 149, 218, 236, 238
Ohsaka, T., 90, 130, 132, 136, 137, 138, 172, 182, 183, 184, 210, 214, 233, 237, 238, 240
Ohsawa, Y., 29, 31, 41, 42, 43, 45, 56, 83, 84
Ohta, M., 54, 86
Oikonomou, A., 90, 130, 232
Okano, T., 79, 88
Oldham, K. B., 285, 291, 307, 329, 371, 372

Opallo, M., 327, <u>372</u>
Oppenheimer, L. E., 71, <u>87</u>
Orata, D., 90, 210, 211, <u>233</u>
Osaka, T., 90, 122, 123, 186, 192, 201, <u>232</u>, <u>233</u>
Osteryoung, J., <u>83</u>
Ouyang, J., <u>86</u>
Overall, B. L., 244, <u>370</u>
Owlia, A., 4, 29, 31, 32, 33, 36, 39, 56, 61, 68, 69, 71, 72, 73, 74, <u>81</u>, <u>83</u>, <u>84</u>, <u>87</u>
Oyama, N., 90, 100, 130, 132, 133, 136, 137, 138, 139, 172, 182, 183, 184, 210, 214, <u>233</u>, <u>235</u>, <u>237</u>, <u>238</u>, <u>240</u>

Paik, J. H., 29, 31, 38, 40, 41, 42, 53, <u>84</u>
Palazzotto, M. C., 90, <u>231</u>
Panero, S., 90, 122, 171, 186, 191, 192, <u>232</u>
Park, J. W., 29, 31, 38, 40, 41, 42, 53, <u>84</u>
Parsons, R., 19, <u>83</u>, 90, 190, <u>232</u>, 250
Passerini, S., 90, 122, 171, 186, 191, 192, <u>232</u>
Paulse, C. D., 100, 122, 186, 192, 193, <u>236</u>, <u>237</u>
Peaceman, D. W., 270, <u>371</u>
Peerce, P. J., 95, 96, 147, 148, 181, 220, <u>235</u>
Pelizzetti, E., 4, 45, 56, <u>81</u>
Penneau, J. F., 121, 213, <u>236</u>
Penner, R. M., 90, 100, 122, 171, 186, 190, <u>232</u>, 262, 307, 327, <u>371</u>
Pergola, F., 52, <u>85</u>

Perlmutter, D. D., 90, 122, 171, 186, 191, 192, <u>232</u>
Pernaut, Jm., 122, <u>237</u>
Peter, L. M., 90, 121, 122, 123, 124, 203, 205, 208, 209, <u>233</u>, <u>234</u>, 287, 348, <u>371</u>
Pfeiffer, B., 209, <u>240</u>
Pfluger, P., 121, 185, <u>237</u>
Phillips, J. B., 314, 361, <u>372</u>
Phipps, J. B., 361, 363, 364, <u>373</u>
Pickup, P. G., 90, 100, 122, 144, 145, 146, 186, 192, 193, <u>234</u>, <u>236</u>, <u>237</u>
Pierce, D. T., 90, <u>234</u>, 278, 280, 287, 289, 290, 291, 312, 314, 316, 320, 348, 349, 350, 351, <u>371</u>, <u>372</u>
Pincus, P. A., 186, <u>240</u>
Pingarron Carrazon, J. M., 52, 55, <u>85</u>, <u>86</u>
Pinkerton, M. J., 90, 113, 145, 220, 221, 222, 223, <u>234</u>, <u>241</u>
Pistoia, G., 90, 121, 122, 123, 171, 201, <u>232</u>
Pletcher, D., 42, <u>84</u>, 323, <u>372</u>
Plichon, V., 173, <u>240</u>
Polo Diez, L. M., 52, 55, <u>85</u>, <u>86</u>
Pons, S., 39, <u>83</u>, 244, 256, 288, <u>370</u>, <u>371</u>
Porter, M. D., 17, <u>82</u>
Pouget, J. P., 121, 122, <u>237</u>
Pouillen, P., 48, 49, 50, <u>84</u>, <u>85</u>
Pourcelly, G., 90, 130, <u>232</u>
Pramauro, E., 4, 45, 56, <u>81</u>
Prins, W., 169, 170, <u>240</u>
Proske, G. E. O., 4, <u>81</u>
Prospieri, P., 90, 122, 171, 186, 191, 192, <u>232</u>
Prud'homme, R. K., 18, 27, <u>83</u>

Author Index

Puglisi, J. V., 334, 335, <u>372</u>
Puyal, M.-C., 36, 38, <u>84</u>

Qiu, Y.-J., 90, 196, <u>234</u>
Queirós, M. A., 244, 327, <u>370</u>
Quintela, P. A., 53, <u>85</u>
Qutubuddin, S., 10, 31, 33, 38, 40, 66, 67, 68, <u>82</u>, <u>84</u>, <u>87</u>

Rabani, J., 129, 130, 182, <u>238</u>
Rachford, H. H., 270, <u>371</u>
Raistrick, I. D., 90, 190, <u>232</u>
Ramaswamy, D., 31, 33, 40, <u>84</u>
Raous, H., 17, 18, 27, <u>82-83</u>
Rauniyar, G., 55, 56, <u>86</u>
Rechnitz, G. A., 78, 79, <u>88</u>
Redondo, A., 90, 123, 203, 211, <u>233</u>, <u>237</u>
Ren, B., 344, <u>373</u>
Rennie, A. R., 15, <u>82</u>
Reverdy, G., 59, <u>86</u>
Reviejo Garcia, A. J., 52, 55, <u>85</u>, <u>86</u>
Reynolds, J. R., 90, 196, <u>234</u>
Richards, T. C., 322, 323, 324, 329, <u>372</u>
Richter, B., 42, 58, <u>82</u>
Richtering, W., 100, <u>236</u>
Risphon, J., 90, 100, 121, 123, 127, 137, 187, 201, 211, <u>232</u>, <u>233</u>
Robinson, L. R., 42, 58, <u>82</u>
Rodriguez, R., 54, <u>85</u>
Rolison, D. R., 39, <u>83</u>, 147, 148, 218, <u>238</u>, 244, 256, <u>370</u>
Roncali, J., 197, <u>240</u>
Rosano, H. L., 28, <u>83</u>
Roullier, L., 100, 161, <u>236</u>, <u>239</u>
Rubinson, J. F., 90, 120, 185, <u>234</u>

Rubinstein, I., 42, 58, <u>82</u>, 90, 100, 102, 121, 123, 127, 128, 129, 137, 187, 201, 203, <u>232</u>, <u>236</u>, <u>237</u>, <u>238</u>, <u>240</u>
Rudnicki, J., 139, 202, <u>238</u>
Ruff, I., 94, 97, 98, 99, 130, <u>235</u>
Ruiz Barrio, A., 52, <u>85</u>
Rusling, J. F., 4, 7, 9, 10, 11, 18, 19, 20, 22, 24, 28, 29, 31, 32, 33, 35, 36, 37, 38, 39, 42, 43, 44, 45, 53, 56, 57, 58, 59, 60, 61, 62, 63, 64, 68, 69, 70, 71, 72, 73, 74, 76, <u>81</u>, <u>82</u>, <u>83</u>, <u>84</u>, <u>86</u>, <u>87</u>
Rydzewski, R., 227, 228, <u>241</u>
Rymden, R., 36, 38, <u>84</u>

Sabatini, E., 42, 58, <u>82</u>, 90, 121, 123, 187, 201, <u>232</u>
Sagi, T., <u>86</u>
Sagiv, J., 42, 58, <u>82</u>
Saji, T., 56, <u>86</u>
Sakai, M., 292, <u>371</u>
Salmon, M., 122, 185, <u>237</u>
Sanford, L. K., 223, <u>241</u>
Sangaranarayanan, M. V., 116, <u>236</u>
Sanui, K., 99, 102, 224, <u>235</u>
Sasaki, K., 121, 185, 199, <u>237</u>
Sasaki, M., 338, <u>372</u>
Sasse, W. H. F., 129, 130, 182, <u>238</u>
Sato, K., 130, 132, 136, 137, 138, 172, 182, 183, 184, <u>238</u>
Saveant, J. M., 53, 61, <u>85</u>, <u>87</u>
Savéant, J.-M., 95, 96, 99, 102, 105, 110, 111, 115, 116, 117, 130, 132, 136, 137, 159, 183, 188, <u>235</u>, <u>236</u>, <u>238</u>, <u>239</u>, 360, <u>373</u>
Savy, M., 90, 190, <u>232</u>

388 Author Index

Scamehorn, J. F., 15, 82
Schecter, R. S., 14, 15, 16, 57, 82
Schiffrin, D. J., 173, 240
Schlenoff, J. B., 193, 219, 240
Schmehl, R. H., 111, 112, 138, 236
Schmickler, W., 108, 236
Schmidt, P. P., 39, 83, 244, 256, 370
Schroeder, A. H., 157, 173, 239
Schuhman, D., 17, 18, 27, 82-83
Schultz, F. A., 90, 231
Schultze, J. W., 209, 240
Schumacher, R., 90, 233
Scott, E. R., 314, 361, 363, 364, 372, 373
Scott, J. C., 121, 185, 237, 240
Scrosati, B., 90, 122, 171, 186, 191, 192, 232
Seiders, R. P., 36, 38, 39, 65, 66, 84
Servagent, S., 90, 122, 197, 233
Sharp, M., 90, 105, 106, 113, 114, 127, 128, 132, 135, 136, 137, 138, 172, 232, 236, 360, 373
Shay, M., 90, 233
Shi, C.-N., 4, 7, 19, 20, 28, 29, 31, 32, 33, 35, 36, 37, 38, 39, 45, 53, 56, 58, 59, 61, 62, 64, 68, 81, 83, 86
Shigehara, K., 100, 130, 132, 136, 137, 138, 139, 235, 238
Shimazaki, Y., 41, 42, 43, 45, 84
Shimazu, K., 209, 210, 211, 240
Shinoda, K., 10, 82
Shinozuka, N., 4, 16, 18, 19, 27, 28, 31, 43, 81, 82
Shiota, K., 90, 122, 123, 201, 233
Shlepakov, A. V., 90, 121, 209, 211, 234

Shu, C.-F., 99, 102, 116, 137, 235, 360, 373
Shukla, S. S., 9, 19, 20, 28, 32, 38, 43, 44, 45, 53, 56, 57, 61, 62, 63, 81, 83
Sidarous, L., 53, 54, 85
Silver, M., 99, 102, 235
Simister, E. A., 15, 82
Skotheim, T. A., 121, 122, 237
Small, B., 330, 372
Smoluchowksi, M., 101, 236
Smyrl, W. H., 90, 100, 122, 156, 171, 186, 192, 196, 232, 233
Sofue, S., 79, 88
Solouki, T., 58, 86
Somasiri, N. L. D., 198, 240
Somasundaran, P., 14, 82
Soriaga, M. P., 19, 83
Sosnoff, C. S., 99, 102, 235
Spiegel, D., 186, 240
Spytsin, M. A., 90, 121, 209, 211, 234
Srinivasa Mohan, L., 116, 236
Stegun, I., 273, 282, 287, 371
Stickney, J. L., 19, 83
Stilbs, P., 36, 38, 84
Storck, W., 130, 132, 238
Street, J.B., 90, 121, 122, 185, 191, 209, 233, 237, 240
Sucheta, A., 69, 70, 71, 87
Suga, K., 41, 42, 43, 45, 84, 86
Sugawara, S., 79, 88
Sugiyama, K., 62, 63, 87
Suib, S. L., 58, 59, 64, 86
Sun, S., 23, 24, 26, 28, 83
Surridge, N. A., 90, 99, 102, 113, 145, 220, 234, 235
Swann, M. J., 90, 122, 148, 197, 225, 233, 239
Syed, A. A., 121, 237

Author Index

Szabó, L., 99, 100, 148, 149, 152, 153, 158, 161, 165, 218, 224, 235, 236, 238, 239
Szentrimay, R., 250
Szulbinski, W., 53, 85

Tabakovic, I., 50, 85
Takamoto, Y., 79, 88
Takenouti, H., 90, 100, 141, 190, 210, 232, 233
Tamura, H., 121, 198, 199, 236, 240
Tanaka, M., 4, 81
Tanford, C., 7, 27, 81
Tang, X., 121, 122, 237
Tanguy, J., 90, 121, 123, 186, 232, 237
Taylor, D. M., 17, 82
Tebbutt, P., 250
Texter, J., 38, 40, 67, 68, 71, 84, 87
Thomalla, M., 53, 55, 56, 70, 85, 86
Thomas, J. K., 58, 86
Thomas, R. K., 15, 82
Thompson, D. H. P., 76, 87
Thormann, W., 323, 372
Thundat, T., 259, 261, 371
Thyssen, A., 209, 240
Tian, Z., 344, 373
Tiddy, G. J. T., 36, 84
Tokuda, K., 100, 132, 133, 137, 138, 235
Tong, L. K. J., 331, 372
Tople, R., 247, 291, 327, 331, 333, 370
Torresi, R., 90, 210, 233
Toth, K., 280, 287, 289, 290, 291, 348, 349, 350, 351, 371
Tran, N., 58, 86
Tribollet, B., 90, 121, 123, 232

Tricot, Y.-M., 78, 88
Trise Frank, M. H., 287, 348, 371
Tronel-Peyroz, E., 17, 18, 27, 82-83
Tsou, Y.-M., 130, 131, 132, 184, 238, 240
Tsukada, A., 90, 100, 141, 233
Tuckerman, L. S., 269, 270, 271, 272, 371
Turro, N., 14, 82
Tyler, P. S., 90, 100, 156, 171, 232

Ueyama, K., 90, 122, 186, 192, 232
Umana, M., 90, 231
Umezawa, Y., 79, 88
Underwood, A. L., 4, 81
Unwin, P. R., 90, 234, 248, 269, 270, 272, 274, 275, 276, 277, 278, 280, 281, 291, 292, 293, 294, 296, 297, 298, 299, 300, 301, 302, 304, 306, 312, 314, 316, 320, 321, 322, 331, 333, 334, 336, 337, 338, 339, 340, 370, 371, 372

Van De Mark, M. R., 90, 231
Vanel, P., 17, 18, 27, 82-83
Van Woert, H. C., 68, 87
Varineau, P. T., 90, 148, 233
Verniette, M., 50, 85
Verrall, R. E., 38, 40, 84
Vieil, E., 90, 121, 122, 197, 213, 233, 236, 237
Villeret, B., 122, 237
Vining, W. J., 105, 236
Vork, F. T. A., 122, 193, 237
Vos, J. G., 90, 139, 141, 172, 218, 233, 238

Wade, W. H., 14, 15, 16, 57, 82
Waldner, E., 161, 239
Waller, A. M., 90, 121, 122, 185, 186, 187, 188, 232
Waltman, R. J., 197, 240
Wang, Z., 4, 29, 31, 32, 33, 36, 39, 56, 59, 64, 68, 69, 71, 72, 73, 81, 83, 84, 86
Watanabe, M., 90, 99, 100, 102, 113, 145, 220, 221, 222, 223, 224, 234, 235, 241
Weaver, M. J., 223, 241
Weber, M., 247, 291, 331, 370
Weinheimer, R. M., 36, 38, 84
Westmoreland, P. G., 4, 81
Whitaker, R. G., 250
White, H. S., 90, 100, 128, 137, 156, 171, 232, 238, 314, 360, 361, 363, 364, 372, 373
Whitlow, S. J., 220, 241
Wichramasinghe, H. K., 244, 370
Widrig, C. A., 27, 28, 83
Wightman, R. M., 39, 84, 244, 247, 256, 277, 291, 292, 323, 327, 331, 333, 370, 371, 372
Wilde, C. P., 90, 122, 148, 197, 225, 233, 239
Wilkinson, M. C., 17, 82
Willis, W. S., 42, 58, 59, 64, 82, 84, 86
Wilson, P. J., 90, 121, 122, 185, 186, 187, 188, 189, 201, 232
Winiecki, A. M., 59, 64, 86
Winkler, K., 323, 372
Winquist, S., 247, 291, 331, 370
Wipf, D. O., 90, 234, 244, 245, 246, 251, 253, 254, 255,
256, 269, 270, 272, 275, 276, 277, 278, 279, 280, 281, 311, 312, 313, 314, 315, 316, 317, 319, 320, 321, 322, 323, 327, 330, 345, 354, 370, 371, 372
Wöhrle, D., 90, 185, 219, 234
Wool, R. P., 220, 241
Wooster, T. T., 90, 99, 100, 102, 113, 145, 220, 224, 234, 235
Wrighton, M. S., 90, 182, 231, 240
Wu, W., 198, 240
Wunder, D. J., 247, 291, 331, 370
Wuu, Y.-M., 343, 372

Xie, Z., 344, 373

Yamada, A., 183, 184, 240
Yamaguchi, S., 100, 132, 133, 137, 138, 235
Yamamoto, H., 182, 183, 184, 240
Yamamoto, N., 90, 210, 233
Yamashita, K., 56, 86
Yang, H., 209, 211, 233, 240, 241, 332, 335, 372
Yang, S. C., 199, 205, 240
Yaniger, S. T., 198, 240
Yap, W. T., 100, 236
Yasunaga, T., 338, 372
Yeh, P., 60, 86-87, 250
Yeskie, M. A., 14, 15, 16, 57, 82
Yoneyama, H., 121, 156, 198, 199, 236, 239, 240

Zaba, B. N., 17, 82
Zana, R., 8, 11, 28, 31, 38, 39, 40, 81, 83, 84
Zembala, M., 19, 83

Author Index

Zhang, H., 42, 58, 82, 84, 90, 99, 102, 113, 145, 220, 221, 222, 223, 224, 234, 235, 241
Zhang, S., 73, 74, 76, 87
Zhang, X., 223, 241, 373
Zhou, F., 90, 234, 278, 291, 292, 293, 294, 297, 298, 299, 300, 312, 314, 316, 334, 336, 337, 338, 371, 372
Zhou, X., 344, 373
Zhu, Y., 24, 28, 83
Zoski, C. G., 285, 291, 323, 371, 372
Zvanut, M. E., 99, 102, 235

SUBJECT INDEX

Activated olefins, reduction of, 50–52
Adiponitrile, electrolytic production of, 3
Adsorption process, SECM in study of, 298–303, 337–339
Adsorption of surfactants, 16–28
　of electroactive surfactants, 25–27
　nonelectroactive surfactants
　　electrochemcial studies, 18–21
　　in situ surface spectroscopy, 21–25
　on solid surfaces, 13–16
　surfactants adsorbed as electrodes, 27–28
Anions, inorganic, influence of surfactants on, 56–58
Applications of SECM, 310–364
　characterization of thin film and membranes, 349–364
　chronoamperometry and chronopotentiometry, 354–355
　cyclic voltammetry, 350–354
　direct electrochemical measurements, 355–361
　SECM imaging of films and membranes, 361–364
　fabrication, 339–348

　heterogeneous electron transfer and reaction rate imaging, 319–331
　studies of heterogeneous kinetics, 320–331
　modes of, 247–250
　　as an electrochemical tool, 247–248
　　as a fabrication tool, 249–250
　　as a imaging tool, 249
　potentiometric and other tips, 348–349
　SECM images, 310–319
　studies of homogeneous chemical reactions and adsorption/desorption processes, 331–339
　　adsorption/desorption processes, 337–339
　　first-order homogeneous reactions, 331–334
　　second-order following reactions, 334–337
Aromatic ketones, reduction of, 50–51
Aromatic nitro compounds, reduction of, 50–52
Atomic force microscopy (ATM), 244

Bicontinuous microemulsions, 11

Subject Index

Blauch–Saveant theory, 117–118

Cations, inorganic, influence of surfactants on, 56–58
Cetyltrimethylammonium bromide (CTAB), 15
Charge transport in polymer-modified electrodes, 89–241
   comparison of traditional and polymer film electrodes, 92–95
   definition and classification, 91–92
   effect of film morphology on charge transport in polymers, 218–228
   mechanical and electrochemical equilibria in polymer layers, 225–228
   polymer self-diffusion, 221–225
   theories of de Gennes, 219–221
   future research goals, 229–231
   results on charge transport in polymer films, 124–218
     conducting polymer films, 184–218
     fixed-site redox polymers, 138–184
     ion-exchange polymers containing electrostatically bound redox centers, 124–138
   theories of electron transport in polymer film electrodes, 95–124
     early models of charge propagation, 95–97
     models of charge transport in electronically conducting polymer films, 120–124
     new theories predicting nonlinear $D(c)$ function, 103–118
     potential dependence of diffusion coefficient, 118–119
     theory of the electron exchange reaction, 97–103
Chronoamperometry, 354–355
Chronopotentiometry, 354–355
Conducting polymer films, 92, 184–218
   poly(aniline), 198–218
   poly(pyrrole), 184–196
   poly(thiophene), 197–198
Conductive microemulsions, diffusion in, 65–68
Coupled electron hopping-ion displacement, 109–116
Critical micelle concentration (CMC), 2, 5–6
   diffusion coefficients at, 39–40
Cyclic voltammetry, 350–354
   fast scan rate on ultramicroelectrode, 90

de Gennes theories on the adsorption, diffusion, and charge transport of polymers, 219–221
Desorption process, SECM in the study of, 298–303, 337–339
Diffusion coefficient, potential dependence of, 118–119
Diffusion in micellar solutions, 28–41
   consequence of solute binding, 40–41

Subject Index

detection of micelles, 40
diffusion coefficients in CMC, 39–40
microelectrode voltammetry, 39
model for two-micelle distribution, 35–39
models for diffusion, 29–31
single size distribution of micelles, 31–35
Diffusion-migration model for long-distance electron hopping, 116–117

Electroactive surfactants, adsorption of, 25–27
Electochemical catalysis
  in micellar solutions, 59–65
  catalytic reduction of organohalides, 61–65
  in microemulsions, 71–74
Electrochemical equilibrium in polymer layers, 225–228
Electrochemical impedance spectroscopy (EIS), 90
Electrochemical quartz crystal microbalance (EQCM), 90, 148
Electrode reactions, SECM applied to the study of, 247–248
Electron spin resonance (ESR), 3
Electron transport in polymer film electrodes, 95–124
  early models of charge propagation, 95–97
  models of charge transport in electronically conducting polymer films, 120–124
  new theories predicting nonlinear $D(c)$ function, 103–118

Blauch–Saveant theory, 117–118
diffusion-migration model for long-distance electron hopping, 116–117
Fritsch-Faules-Faulkner model, 105–108
He–Chen model, 103–105
ion association and electric field effects, 116
percolation theory, 108–109
theories of coupled electron hopping-ion displacement, 109–116
potential dependence of diffusion coefficient, 118–119
theory of the electron exchange reaction, 97–103
Electrostatically bound redox centers, ion-exchange polymers containing, 124–138

Fabrication of microstructures on surfactants, SECM in the study of, 249–250, 339–348
Fast scan-rate cyclic voltammetry on ultramicroelectrodes, 90
Film morphology effect on charge transport in polymers, 218–228
  mechanical and electrochemical equilibria in polymer layers, 225–228
  polymer self-diffusion, 221–225
  theories of de Gennes, 219–221
Fixed-site redox polymers, 138–184

[Fixed-site redox polymers]
  containing inorganic complex ions, 138–146
  organic redox polymers, 157–184
    poly(tetracyanoquinodimethane), 158–182
    poly(tetrathiafulvalene), 157–158
    quinone polymers, 158
    viologen, 182–184
  organometallic redox polymers, 146–157
Fritsch-Faules-Faulkner model (of nonlinear function), 105–108

Generation/collection (G/C) mode of SECM operation, 287–293
  G/C mode with amperometric tip, 291–293
  potentiometric G/C mode, 287–291

He–Chen model (of nonlinear function), 103–105
Heterogeneous electron transfer in micellar solutions, 41–46
  electron transfer processes, 41–42
  heterogeneous electron transfer kinetics, 43–45
  question of reactants exchanging electrons with electrodes while in micelles, 45–46
Heterogeneous kinetics, SECM in the studies of, 320–331
Hexamethylenediamine, 3

Historical survey of surfactants in electrochemistry, 2–4
Homogeneous chemical reactions, SECM in the study of, 331–336
  first-order homogeneous reactions, 331–334
  second-order following reactions, 334–336
Hyamine, organic oxidations using, 53–54

Imaging as an SECM application, 249, 310–319
Inorganic ions
  fixed-rate redox polymers containing, 138–146
  influence of surfactants on, 56–58
Instrumentation for SECM, 251–268
  basic apparatus, 251–253
  tip position modulation, 253–255
  tip preparation, 255–268
    Apiezon wax coated ultramicroelectrode tip, 259–262
    characterization of the tip, 262–268
    construction of submicrometer with electrode beam lithography, 259
    disk-in-glass microelectrodes, 255–258
Interfacial tension, lowering of, 2
Ion association effect within the polymer field, 116

Subject Index

Ion-exchange polymeric systems, 92
  containing electrostatically bound redox clusters, 124–138

Ketones, aromatic, reduction of, 50–52
Kinetics of the heterogeneous electron transfer, 43–45
Kinetics in micellar systems, 9–10

Lamellar dispersions, 11–12, 74–76
Liposome marker release method, 78–79
Long-distance electron hopping, diffusion–migration model for, 116–117

Mechanical equilibria in polymer layers, 225–228
Membranes and thin films, SECM in the study of, 349–364
Micelles and microemulsions, 1–88
  electrochemistry in lamellar and vesicle dispersions, 74–79
  detecting photochemistry at CdS particles in vesicles, 77–78
  immunological detection marker release from liposomes, 78–79
  lamellar dispersions, 74–76
  vesicle dispersions, 76–77
  electrochemistry in micellar solutions, 16–65
  adsorption of surfactants, 16–28
  diffusion in micellar solutions, 28–41
  electrochemical catalysis in micellar solutions, 59–65
  electrochemical reactions in micelles, 46–59
  heterogeneous electron transfer in micellar solutions, 41–46
  electrochemistry in microemulsions, 65–74
  diffusion in conductive microemulsions, 65–68
  diffusion studies with microelectrodes, 68–69
  electrochemical catalysis, 71–74
  electrochemical reactions in microemulsions, 69–71
  historical survey, 2–4
  surfactant microstructures, 4–16
  adsorption of surfactants on solid surfaces, 13–16
  microemulsions, 10–11
  surfactant structure controls and system architecture, 12–13
  surfactants and micelles, 4–10
  vesicles and lamellar dispersions, 11–12
Microelectrode voltammetry, 39
Models of charge propagation, 95–97
Models of charge transport in electronically conducting polymer films, 120–124

Nitro compounds, aromatic, reduction of, 46–50

Non-disk-shaped microtips in
SECM experiments,
303–310
Nonelectroactive surfactants
electrochemical studies, 18–21
in situ surface spectroscopy,
21–25
Non-steady-state SECM
measurements, 268–274

Oil-in-water (o/w)
microemulsions, 10
Olefins, activated, reduction of,
50–52
Organic oxidations, 54–55
using hyamine, 53–54
Organic radicals, micellar effects
on, 52–53
Organic redox polymers, 157–184
Organohalides, catalytic
reductions of, 61–65
Organometallic(s)
influence of surfactants on,
56–58
redox polymers, 146–157

Percolation theory, 108–109
transition between diffusion
behavior and percolation
behavior, 117–118
Photochemical studies of CdS
particles in vesicles, 77–78
Poly(aniline), 92, 198–218
Poly(pyrrole), 92, 184–196
Poly(tetracyanoquinodimethane),
158–162
Poly(tetrathiafulvalene), 157–158
Poly(thiophene) film, 197–198
Polymer-modified electrodes,
charge transfer in, 89–241
comparison of traditional and
polymer film electrodes,
92–95
definition and classification,
91–92
effect of film morphology on
charge transport in
polymers, 218–228
mechanical and electrochemical
equilibria in polymer
layers, 225–228
polymer self-diffusion,
221–225
theories of de Gennes,
219–221
future research goals, 229–231
results on charge transport in
polymer films, 124–218
conducting polymer films,
184–218
fixed-site redox polymers,
138–184
ion-exchange polymers
containing electrostatically
bound redox centers,
124–138
theories of electron transport in
polymer film electrodes,
95–124
early models of charge
propagation, 95–97
models of charge transport in
electronically conducting
polymer films, 120–124
new theories predicting
nonlinear D(c) function,
103–118
potential dependence of
diffusion coefficient,
118–119
theory of the electron
exchange reaction, 97–103

Subject Index

Potentiometric G/C mode of SECM, 287–291

Quinone polymers, 158

Reaction rate imaging by SECM, 314–316
Redox polymers, 92
 fixed-site, 138–184
 organic, 157–184
 organometallic, 146–157
Reverse micelles, 9

Scanning electrochemical microscopy (SECM), 90, 243–373
 abbreviations, 365–366
 applications, 310–364
  characterization of thin films and membranes, 349–364
  fabrication, 339–348
  heterogeneous electron transfer and reaction rate imaging, 319–331
  potentiometric and other tips, 348–349
  SECM images, 310–319
  studies of homogeneous chemical reactions and adsorption/desorption processes, 331–339
 instrumentation, 251–268
  basic apparatus, 251–253
  tip position modulation, 253–255
  tip preparation, 255–268
 list of symbols, 366–370
 modes of application, 247–250
  as an electrochemical tool, 247–248
  as a fabrication tool, 249–250
  as an imaging tool, 249
 principles, 244–247
 theory, 268–310
  adsorption/desorption processes, 296–303
  generation/collection (G/C) mode of operation, 287–293
  processes with first- and second-order homogeneous reactions in the gap, 293–298
  SECM with disk-shaped tip in the feedback mode of operation, 268–287
  SECM with a nondisk tip and tip shape characterization, 303–310
Scanning tunneling microscopy (STM), 244
Self-diffusion of polymers, 221–225
Steady-state SECM measurements, 274–287
Structure of micelles, 5–7
Surfactant microstructures, 4–16
 adsorption of surfactants on solid surfaces, 13–16
 microemulsions, 10–11
 surfactant structure controls system architecture, 12–13
 surfactants and micelles, 4–10
  kinetics in micellar systems, 9–10
  micellar dynamics, 7–9
  reverse micelles, 9
  structure of micelles, 5–7
 vesicles and lamellar dispersions, 11–12

Subject Index

Surfactants, historical survey of, 2–4

Theories predicting nonlinear D(c) function, 103–118
  Blauch–Saveant theory, 117–118
  diffusion-migration model for long-distance electron hopping, 116–117
  Fritsch-Faules–Faulkner model, 105–108
  He–Chen model, 103–105
  ion association and electric field effects, 116
  percolation theory, 108–109
  theories of coupled electron hopping-ion displacement, 109–116
Theory of the electron exchange reaction, 97–103
Thin films and membranes, SECM in the study of, 349–364
  chronoamperometry and chronopotentiometry, 354–355
  cyclic voltammetry, 350–354
  direct electrochemical measurements, 355–361
  SECM imaging of films and membranes, 361–364
Traditional electrodes, comparison of polymer film electrodes and, 92–95
Two-cell micelle distribution, model for, 35–39

Ultramicroelectrode (UME), 244–247
  fast scan rate cyclic voltammetry on, 90
Uncomplicated non-steady-state SECM measurements, 268–274

Vesicle(s), 11–12
  detecting photochemistry at CdS particles in, 77–78
  dispersions, 76–77
Viologens (organic redox polymers), 182–184

Water–in–oil (w/o) microemulsions, 10–11